T0348465

ADVANCES IN
ORGAN BIOLOGY

Volume 5A • 1998

MOLECULAR AND CELLULAR BIOLOGY OF BONE

MOLECULAR AND CELLULAR BIOLOGY OF BONE

Guest Editor

MONE ZAIDI
Medical College of Pennsylvania
School of Medicine and
Veterans Affairs Medical Center
Philadelphia, Pennsylvania

Associate Guest Editors

OLUGBENGA A. ADEBANJO
Medical College of Pennsylvania
School of Medicine and
Veterans Affairs Medical Center
Philadelphia, Pennsylvania

CHRISTOPHER L. -H. HUANG
Department of Physiology
University of Cambridge
Cambridge, England

ADVANCES IN ORGAN BIOLOGY

MOLECULAR AND CELLULAR BIOLOGY OF BONE

Guest Editor: MONE ZAIDI
Veterans Affairs Medical Center

Series Editor: E. EDWARD BITTAR
Department of Physiology
University of Wisconsin-Madison

Associate Guest Editors: OLUGBENGA A. ADEBANJO
Veterans Affairs Medical Center

CHRISTOPHER L. -H. HUANG
Department of Physiology
University of Cambridge

VOLUME 5A • 1998

Stamford, Connecticut *London, England*

100 Prospect Street
Stamford, Connecticut 06901

JAI PRESS LTD.
38 Tavistock Street
Covent Garden
London WC2E 7PB
England

ISBN: 0-7623-0390-5

Printed and bound by CPI Antony Rowe, Eastbourne

CONTENTS (Volume 5A)

LIST OF CONTRIBUTORS xi

FOREWORD
Iain MacIntyre xxiii

PREFACE
Mone Zaidi xxvii

SECTION I. GENERAL PRINCIPLES

ASPECTS OF ANATOMY AND DEVELOPMENT OF BONE: THE NM, μM, AND MM HIERARCHY
Alan Boyde and Sheila Jones 3

PHYSIOLOGY OF BONE REMODELING
Chantal Chenu and Pierre Dominique Delmas 45

HORMONAL REGULATION OF BONE REMODELING
Kong Wah Ng and T. John Martin 65

COUPLING OF BONE FORMATION AND BONE RESORPTION: A MODEL
James T. Ryaby, Robert J. Fitzsimmons, Subburaman Mohan, and David J. Baylink 101

MECHANOTRANSDUCTION IN BONE
Elisabeth H. Burger, Jenneke Klein-Nulend, and Stephen C. Cowin 123

VASCULAR CONTROL OF BONE REMODELING
Ted S. Gross and Thomas L. Clemens 137

PARATHYROID HORMONE AND ITS RECEPTORS
Abdul B. Abou-Samra 161

THE STRUCTURE AND MOLECULAR BIOLOGY OF THE CALCITONIN RECEPTOR
Steven R. Goldring 187

THE VITAMIN D RECEPTOR: DISCOVERY, STRUCTURE, AND FUNCTION
J. Wesley Pike 213

MOLECULAR PHYSIOLOGY OF AVIAN BONE
Christopher G. Dacke 243

CONTENTS (Volume 5B)

LIST OF CONTRIBUTORS xi

FOREWORD
Iain MacIntyre xxiii

PREFACE
Mone Zaidi xxvii

SECTION II. BONE RESORPTION

OSTEOCLASTOGENESIS, ITS CONTROL, AND ITS DEFECTS
Etsuko Abe, Tomoo Yamate, Hanna Mocharla,
Yasuto Taguchi, and Matsuo Yamamoto 289

OSTEOCLAST INTEGRINS: ADHESION AND SIGNALING
Geetha Shankar and Michael Horton 315

HORMONAL REGULATION OF FUNCTIONAL
OSTEOCLAST PROTEINS
F. Patrick Ross 331

THE OSTEOCLAST CYTOSKELETON
Alberta Zambonin Zallone, Maria Grano, and Silvia Colucci 347

ROLE OF PROTEASES IN OSTEOCLASTIC RESORPTION
Toshio Kokubo, Osamu Ishibashi, and Masayoshi Kumegawa 359

EXTRACELLULAR CALCIUM ION SENSING IN
THE OSTEOCLASTS
Olugbenga Adebanjo and Mone Zaidi 371

THE OSTEOCLAST MOLECULAR PHENOTYPE
Dennis Sakai and Cedric Minkin 385

ION CHANNELS IN OSTEOCLASTS
A. Frederik Weidema, S. Jeffrey Dixon, and Stephen M. Sims 423

SECTION III. BONE FORMATION

OSTEOBLASTS AND BONE FORMATION
Pierre J. Marie 445

OSTEOBLAST LINEAGE
James T. Triffitt and Richard O. C. Oreffo 475

OSTEOBLAST RECEPTORS
Janet E. Henderson and David Goltzman 499

COLLAGENASE AND OTHER OSTEOBLAST ENZYMES
Anthony Vernillo and Barry Rifkin 513

BIOLOGY OF OSTEOCYTES
P.J. Nijweide, N.E. Ajubi, E.M. Aarden, and A. Van der Plas 529

CELL–CELL COMMUNICATION IN BONE
Roberto Civitelli 543

THE COLLAGENOUS AND NONCOLLAGENOUS PROTEINS OF CELLS IN THE OSTEOBLASTIC LINEAGE
Pamela Gehron Robey and Paolo Bianco 565

THE ROLE OF GROWTH FACTORS IN BONE FORMATION
Lynda F. Bonewald and Sarah L. Dallas 591

SYSTEMIC CONTROL OF BONE FORMATION
Toshio Matsumoto and Yasuhiro Takeuchi 615

THE DIRECT AND INDIRECT EFFECTS OF ESTROGEN ON BONE FORMATION
Timothy J. Chambers 627

CONTENTS (Volume 5C)

LIST OF CONTRIBUTORS xi

FOREWORD
Iain MacIntyre xxiii

PREFACE
Mone Zaidi xxvii

SECTION IV. MOLECULAR BASIS OF BONE DISEASE

ESTROGEN AND BONE LOSS
Roberto Pacifici 641

PAGET'S DISEASE OF BONE
G. David Roodman 661

INHERITED AND ACQUIRED DISORDERS OF THE EXTRACELLULAR CA^{2+}_{o}-SENSING RECEPTOR
Edward M. Brown and Steven C. Hebert 677

BONE DISEASE IN MALIGNANCY
Brendan F. Boyce, Toshiyuki Yoneda, and Theresa A. Guise 709

MECHANISMS OF IMMUNOSUPRESSANT-INDUCED BONE DISEASE
Grant R. Goodman and Solomon Epstein 739

DEVELOPMENTAL DISORDERS OF BONE
Jay R. Shapiro 765

SKELETAL FLUOROSIS: MOLECULAR ASPECTS
Ambrish Mithal 797

MOLECULAR PHARMACOLOGY OF ANTIRESORPTIVE THERAPY FOR OSTEOPOROSIS
Olugbenga A. Adebanjo, Edna Schwab, Li Sun, Michael Pazianas, Baljit Moonga, and Mone Zaidi 809

BISPHOSPHONATES: MECHANISMS OF ACTION
Herbert Fleisch 835

NOVEL BONE-FORMING AGENTS
Ian R. Reid 851

TRANSGENIC MODELS FOR BONE DISEASE
Pietro De Togni 869

INDEX 891

LIST OF CONTRIBUTORS

E.M. Aarden — Research Scientist
Department of Cell Biology
Faculty of Medicine, Leiden University
Leiden, The Netherlands

Etsuko Abe, PhD — Research Professor of Medicine
Department of Medicine
University of Arkansas for Medical Sciences
Little Rock, Arkansas

A.B. Abou-Samra, MD — Associate Professor of Medicine
Endocrine Unit, Department of Medicine
Massachusetts General Hospital
Harvard Medical School
Boston, Massachusetts

Olugbenga A. Adebanjo, MD — Assistant Professor of Medicine
Department of Medicine
Medical College of Pennsylvania
School of Medicine and Veterans Affairs Medical Center
Philadelphia, Pennsylvania

N.E. Ajubi — Research Scientist
Department of Cell Biology
Faculty of Medicine
Leiden University
Leiden, The Netherlands

David J. Baylink, MD

Distinguished Professor of Medicine
Loma Linda University
and Associate Vice President
for Medical Affairs for Research
J.L. Pettis Veterans Affairs Medical Center
Loma Linda, California

Paolo Bianco, MD

Dipartmento di Biopatologia Umana
Universita La Sapienza
Rome, Italy

L.F. Bonewald, PhD

Associate Professor of Medicine
Department of Medicine
University of Texas Health Science Center
San Antonio, Texas

Brendan F. Boyce

Professor of Pathology
Department of Medicine
Division of Endocrinology and Metabolism
University of Texas Health Science Center
San Antonio, Texas

Alan Boyde, PhD

Professor of Mineralized Tissue Biology
Department of Anatomy and
Developmental Biology
University College London
London, England

Edward M. Brown, MD

Professor of Medicine
Endocrine-Hypertension and Renal
Divisions
Brigham and Women's Hospital
Boston, Massachusetts

Elisabeth H. Burger, PhD

Professor
Department of Oral Cell Biology
ACTA-Vrije Universiteit
Amsterdam, The Netherlands

T.J Chambers, PhD, MBBS, MRCPath

Professor and Chairman
Department of Histopathology
St. George's Hospital Medical School
London, England

Chantal Chenu, PhD | Staff Research Fellow
INSERM
Hôpital Edouard Herriot
Lyon, France

Roberto Civitelli, MD | Associate Professor of Medicine and Orthopedic Surgery and Assistant Professor of Cell Biology and Physiology
Division of Bone and Mineral Diseases
Washington University School of Medicine
St. Louis, Missouri

Thomas L. Clemens, PhD | Professor of Medicine
Department of Molecular and Cellular Physiology and Orthopedic Surgery
University of Cincinnati Medical Center
Cincinnati, Ohio

Silvia Colucci, PhD | Assistant Professor of Histology
Institute of Human Anatomy
University of Bari
Bari, Italy

Stephen C. Cowin | Department of Mechanical Engineering
City University of New York
New York, New York

C.G. Dacke, B.Tech, PhD, FIBiol | Reader and Head, Pharmacology Division
School of Pharmacy and Biomedical Science
University of Portsmouth
Portsmouth, England

Sarah L. Dallas, PhD | Assistant Professor of Medicine
Department of Medicine
University of Texas Health Science Center
San Antonio, Texas

Pietro De Togni, MD | Assistant Professor of Pathology
Immunogenetics and Transplantation Laboratory
University of Arkansas for Medical Sciences
Little Rock, Arkansas

P.D. Delmas, MD, PhD	Professor of Medicine INSERM Hôpital Edouard Herriot Lyon, France
S.J. Dixon, DDS, PhD	Associate Professor of Physiology and Oral Biology Department of Physiology Faculty of Dentistry The University of Western Ontario London, Ontario, Canada
S. Epstein, MD, FRCP	Professor of Medicine and Chief Division of Endocrinology Medical College of Pennsylvania Hahnemann School of Medicine Philadelphia, Pennsylvania
R. J. Fitzsimmons, PhD	Assistant Research Professor of Medicine and Director Mineral Metabolism Jerry L. Pettis Veterans Affairs Medical Center Loma Linda University Loma Linda, California
Herbert Fleisch, MD	Professor and Chairman Department of Pathophysiology University of Berne Berne, Switzerland
Steven R. Goldring, MD	Associate Professor of Medicine and Chief of Rhematology Beth Israel-Deaconess Hospital Harvard Medical School Boston, Massachusetts
David Goltzman, MD	Professor and Chairman Department of Medicine McGill University,Royal Victoria Hospital Montréal, Québec, Canada
Grant R. Goodman, MD	Research Associate Department of Medicine Albert Einstein Medical Center Philadelphia, Pennsylvania

Maria Grano, PhD	Assistant Professor of Histology Institute of Human Anatomy University of Bari Bari, Italy
Ted S. Gross, PhD	Assistant Professor Departments of Medicine and Molecular and Cellular Physiology and Orthopedic Surgery University of Cincinnati Medical Center Cincinnati, Ohio
Theresa A. Guise, MD	Assistant Professor of Medicine Division of Endocrinology and Metabolism University of Texas Health Science Center San Antonio, Texas
Steven C. Hebert, MD	Professor of Medicine and Chief, Division of Nephrology Vanderbelt University Nashville, Tennessee
Janet E. Henderson, PhD	Assistant Professor of Medicine Department of Medicine McGill University Montréal, Québec, Canada
M. Horton, MD, FRCP, FRCPath	Professor Rayne Institute Bone and Mineral Center University of London London, England
Osamu Ishibashi, MS	Scientist Ciba-Geigy Japan Limited International Research Laboratories Takarazuka, Japan
Sheila Jones, PhD	Professor of Anatomy Department of Anatomy and Developmental Biology University College London London, England

J. Klein-Nulend, PhD
Assistant Professor
Department of Oral Cell Biology
ACTA-Vrije Universiteit
Amsterdam, The Netherlands

Toshio Kokubo, PhD
Group Leader
International Research Laboratories
Ciba Geigy Japan Limited
Takarazuka, Japan

Masayoshi Kumegawa, DDS
Professor
Department of Oral Anatomy
Meikai University School of Dentistry
Saitama, Japan

Pierre J. Marie, PhD
Professor
Cell and Molecular Biology of Bone and Cartilage
Lariboisière Hospital
Paris, France

T.J. Martin, MD, DSC, FRCPA, FRACP
Professor of Medicine
St. Vincent's Institute of Medical Research
University of Melbourne
Fitzroy, Victoria, Australia

Toshio Matsumoto, MD
Professor and Chairman
First Department of Medicine
Tokushima University School of Medicine
Tokushima, Japan

Cedric Minkin, PhD
Professor
Department of Basic Sciences
University of Southern California School of Dentistry
Los Angeles, California

Ambrish Mithal, MD, DM
Professor
Department of Medical Endocrinology
Sanjay Gandhi Post Graduate Institute of Medical Sciences
Lucknow, India

Hanna Mocharla, PhD
Research Instructor
Department of Medicine
University of Arkansas for Medical Sciences
Little Rock, Arkansas

S. Mohan, PhD
Research Professor of Medicine, Biochemistry, and Physiology
J.L. Pettis Veterans Affairs Medical Center
Loma Linda University
Loma Linda, California

Baljit Moonga, PhD
Assistant Professor of Medicine
Medical College of Pennsylvania
School of Medicine and
Veterans Affairs Medical Center
Philadelphia, Pennsylvania

K.W. Ng, MBBS, MD, FRACP
Associate Professor
Department of Medicine
The University of Melbourne
St. Vincent's Hospital
Fitzroy, Victoria, Australia

Peter J. Nijweide
Professor
Department of Cell Biology
Faculty of Medicine
Leiden University
Leiden, The Netherlands

Richard O.C. Oreffo, D. Phil.
MRC Research Fellows
MRC Bone Research Laboratory
Nuffield Orthopedic Center
University of Oxford
Headington
Oxford, England

Roberto Pacifici, MD
Associate Professor of Medicine
Division of Bone and Mineral Diseases
Washington University Medical Center
St. Louis, Missouri

Michael Pazianas, MD
Associate Professor of Medicine
Division of Geriatric Medicine and Institute on Aging
University of Pennsylvania
Philadelphia, Pennsylvania

J. Wesley Pike, PhD
Professor of Medicine
Department of Molecular and Cellular Physiology
University of Cincinnati Medical Center
Cincinnati, Ohio

James T. Ryaby, PhD
Director of Research
Orthologic Corporation
Phoenix, Arizona

Ian R. Reid, MD
Associate Professor of Medicine
Department of Medicine
University of Auckland
Auckland, New Zealand

Barry Rifkin, DDS, PhD
Professor and Dean
State University of New York Dental School
Stony Brook, New York

Pamela Gehron Robey, PhD
Chief
Craniofacial and Skeletal Diseases
National Institute of Dental Research
National Institutes of Health
Bethesda, Maryland

G. David Roodman, MD
Professor of Medicine and Chief of Hematology
Audie Murphy Veterans Affairs Medical Center
University of Texas Health Science Center
San Antonio, Texas

F. Patrick Ross, PhD
Associate Professor of Pathology
Department of Pathology
Barnes-Jewish Hospital
St. Louis, Missouri

Dennis Sakai, PhD
Research Professor
Department of Basic Sciences
University of Southern California School of Dentistry
Los Angeles, California

Edna Schwab, MD
Assistant Professor of Medicine
Division of Geriatric Medicine and Institution Aging
University of Pennsylvania
Philadelphia, Pennsylvania

Geetha Shankar, PhD
Scientist
NPS Pharmaceuticals Inc.
Salt Lake City, Utah

Jay Shapiro, MD
Professor
Department of Medicine
Walter Reed Army Medical Center
Bethesda, Maryland

Stephen M. Sims, PhD
Associate Professor
Department of Physiology
Faculty of Medicine and Dentistry
The University of Western Ontario
London, Ontario, Canada

Li Sun, MD, PhD
Research Fellow
Medical College of Pennsylvania
School of Medicine and Veterans Affairs Medical Center
Philadelphia, Pennsylvania

Yasuto Taguchi, MD
Research Fellow
Department of Medicine
University of Arkansas for Medical Sciences
Little Rock, Arkansas

Yasuhiro Takeuchi, MD
Assistant Professor
Fourth Department of Internal Medicine
University of Tokyo School of Medicine
Tokyo, Japan

James T. Triffitt, PhD
Head of Department
MRC Bone Research Laboratory
Nuffield Orthopedic Center
University of Oxford
Headington
Oxford, England

A. Van der Plas
Head of Technical Staff
Department of Cell Biology
Faculty of Medicine
Leiden University
Leiden, The Netherlands

Anthony Vernillo, PhD DDS
Associate Professor
Department of Oral Medicine and Pathology
New York University College of Dentistry
New York, New York

A. Frederik Weidema, PhD
Research Associate
Laboratorium voor Fysiologie
Katholieke Universiteit Leuven
Herestraat, Leuven, Belgium

Matsuo Yamamoto, PhD
Research Fellow
Department of Medicine
University of Arkansas for Medical Sciences
Little Rock, Arkansas

Tomoo Yamate, MD, PhD
Instructor
Department of Medicine
University of Arkansas for Medical Sciences
Little Rock Arkansas

Toshiyuki Yoneda, DDS, PhD
Professor of Medicine
Department of Medicine
Division of Endocrinology and Metabolism
University of Texas Health Science Center
San Antonio, Texas

Alberta Zambonin Zallone, PhD
Professor of Histology
Institute of Human Anatomy
University of Bari
Bari, Italy

M. Zaidi, MD, PhD, FRCP, FRCPath
Professor of Medicine and Associate Dean
Medical College of Pennsylvania
School of Medicine
Associate Chief of Staff and Chief, Geriatrics and Extended Care
Veterans Affairs Medical Center
Philadelphia, Pennsylvania

FOREWORD

These volumes differ from the current conventional texts on bone cell biology. Biology itself is advancing at breakneck speed and many presentations completely fail to present the field n a truly modern context. This text does not attempt to present detailed clinical descriptions. Rather, after discussion of basic concepts, there is a concentration on recently developed findings equally relevant to basic research and a modern understanding of metabolic bone disease. The book will afford productive new insights into the intimate inter-relation of experimental findings and clinical understanding. Modern medicine is founded in the laboratory and demands of its practitioners a broad scientific understanding: these volumes are written to exemplify this approach. This book is likely to become essential reading equally for laboratory and clinical scientists.

Ian MacIntyre, FRS
Research Director
William Harvey Research Institute
London, England

DEDICATION

To Professor Iain MacIntyre,
MBChB, PhD, Hon MD, FRCP, FRCPath, DSc, FRS

In admiration of his seminal contributions to bone and mineral research that have spanned over more than four decades, and

In gratitude for introducing us into the field of bone metabolism and for his continued encouragement, assistance, and friendship over many years

PREFACE

The intention of putting this book together has been not to develop a full reference text for bone biology and bone disease, but to allow for an effective dissemination of recent knowledge within critical areas in the field. We have therefore invited experts from all over the world to contribute in a way that could result in a complete, but easily readable text. We believe that the volume should not only aid our understanding of basic concepts, but should also guide the more provocative reader toward searching recent developments in metabolic bone disease.

For easy reading and reference, we have divided the text into three subvolumes. Volume 5A contains chapters outlining basic concepts stretching from structural anatomy to molecular physiology. Section I in Volume 5B is devoted to understanding concepts of bone resorption, particularly in reference to the biology of the resorptive cell, the osteoclast. Section II in Volume 5B contains chapters relating to the formation of bone with particular emphasis on regulation. Volume 5C introduces some key concepts relating to metabolic bone disease. These latter chapters are not meant to augment clinical knowledge; nevertheless, these do emphasize the molecular and cellular pathophysiology of clinical correlates. We do hope that the three subvolumes, when read in conjunction, will provide interesting reading for those dedicated to the fast emerging field of bone biology.

We are indebted to the authors for their significant and timely contributions to the field of bone metabolism. We are also grateful to Christian Costeines (JAI Press) and Michael Pazianas (University of Pennsylvania) for their efforts in ensuring the creation of quality publication. The editors also acknowledge the support and perseverance of their families during the long hours of editing.

Mone Zaidi
Guest Editor
Olugbenga A. Adebanjo
Christopher L.-H. Huang
Associate Guest Editors

SECTION I

GENERAL PRINCIPLES

ASPECTS OF ANATOMY AND DEVELOPMENT OF BONE

THE nm, μm AND mm HIERARCHY

Alan Boyde and Sheila Jones

I. Introduction 4
II. The State of the Mineral Phase and its Packing in Bone 5
III. Collagen Order, Matrix Pattern and Fiber Dimensions and Range, Lamellae 6
IV. Compact Bone 9
A. The Advantages of the Use of Circularly Polarized Light 9
B. Collagen Versus Mineral Contributions to the Polarized Light Image 10
C. Preferred Orientation of Collagen in Compact Bone 11
V. Spongy Bone 11
A. Metaphyseal Trabeculation in Laboratory Rodents 11
B. Architectural Changes in the Aging of Human Lumbar Vertebral Body Trabecular Bone 13
C. Resorption-Repair Coupling is not Site Specific 15
D. Identification of Bone Surface States by SEM 17
E. Bone Quality and Bad Joins? 19
F. Is Bone Put Back on Unloaded Struts? 20

Advances in Organ Biology
Volume 5A, pages 3-44.

ISBN: 0-7623-0390-5

G. The Iliac Crest 21
H. The Femoral Neck 22
I. The Distal Femur. 23
VI. Better Methods for Investigating the Structure of Trabecular Bone. 23
A. Complete 3D Data Sets From μCT, MRI, or Serial Sectioning 23
B. Tilting Beams or Samples: Stereo and Stereophotogrammetry. 24
C. Deep Field Microscopy with Rotating Samples:
Continuous Motion Parallax 25
D. Internal Casting. 26
E. Frequency Domain Analysis of Trabecular Bone Structure 27
F. TBV from Digital Processing of X-rays of Parallel Slices 27
G. Giving a Correct Impression of a Complicated 3D Structure
From a Single Projection: Advances in SEM Technique 29
VII. Qualitative Variations in Mineralization Pattern and Degree. 32
A. Woven Bone 32
B. Lamellar Bone. 32
C. Extrinsic Fibers in Bone 33
D. Endochondral Mineralization 33
E. Calcified Fibrocartilage 35
F. Reversal and Cement Lines. 35
G. Osteocytic Death and Mineralization 35
VIII. Quantitative Study of Mineralization Degree 37
A. The Meaning of Bone Density. 37
B. Determination of Bone Mineralization Density Using BSE-SEM 37
IX. Summary. 39
Acknowledgments. 39

I. INTRODUCTION

Bone is a tissue, but bones are complex organs, largely made of bone and cartilage (both uncalcified and calcified), and enclosed red (hemopoietic) and yellow (fatty) marrow that contain osteoprogenitor cells. Bone, as a calcified connective tissue, contains cells and extracellular matrix constituents that vary in relative proportions according to the age and health of the individual, just as the microanatomical and anatomical features change, rapidly or gradually, with time. Information about changing constitution and structure is therefore embedded within the bones as well as apparent at its surfaces. We can read the bone at several hierarchical levels and relate architectural changes to histological variation. Although the architecture is obviously three-dimensional (3D) it is often neglected: that the constitution of the tissue is also 3D is equally disregarded because we are accustomed to section bone to appraise it microscopically. The two-dimensional (2D) views are notoriously inadequate for the understanding of varied phases within an irregular solid and often misrepresentative of the functional tissue even at the architectural level.

Ham and Leeson (1961) wrote:
They were like young microscopists
Who study single sections
And picture "wholes" from single parts
With many misconceptions -
Especially if they never learn
To think in three dimensions.

In this chapter we highlight some of the ways that can be used to achieve an integrated approach to the constitution and structure of bone during growth, maturity, and aging, and some of the findings made possible through these means. We shall center our commentary on those of our observations which are either controversial in being in conflict with current accepted opinion, or develop known themes, but using novel techniques. We have chosen to focus on a small number of particularly common topics where we would feel it fruitful to redirect thought in bone research. All our topics address bone density in one sense or another.

II. THE STATE OF THE MINERAL PHASE AND ITS PACKING IN BONE

Anorganic bone hangs together: and it fizzes in acid.

This currently widely accepted and quoted view concerning the state of subdivision and the degree of admixture of bone mineral with collagen (Katz et al., 1989; Weiner and Traub, 1992; review by Currey, 1996) is flawed to the extent that it ignores the fact that deproteinized (anorganic) bone has a continuity of its own. If bone salt crystals were the separated, short elements commonly depicted, the tissue would fall apart when the organic matrix is dissolved or oxidized, but it does not, and we profit from this fact in preparing bone samples for scanning electron microscopy (SEM).

When examined by transmission electron microscopy (TEM), sections prepared by ion beam thinning show the mineral as a natural electron dense phase outlining the collagen fibrils of the matrix within which it is deposited: this preparation method avoids the contact of an aqueous phase with the finely divided bone mineral in the section and its physical disruption during the preparation of ultra-thin sections for TEM. Collagen encapsulated within the mineral phase is negatively stained by the electron dense mineral phase, which demonstrates a great length for its individual elements (Boyde and Pawley, 1975). The particles of which it is composed are maximally a few unit cells thick (such narrow structures would be flexible). They are not like the short, thin, flat bricks so commonly seen in figurative illustrations of bone mineral packing in collagen. That mineral which forms within fibrils (70% of the volume of the collagen fibrils is water) is organized as long thin plates or tubes which effectively envelop the collagen microstructure: they cannot not be broken in ultramicrotomy. This is in agreement with the fact that bone salt inside collagen fibrils so closely shields the collagen that it prevents enzymatic access during osteoclasis unless demineralization occurs first.

Matrix constituents other than collagen may be lost or gained during mineralization, so that osteoid is easily distinguished from previously mineralized matrix after the mineral has been removed in the preparation of stained, demineralized histological sections.

A second misconception is widespread in describing the chemistry of the mineral phase. Osteoid is substantially dehydrated, under water, by replacement of matrix water by a carbonated apatite phase (akin to the mineral Dahllite). It differs from hydroxyapatite ($Ca_{10}(PO_4)_6(OH)_2$) in having variable and sometimes large Mg and carbonate fractions. The variation in the composition of bone mineral and its non-apatitic nature should have an influence on the computation of bone mineral content by x-ray photon-absorption and scattering methods, but these usually assume wrongly that bone mineral is hydroxyapatite.

III. COLLAGEN ORDER: MATRIX PATTERN AND FIBER DIMENSIONS AND RANGE, LAMELLAE

Immature or woven bone contains collagen fibrils/fibers which have a wide range of diameters (< 0.1 to >3 μm). The fibers form a random felt-work: the inappropriate name stems from the fact that only the larger fibers can be resolved using crossed linear polarizers (LPL): since these are seen in the 45° sectors, they may look like the warp and woof of a woven fabric. Cells are included in this matrix in an apparently haphazard fashion.

In lamellar bone, there is a somewhat contrasting orientation between successive 2–3 μm thick layers of 2–3 μm wide branching bundles (Boyde, 1972): the layered arrangement gives rise to sufficient contrast to be seen well in LPL, and the name is appropriate. At an intermediate scale of organization, bone's mechanical properties result from this layered structure of the matrix organization which is favorably exposed to view by the activities of osteoclasts (Reid, 1986). The osteocyte lacunae in lamellar bone conform to the microstructure of the matrix. They are commonly three-axis ellipsoids (like plumstones), with the longest axis parallel to the collagen domain in which they lie, and the shortest perpendicular to the forming surface. The cell orientation layering in depth is nicely seen in 3D, either directly in a real time 3D microscope or in a reconstruction from serial optical sections recorded with a reflection confocal microscope (Boyde, 1987).

Osteocytes may occupy a third to a half of the matrix volume in woven bone, but only about 2% in young and 1% in adult human lamellar bone. It may be possible to infer the rate of matrix formation from the size and shape of the osteocyte lacunae, as these reflect to some measure the plumpness of the osteoblast just prior to its change in status.

Some bone shows little change in collagen orientation between successive (erstwhile) lamellae, and is often called parallel fibered. It is probably better to regard this type of bone as lying at one end of the range of possibilities for contrasting orientations between lamellae.

How does the collagen become oriented? Osteoblasts constitute a pavement of fully differentiated cells that is one cell layer thick. The cells act together in groups to make patches of bone that show structural conformity (Jones et al., 1975). The orientation of the collagen within the surface layer of osteoid reflects the orientation of the osteoblasts immediately in contact with it, rather than the ordering of the collagen dictating the lie of the cells, although it may be that both the cells and the assembling collagen can respond to the same strains.

The intermittent activity of osteoblasts, which is evidenced in the layered structure in bone, may possibly be related to the circadian rhythm of parathyroid hormone (PTH) (Logue et al., 1990) and other hormones; osteoblasts have receptors for PTH/ parathyroid hormone-related protein (PTHrP), and PTH is a mitogen for cultured osteoblastic cells in pulsed doses.

The standard lamellar model of lamellar bone grew in stature following the introduction of SEM. Not only did this lead to the knowledge of the domain organization of bone collagen but also to that of the osteoblasts (Boyde, 1972; Jones, 1973). SEM-based information also led to detailed modification of the criss-cross layer concept by showing some change in orientation within each layer, and particularly at the layer boundaries (Boyde, 1972). The twisted or spiral plywood aspects of the structure have been particularly attractive to certain observers (Bouligand, 1972). Such minutiae have been confirmed in our unpublished studies using high resolution, video rate scanning, reflection mode, confocal optical microscopy (see below).

In the conventional light microscopic preparation, the sample is at least translucent, if not nearly transparent. Image contrast may be obtained by selective staining of tissue components or by interference optical effects, as, for example, in the utilization of polarized light. To make specimens transparent, the strong differences in the refractive index (RI) between mineralized bone matrix and cellular material must be reduced: this is conventionally done by fixing the tissue to render cells both stable and permeable and by substituting water with high RI substances to match that of the bone matrix. Such preparation minimizes optical scatter in the sample, and makes it possible to obtain confocal images at greater depths, but it is not possible to make bone optically homogeneous. The small differences in RI can be used to advantage in reflection imaging in the confocal mode, which can be used to study fine detail of fiber and fibril arrangements in bone matrix. This is done by rapid through focusing under the optimal confocal optical-sectioning conditions: we use a video rate confocal laser scanning microscope (Noran Odyssey) with 1.4NA objective lenses. Since the evidence is dynamic, it is not possible to publish it in print.

The lamellar organization of bone was discovered through the application of linearly polarized light microscopy (PLM/LPL; Gebhardt, 1905). Gebhardt theorized that the differing arrangements of lamellar spirals in osteons would relate to specialized function. Direct evidence for the role of lamellar orientation in determining the physical properties of bone comes from the patent relationship to microhardness. Lamellae polish (and etch) at different rates dependent on their orientation (Boyde, 1984b) and soft-lap polishing is an excellent means of demonstrating their existence in surface preparations: more mature (better mineralized) tissue is both relatively harder and less prone to develop polishing relief. Wet collagen arrays also shrink anisotropically, such that nearest neighbor fibers aggregate parallel with their long axes. Polished and then demineralized section surfaces of lamellar bone therefore exhibit layers of seemingly more and less dense collagen packing, irrespective of the direction of viewing. These considerations, when added to the known spiraling of fibrils within bundles and the sine/cosine relation of apparent to real length of rodlike structures, probably account for the appearances highlighted by Marotti et al. (1994) and which they believe to shed real doubt on the classical view of bone lamellation.

The micron-range variation in electron backscattering is so dependent on local topography that the magnitude of surface relief developed on polished bone samples will dominate any probable local change in mineral content (Howell and Boyde, 1994). Samples micro-milled to a surface relief of 100 nm or better (unachievable by polishing) do not demonstrate lamellation in backscattered electron (BSE) images (Boyde and Jones, 1996). Evidence has not yet been produced which successfully challenges the basic precepts of the domain model (Boyde, 1972) of lamellar bone micro-architecture.

IV. COMPACT BONE

Bone structure makes sense—in compact as well as trabecular lamellar bone.

Wolff's (1892) law referred to trabecular bone and the distribution of bone bulk. However, bone structural organization makes good sense over many different hierarchical scales. In this section we shall consider the statistics of collagen fiber orientation and variations in lamellation in compact bone.

A. The Advantages of the Use of Circularly Polarized Light

In standard PLM, using crossed linear polarizers, uniaxially birefringent materials with a significant component of their orientation lying in the plane of section (i.e., perpendicular to the optic, observation axis) appear maximally bright when furthest from being parallel to either filter (i.e., at 45°). Circularly polarized light (CPL) conditions are achieved by using four filters: (a) a polar; (b) a 45° quarter-wave retardation plate for the wavelength of observation, then the specimen; (c) a 135° quarter-wave plate; and (d) a 90° polarizer, the analyzer. With these correctly aligned, all birefringent material (and here we are mostly concerned with collagen) appears bright if lying in the plane of section, or dark if lying in the optic axis. Intermediate orientations in the angle of orientation with respect to the optic axis give rise to the observed variations in brightness, which, for a constant optical path length (section thickness), depend upon the vertical element of orientation in a sine/cosine relationship. Thus CPL is unbiased by the direction of the birefringent material in the section plane: its use radically improves the chances for quantitation (Boyde et al., 1984; Boyde and Riggs, 1990; Riggs et al., 1993a,b). For quantitative PLM, it is particularly important that the section be exceptionally well cleared and it should be plane parallel because all interference optical effects are directly proportional to path length. If direct comparisons are to be made from section to section according to some standard scheme, then all sections should have the same thickness. It is difficult to satisfy the uniform thickness criterion for much of trabecular bone and most studies have used compact tissue, since this which will hold together in, for example, 100 μm thick sawn sections.

B. Collagen Versus Mineral Contributions to the Polarized Light Image

The birefringence of mineralized bone is of positive sign. Both collagen and the mineral phase contribute to the measured birefringence, but their contributions oppose each other. Although the positive form birefringence of collagen always strongly dominates the negative intrinsic birefringence of the bone salt (apatite), the latter is important, and removing it, by decalcifying the section, notably enhances the total birefringence. This shows either that the higher RI mineral (replacing water, which has a lower RI than protein) induces a negative form birefringence which would result even if the material itself were not optically anisotropic, and/or that the contribution of any intrinsic birefringence of the mineral was such as to exactly oppose the collagen form birefringence. This requires that the apatitic crystals be aligned parallel with the protein-rodlet/water-space array which generates the form. Electron microscopic evidence shows that the axes of collagen and the mineral within it are parallel to within a few degrees; certainly parallel for the purpose of this discussion, and more parallel than one fibril is with another within a single bundle or lamella in one domain. However, we note here for the first time that mineral will reduce the collagen birefringence irrespective of its orientation, and that extrafibrillar mineral may conceivably have any orientation. Here we deal with lamellated bone, but PLM evidence shows that dentine contains a substantial extracollagenous mineral phase which orients perpendicular to the local mineralizing front. The same occurs in calcospheritic cartilage calcification, and probably also in woven bone, though the growing clusters here are too small for equivalent PLM studies.

The discussion above matters because in all transverse sections of secondary osteons the most peripheral lamellae which mate to the reversal line are brighter in CPL (Riggs et al., 1993a,b). BSE imaging and microradiography show that these regions are also the least well mineralized within osteons. Thus some of the enhanced CPL brightness may be explained by mineralization density rather than collagen orientation.

C. Preferred Orientation of Collagen in Compact Bone

Studies in a bone in which the pattern of *in vivo* strain had been unambiguously determined (Biewener et al., 1983) showed that within secondarily remodeled compact bone there are important differences in the orientation of collagen lamellae in volumes subjected to tension as against compressive loading (Boyde and Riggs, 1990; Riggs et al., 1993a,b). In sites loaded more in compression, far more reworking of the structure occurs and the mean collagen orientation is at a larger angle to the longitudinal axis, compared with the more nearly longitudinal slant in tension sites.

A more obscure relationship was found for both mean collagen orientation (McMahon et al., 1995) and turnover history (Skedros et al., 1994a,b) in the ovine (artiodactyl) calcaneus, another bone for which strain patterns had been determined by Lanyon (1973). In this case, little primary osteonal structure is retained and all cortices are subjected to remodeling. The *in vivo* strain observations had shown that compression was suffered by all regions, but some regions experienced more tension than others. McMahon et al. (1995) suggest that the most important element of strain experience determining a more LS collagen orientation is the relative magnitude of tensile loading. Purer tensile loading seems to protect against the requirement to replace compact bone. More active replacement cycling is seen as the consequence of compression or the overlay of compression on tension experience. It is clear that not all osteons are mechanically equal, and we now have a better basis for understanding differences in microstructural layup.

Excluding surface parallel, circumferential lamellae at periosteal and endosteal surfaces, compact bone in man is secondary osteonal in the bulk. In the young and in other mammals, however, we may not safely assume that this is the case. In some equine radii, primary osteonal systems may persist lifelong throughout most of a midshaft tension cortex.

V. SPONGY BONE

It is important that the variations in architectural structure that occur in young, adult, and aged bones of different mammals are taken into account in assessing normal, experimental, or pathological material.

A. Metaphyseal Trabeculation in Laboratory Rodents

Cartilage bones are preceded by and initially form within a cartilage model. A hypertrophic zone develops in the middle of the model and the central cells hypertrophy so that longitudinal growth is favored. The oldest, most central, hypertrophied cartilage calcifies and is invaded by periosteal vascular tissue, which partly destroys the calcified cartilage, before osteoblasts differentiate to lay down bone to encase the remnants of calcified cartilage. This endochondral ossification leads to the formation of initial fine "trabeculae" (partition walls) of bone containing calcified cartilage. The 3D organization of these is extremely difficult to interpret from the usual 2D view given by longitudinal sections through the growth plate, yet it may be crucial to obtaining an understanding of the morphological disruption of normal structure brought about by the effects of experimental genetic manipulations manifesting in bones (Hayman et al., 1996). The metaphyseal trabeculae can have many 3D forms and are normally subject to rapid and highly significant architectural changes soon after their formation: generally, we would anticipate the initial structure to be a honeycomb structure with thick cell walls. Metaphyseal regions in still-growing animals have also been adopted by many researchers as a convenient model system in which to investigate hormonal or therapeutic agents which might affect trabecular bone turnover. For example, rats or mice are ovariectomized prior to an overlay treatment which may or may not influence the outcome of the estrogen withdrawal, attention being paid to proximal femoral or tibial metaphyseal regions or to lumbar vertebrae. They have also been used in immobilization osteoporosis experimentation.

Rat lumbar vertebrae have also been used in compressive strength testing correlated with histomorphometry (Ejersted et al., 1995). The rat bones have an anatomy which differs importantly from that of man at any age. The rat vertebral body has a substantial cortex and a denser distribution of trabeculae towards its ends and the sides of the central part of the shaft. Thus both appearances and measurements will be strongly affected by the exact location of a histological section.

Rodent caudal vertebrae may be advantageously substituted for lumbar vertebrae or limb bones, since these are long bones with two growth plates. They are straighter and more easily oriented, thus avoiding difficulties in reproducibility in centration and alignment for sectioning. Rat caudal vertebrae have been adopted for use in both *in vitro* and *in vivo* strain-related growth and turnover studies (Lean et al., 1995). They may also be used as a means of biopsy via tail amputation. These bones, however, have a remarkable structure. Very often, trabeculae are largely confined to the ends of the otherwise nearly empty medullary cavity where they fuse to form a perforated plate: this plate may give rise to single or double longitudinal rods which join to the facing plate at the other end of the shaft (Figure 1).

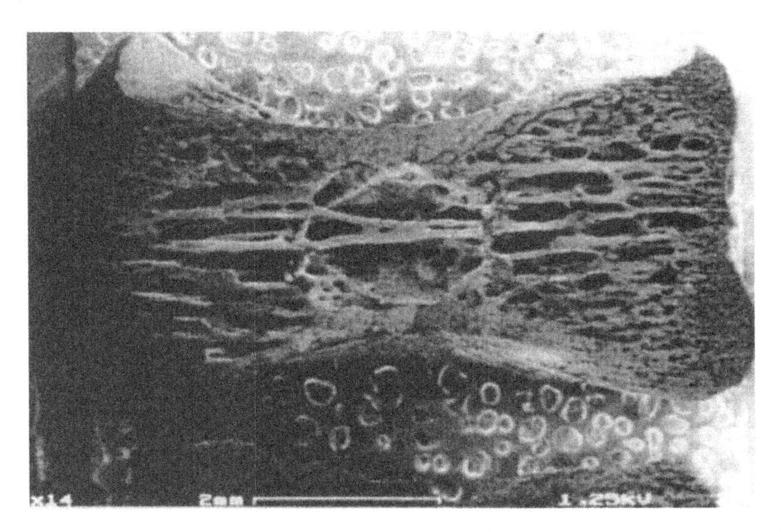

Figure 1. Rat caudal vertebra, cut longitudinally, made anorganic by treatment with NaOCl (which has removed epiphyses) and gold sputter-coated. Note one double-rod trabeculum passing central medullary cavity. 1.25 kV SE SEM.

B. Architectural Changes in the Aging of Human Lumbar Vertebral Body Trabecular Bone

The cancellous bone in the vertebral bodies of subadult humans is spongy: it is a cellular foamlike structure, but with more mass between the foam cells (here meaning the marrow spaces within the bony continuum) than we would normally conceive of when using the term foam, with more of the solid material packed into the linear junctures between three or more facing cells (and thereby often forming short rodlike elements), and with the planar walls between pairs of cells largely perforated or missing. It may be difficult to recognize the plane of section or to determine which is the longitudinal, load-bearing direction in sagittal or coronal sections (Figure 2).

The mature adult vertebral body trabecular bone architecture is very unlike the subadult in these respects (Jayasinghe et al., 1993, 1994). Most importantly, it is always possible for a practised observer to recognize, unambiguously, the load bearing axis in the mature adult (Figure 3). Coronal, sagittal, and longitudinal views in other rotations still appear similar to each other, but quite dissimilar to transversely cut tissue. The bone resembles a honeycomb with less than perfect walls between cells. Thus vertebral body spongiosa in younger, mature adults is evidently porous when prepared as cleaned, macerated thick transverse sections. In either the superior

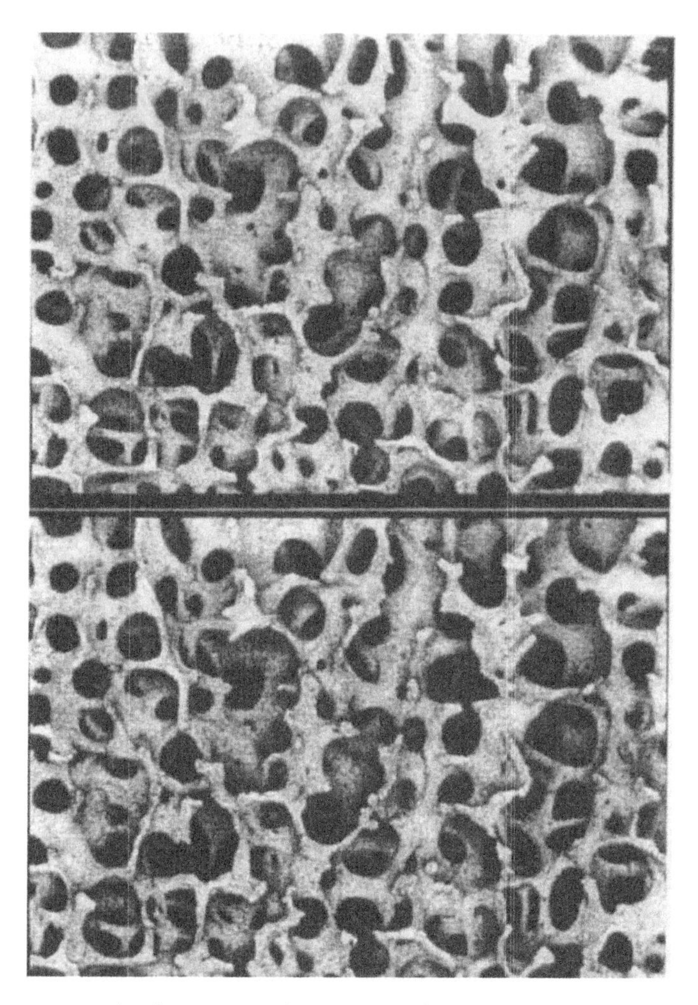

Figure 2. Human lumbar vertebral body cut longitudinally: 4,000 year BP archeological sample (loaned by Dr. M. Kneissel) was determined as female and aged about 20 years from standard anthropological methods. BSE image at 1.25 kV recorded with micro channel plate detector using uncoated sample. Stereo-pair image, tilt angle difference = 10°. Longer field dimension (height) = 8 mm.

or inferior portions, which are more finely divided, or in the more grossly partitioned central section of the body, one will be able to see many large tubular marrow spaces with linear penetration through a 4 mm thick macerated section (Jayasinghe et al., 1994). Longitudinal sections of 4mm will, however, almost entirely obstruct the direct passage of a beam of light or electrons.

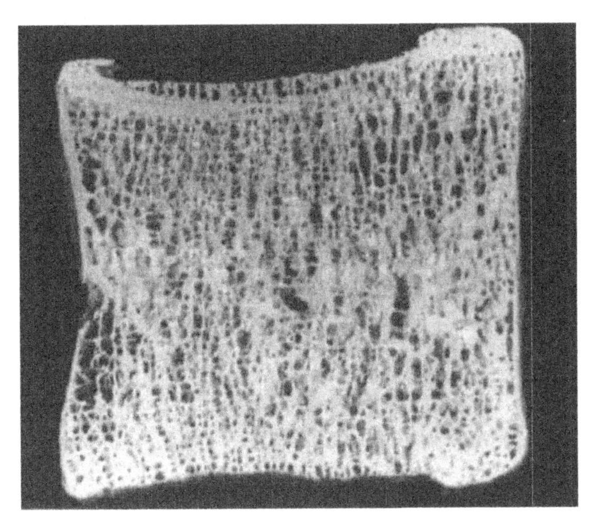

Figure 3. Human fourth lumbar vertebral body: 35-year-old male, photo of 4 mm thick sagittal slice against black background (see mm scale in Figure 4).

The cancellous bone in an aging individual is a latticework of rods, and the original curved plates (the walls of the cells in the honeycomb) are far less extensive. These architectural changes are associated with loss of tissue occupancy by bone, but the increased 3D extent of individual, line-of-sight, pore channels is greater in proportion than the net loss of bone tissue per unit volume (in agreement with findings from the 2D marrow space "star volume" method; Vesterby, 1990). Thus in bone from aged individuals, including those with relatively high bone volume fraction for their age, substantial direct light transmission can be seen through clearings in any direction of view through 4 mm of macerated tissue (Figure 4).

C. Resorption-Repair Coupling is not Site Specific

Current opinion regarding the sequence of events leading to the development of bone porosis is strongly influenced by the idea of the BMU and the activation, resorption, and coupled-formation hypothesis (e.g., Eriksen et al., 1994), which subscribes to the view that porosis is the end-result of a large number of cycles in which a standard packet of resorption is replaced by an inadequate packet of repair: the maintenance of peak bone mass depends upon the packets of bone lost in resorption being balanced by packets of bone added in formative activity. The corresponding ubiquitous diagrams

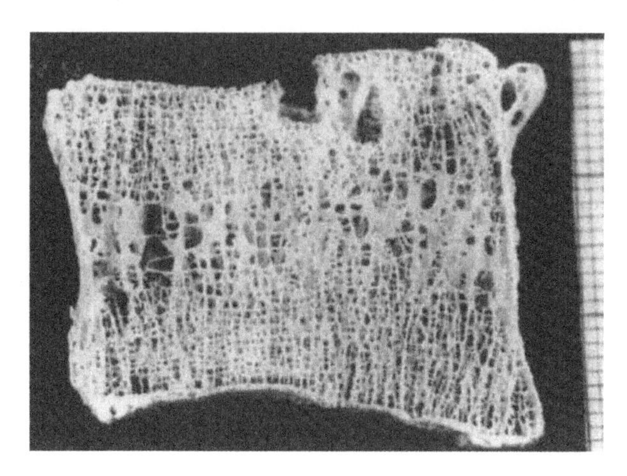

Figure 4. Human fourth lumbar vertebral body: 89-year-old female, photo of 4 mm thick sagittal slice against black background (scale in mm).

in scientific papers and osteoporosis-related advertising literature do a conspicuous disservice to the field in implying site specificity—that new bone packets must lodge only in prior resorption fields in trabecular bone. Even though diagrammatically satisfying, it is not necessary that the new bone exactly replaces the space excavated, only that the sum of activities results in an equally, or at least sufficiently, strong structure.

The sequence of developmental age changes in vertebral spongiosa architecture and apparent porosity occurs in all bones, although the exact details vary between bones and individuals. These modeling changes can only be explained through bone being taken away by osteoclastic resorption and being added back by osteoblasts in different locations. The evidence of bone microarchitectural reorganization alone is sufficient to suggest that spatial coupling is not exact, but this conclusion is confirmed by direct SEM studies of the surface activity condition of surfaces in human cancellous bone. The SEM examination of trabecular surfaces allows us to survey a very high proportion of surfaces more rapidly and efficiently than could be achieved with LM. It shows that new bone packet deposition occurs at sites of prior resorption, but often elsewhere, and that many resorbed sites remain unrepaired over a considerable time. A striking feature, impossible to glean from either 2D histology or popular diagrams is the irregularity and extensiveness of these areas of activity. This conclusion leads to a simpler explanation for osteopenic change than that currently favored and suggests a shift in emphasis of attempts to control bone remodeling. But first we should explain the nature of the evidence which we find so compelling.

D. Identification of Bone Surface States by SEM

Resorbing or Resorbed

Resorbing surfaces are identified in routine histology by the presence of osteoclasts adjacent to a scalloped surface which are resorbed where the scalloping is due to prior activity; it cannot be known how recently such features were vacated by the osteoclasts. In SEM study of samples with cells still present, the lining of osteoclastic resorption pits usually has a thin fringe of demineralized matrix at the time that the osteoclast moves from the lacuna. It is not clear whether this fringe may self-destruct, or whether other cells help to remove it. Resorbed surfaces show released osteocytes, and perhaps macrophages which may be involved in the degradation of the residual demineralized matrix fringe. It is most likely that resorbed surfaces may remain without a special cell covering for some time after resorption; during this time, they may accumulate the material corresponding to (at least a part of) the cement line matrix (McKee and Nanci, 1996). Finally, and usually, according to most accounts, they are then covered with osteoblasts again.

SEM specimens from which both cells and osteoid have been removed (and these are the simplest to prepare) do not allow of the distinction between active and prior resorption. They do, however, provide the best chance for estimating the true total extent of total resorptive activity on an areal basis. They also indicate qualitative differences in the nature of resorption; for example, whether it appears to be aggressive (focal excessive endosteal resorption; Arnold, 1970), with many punctate, deep pits with knife edged boundaries due to intense activity in depth, or whether the process has been a light surface skimming, as in sculpting surfaces to a shallow depth.

Forming Surfaces and Recently Deposited Bone Packets

Forming surfaces, as seen in routine histology, are assumed to be present where osteoid underlies plump (cuboidal or low columnar) osteoblasts. In routine (secondary electron mode, topographic-imaging) SEM, the simplest way to recognize the osteoid-bone junction (mineralizing front), is via the partially discontinuous mineral particles clusters which can be uncovered by dissolving the organic matrix. Proof that osteoid was dissolved can be obtained by examining the same field before and after such treatment (with, for example, an NaOCl solution, or plasma ashing; Boyde, 1972, 1984a).

Recently deposited bone packets can be located by their lower electron backscattering than surrounding surface bone (Figure 5; Boyde and Jones,

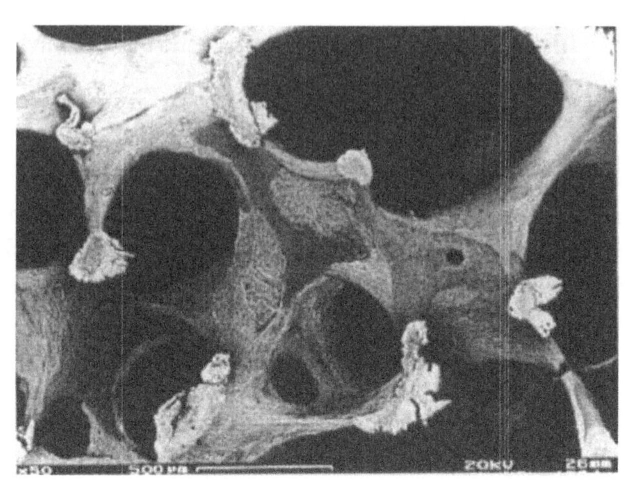

Figure 5. Human fourth lumbar vertebral body: 61-year-old male, 20 kV BSE SEM image showing (recently formed) low density surface packets.

1996). Evidence that an electron-dense mineralizing front lies deep to, rather than at, the intact low-density osteoid matrix surface can be obtained by comparing backscattered electron images at contrasting accelerating potentials.

Resting Surfaces, Fully Mineralized Collagen, and no Osteoid

At a resting surface, matrix formation has ceased, and mineralization catches up to the matrix surface. In routine histology, such surfaces are identified as smooth and covered by thin lining cells. In routine SEM, such a surface appears unaltered after attempts to dissolve or oxidize away the osteoid. The surface collagen fiber bundles are so completely mineralized that their continuity provides the means of recognition after the simplest one-step preparative procedure—making the sample (at least superficially) anorganic.

There remains some uncertainty as to whether a very thin layer of unmineralized material (osteoid) persists at the cell-matrix interface. Chambers (1988) speculates that such a layer is always present to form a barrier to osteoclastic recognition of mineralized bone matrix, and that osteoblasts play a resorptive role in removing it to prepare a surface for recognition. Using SEM, it is possible to find areas of anorganic matrix surface in which a coating of mineralized, collagen-free ground substance obscures the collagen fibers at the surface. These have been described as prolonged-resting surfaces and assumed to be due to the progression of mineralization beyond the collagen of the last layer of the matrix as a maturation phenomenon. As with the finding of matrix surfaces which are essentially unaltered by deproteinization or oxidation, this speaks against the notion of a significant layer of non-mineralized collagen persisting on all bone surfaces. It cannot be surely known when the organic matrix of any non-collagenous material formed. Firstly, it could be an osteopontin- and/or bone sialoprotein-rich (laminae limitantes) matrix which can form on any free surface in bone (McKee and Nanci, 1996). Secondly, it might reflect a rapid influx of mineral into bone as a response to a changed functional status of the osteoblasts due to incompetence in the ionic influx control mechanism of the cell sheet. For example, excess PTH was shown to cause mineralization of the most superficial osteoid in a short-term rat experiment (Jones, 1973). Thirdly, the phenomenon may occur regularly, but represent a pathological age change.

E. Bone Quality and Bad Joins?

Arrested Mineralization Fronts and their Entrapment

The existence of surface osteoid implies the failure to calcify. A high incidence of abnormal, arrested-calcification, mineral fronts is found in the vertebral spongiosa of aged, and particularly in aged osteoporotic, individuals. Trapped, poorly mineralized matrix may be sandwiched between such arrested mineralization fronts and the matrix in new bone packets which cover them without a prior resorption step (Jayasinghe et al., 1993). The occurrence of defective joins of micro-callus to old lamellar bone in repairing crush fractures can be demonstrated in BSE-SEM images of vertebral cancellous bone (Boyde et al., 1992).

Reduction in Stiffness and Strength Following Hydrogen Peroxide Treatment

It would be very difficult to use an imaging method to quantify faults which occupy such a tiny fraction of the total volume of bone, which itself is a small part of the bone organ. We conjectured that the existence of potentially weaker features would be detectable by removing them. To test this idea, we prepared 3 mm thick parasagittal sections of L4 bodies and trimmed them to constant dimensions, prior to applying a 3-line bending test to derive an empirical value for the stiffness of each sample at loads well below any which might have induced damage. The same samples were then treated with hydrogen peroxide, which removes non-mineralized matrix but leaves mineralized regions intact. The samples were then retested along the same loading lines, and the reduction in stiffness noted.

The results of this experiment indicate that planar mineralization faults, undetectable by routine histomorphometric procedures, may be a part of the whole complex of changes which render bone incapable of withstanding loads which would otherwise not be dangerous.

F. Is Bone Put Back on Unloaded Struts?

It is commonly believed that a structural element which is interrupted by resorption will not be repaired; indeed, it makes little teleological sense that osteoblastic activity should be stimulated on a free-ended rod which resulted from the osteoclastic partitioning of a strut. Mosekilde (1990, 1993) reported that she was unable to find such events. We have found that they do occur, albeit rarely (Jayasinghe et al., 1993).

Another means by which bone is put back on unloaded struts concerns the involvement of microcallus in crush fracture repair. Hahn et al. (1995) and our unpublished SEM findings show that the reduced number of vertical trabeculae often show a compensatory increase in girth which is initially achieved through the deposition of microcallus.

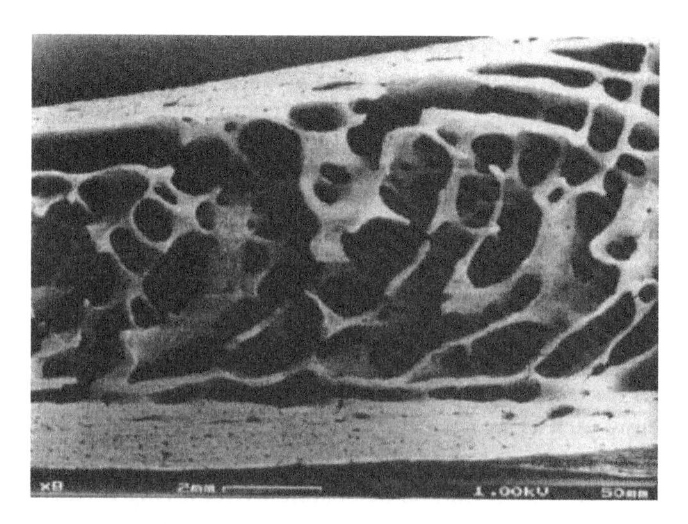

Figure 6. Vertical section of iliac crest, 42y female, gold-palladium sputter coated. Thicker cortex (nearest labelling) is external. Wider end (right) is superior. Region show is that from which trephine core biopsies are normally taken. 1kV SE SEM.

G. The Iliac Crest

The iliac crest is the most studied bone in man, because it alone is routinely biopsied for histomorphometrical studies (Wakamatsu and Sissons, 1969). This bone has a thick external cortex and a thinner internal cortex sandwiching several millimeters of cancellous tissue. The mature adult architecture of the latter consists of extensive intersecting perforated plates. These are closer to each other and more nearly parallel to the top of the crest in the tissue closest to the crest. The plates arc more steeply downwards and incline either externally or internally in the region, 1 to 3 cm below the crest, from which the trephine core biopsy sample is taken in the live patient (Figure 6). The trabeculation is thus highly anisotropic, and radically different from that present in lumbar vertebral bodies. It cannot be properly comprehended from single section views. We do not have sufficient 3D data from age series on architectural change in this bone. Studies of remodeling patch distribution in adult material reveal substantial spatially uncoupled packet formation, which indicates long-term change. The cortex thins by endocortical trabecularization, and cancellous porosity also increases with conversion of plates to rods (cf. Figures 6 and 7).

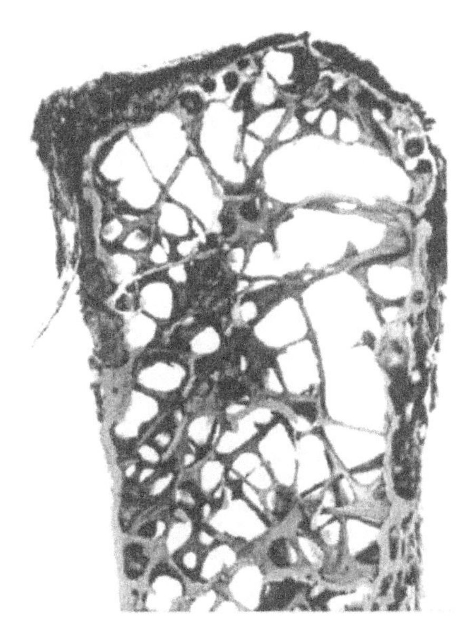

Figure 7. 4-mm thick vertical section of iliac crest, 91y male, gold sputter coated, lying on lead background which gives high BSE signal and appears white in this 20-kV BSE-SEM image. External cortex is below or left. Wider end (top) is superior edge of crest. Lower part of field is that from which trephine core biopsies are normally taken. Widest bone dimension = 12 mm just below crest. Preparation by V.J. Kingsmill.

H. The Femoral Neck

Fracture of the femoral neck as a consequence of osteoporosis is catastrophic. The strength of this unit is held to relate primarily to its gross anatomy, the mass and thickness of its cortices, as well as to the highly organized trabecular component. This again consists mainly of curved, arched, intersecting plates and cannot be understood from single histological images or radiographic projections, yet elements in such images are often interpreted with confidence as tension and compression trabeculae. It was the anteroposterior projection of the trabecular anatomy in the proximal femur which

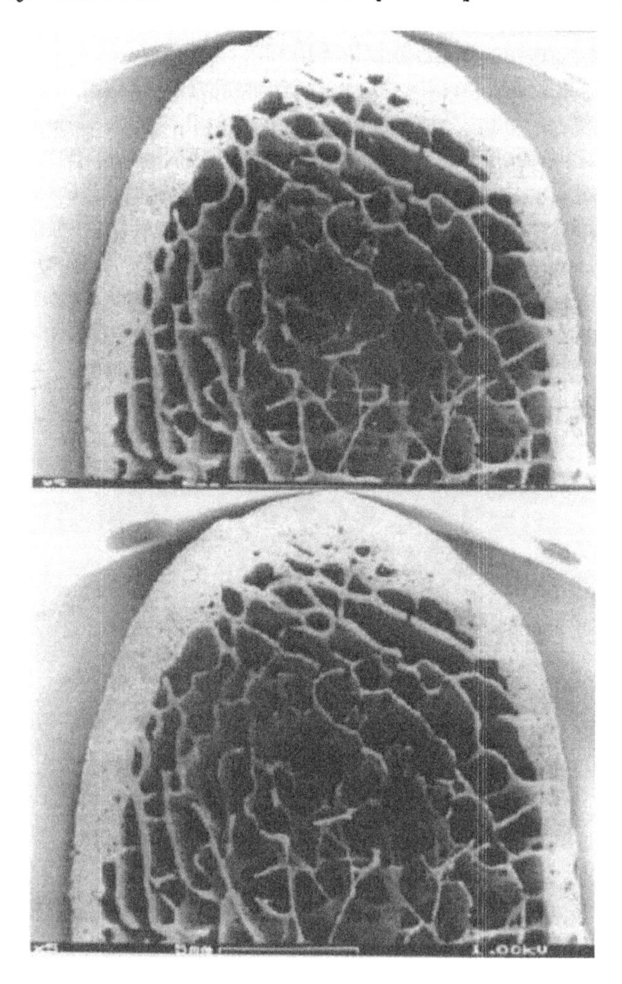

Figure 8. Cross-cut femoral neck, 45y male, Au-Pd sputter coated, 1kV SE SEM. Stereo-pair.

most excited the interest of Wolff (1892) and his predecessors. However, several principal structural axes can be seen in the cross section of the neck (Figure 8). Age changes are known to include the trabecularization of the endocortex and the extensive perforation of the plates to leave more rodlike residues.

I. The Distal Femur

Distal femoral trabecular bone is of considerable interest from both structural and functional aspects. The knee is a hinge joint, and the directions of principal stress and strain are concentrated in the sagittal plane. Three major structural axes are easily conceived and defined. Trabebeculae are plates which are very extensive in the antero-posterior vertical direction, with less significant cross connections (Koch, 1917; Murray, 1936). Such XYZ anisotropy is exceptionally well developed in the equine cannon bone (distal third metacarpal; Boyde et al., 1996).

VI. BETTER METHODS FOR INVESTIGATING THE STRUCTURE OF TRABECULAR BONE

A. Complete 3D Data Sets from μCT, MRI, or Serial Sectioning

It should be clear from the content of the previous section that traditional microscopic methods are not adequate for helping us to understand such complex 3D structures. It is possible to generate complete 3D data sets for meaningful volumes of trabecular bone, which specify whether an XYZ address in a binary voxel array is in or out of bone, and at a satisfactory resolution (i.e., roughly matching that of the unaided eye) with both magnetic resonance imaging (MRI) and μ computed tomography (μCT) imaging or by serial sectioning. Such data sets can be reprocessed to generate any number of views and any angle of view through the structure. They are of great assistance in both perception and in the derivation of structure related measurements, the latter either following those which are routinely made with 2D sections, or being based on new procedures of 3D image and structure analysis.

The μCT data can tell us something about the composition of the bone in the voxel and at a resolution of 5 to 10 μm (which merits the name microscopy), but only at a great cost of time in data acquisition. The output of such elaborate and costly data sets usually involves the production of discrete views or sets of views of the structure, but these can be obtained with far greater economy and efficiency using established and common imaging equipment, and with the advantage that the obtainable resolution extends beyond the range required to visualize the whole structure. Both LM and SEM methods, for example, allow us to discriminate important detail in context and can be exploited dynamically for a better appreciation of structural architecture. These are affordable laboratory methods.

B. Tilting Beams or Samples: Stereo and Stereophotogrammetry

A particular advantage of SEM in examining trabecular bone structures is that the site of most of the action is the surface of the bone and the surface is what we see in the SEM. The SEM works at extremely small aperture, and, relative to the magnification range conventionally employed, it therefore has a large depth of field. Because of the depth of focus, it is not possible to make a correct interpretation of the morphology of a surface recorded in an SEM, any more than it is possible with any other deep field optical system in which only one projection of the object is available. Stereoscopic means of studying the potentially 3D image in an SEM have always been available but are little employed either in practice or in textbooks. However, the reader needs to remember the loss which results (view Figure 2 or 8 in stereo; Boyde, 1972, 1973).

Real-time stereo SEM works by tilting the electron beam rapidly between two principal vantage points, and a 3D video display is generated by standard means such as the anaglyph method employing red versus blue-green filters (Boyde et al., 1972). It is an excellent means of 3D imaging in the fieldwidth range from 5 mm to 50 μm (nominal magnifications of 20X to 2000X), and a delight for the operator who can change the viewpoint by additionally rotating and tilting the sample at the SEM. The output is, however, not conveniently portable.

At very low nominal magnifications of from unity to several times, SEM usually provides less depth of field than conventional light optical systems which are more convenient in providing an overview which is easily interpreted from ordinary, learned experience. Cleaned (macerated) bone samples are conveniently studied with a stereo microscope; there are modern 3D viewing systems which are more appealing and/or handier, but they provide no more information. Stereo only provides a 3D concept of what can be seen from one mean viewing direction and the information is dedicated to the single observer.

However, what can be seen in both members of a stereo pair of views can be measured in 3D if the geometry is known. High quality photographic stereo recordings can be made with 35 mm photography at close to unity magnification at the photo negative scale (Jayasinghe et al., 1994). Shift or tilt can be used to generate the stereoscopic parallax. Simple stereophotogrammetry can be used to measure the true lengths of rodlike trabeculae to provide a measure of porosity in macerated, clean, dry cancellous bone (Jayasinghe, 1991; Kneissel et al., 1994). We currently use a software package developed by our colleague P.G.T. Howell in combination with a Ross Instruments Ltd SFS3 stereocomparator for this purpose.

C. Deep Field Microscopy with Rotating Samples: Continuous Motion Parallax

To study structures as complex as trabecular bone, it is apt to have as many projections as possible and often useful to make these available to multiple observers. We developed a method in which a deep field optical system was used to store the image of the bone sample obtained using a color video camera, moving the object at a constant velocity while recording. We cut standard 4 mm gauge, square cross-section beams which were rotated coaxially with the shaft of a variable speed DC motor; the structure is seen by replaying the video tape. Rotary motion parallax alone gives a powerful 3D effect, but this can be enhanced by placing a suitable neutral density filter in front of one eye, with the effect of creating a time delay in registering that projection on the brain. The delayed eye sees an earlier video frame which is also genuinely a view from a different center (Boyde et al., 1990). The method exploits all our learned experience in visual location. It enables us to interpret and understand bone structure in a unique fashion. It shows clearly the continuum, as well as where there are real discontinuities like free ends. It shows how limited we are with linguistic terms for complex 3D arrangements, and just how dull rod, plate, and node are in the context of structures, which however disarranged by disease, are always beautiful to behold. The method is therefore an effective and efficient approach to study the differences in 3D structure between young, old, and frankly porotic conditions in lumbar vertebral bodies. It can easily be made quantitative for connectivity and feature density. Porosity can be indexed for standard gauge samples by measuring the mean light transmission during the 360° rotation cycle.

Kindred image sets can be obtained with the SEM, single images being first obtained before their sequential replay at video rates. SEM specimen stages are not suited to the generation of large numbers of separate positions with small, precisely separated angular differences. We therefore mounted our samples coaxial with the shaft of a stepping motor, itself mounted on the SEM stage. Similarly, we record x-ray images of stepper motor mounted trabecular bone samples using a digital x-ray imaging device designed for the intra-oral use: a 256 grey level, 768*512 pixel image is stored for each rotation, and we use 96 or 192 images in a 360° cycle (Boyde et al., 1996).

D. Internal Casting

Looking at dry spongy bone with light or electrons tells us whether there is bone in that line of sight. It will not tell us whether there are holes inside the bone, and SEM or deep field optical microscopy of dry samples may be misleading in failing to reveal the presence of blood vessel (Haversian) canal spaces within rod or plate shaped trabeculae (Boyde et al., 1996). Spaces within either compact or spongy bone can be studied by making casts with, for example, poly-methyl-methacrylate (PMMA): the bone is dissolved. Conditions can be varied to replicate, or not, the lacunar and canalicular system, or to dissociate the casts of such small elements and to retain only large features (Figure 9). The latter are particularly valuable for appreciating the extent and the complexity of canals in compact bone, and are excellent at showing the gradation of properties in the border zone with trabecular bone, the scourge of the histomorphometrist. It is possible to inject and cast the entire blood vascular system within bones of small animals, but this is difficult to do and retain much of the context of the vessels within the tissues.

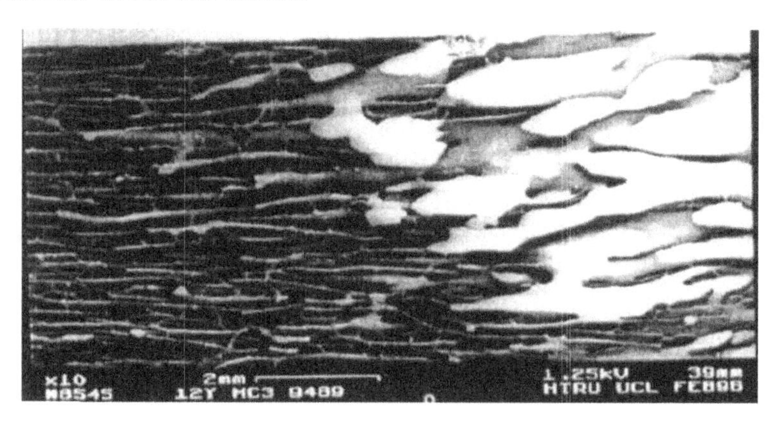

Figure 9. PMMA cast of marrow and Haversian canal space in 12y horse third metacarpal (lateral, distal): solid PMMA shows where no bone was present. Left of field shows compact bone, right shows cancellous region. Note presence of fine casts of canals which were in trabeculae (e.g., bottom right hand corner). Carbon and gold coated.

E. Frequency Domain Analysis of Trabecular Bone Structure

A picture may be worth a thousand words, but many now think that a number is worth a thousand pictures. Much effort is devoted to structure and texture analysis of bone images. It is clearly a simple matter to determine the fraction of tissue which is bone, and this is often considered to be the most important parameter to determine. It is obvious that bone is anisotropic, but how to enumerate this property is not so obvious. With a 2D section, one pragmatic approach is to measure intercept frequency as a function of rotation of the intercept direction (Biewener et al., 1996).

Fast Fourier Transform (2DFFT) methods allow us to quantify the numerical density of image elements as a function of direction by measuring their spatial frequency (Oxnard, 1980, 1986), although the question is open as to the extent to which the image is related in a logical way to the distribution of structural elements in the bone. We have used x-ray images of 3 mm thick sections captured using a fine focus x-ray tube, a long focus to film distance, and fine grained industrial x-ray film. The processed images are digitized to 16 Bits using a cooled CCD camera and a Noran TN8502 image analyzing system. Using a standard commercial software package (Noran, Middleton WI, USA), the power spectrum, equivalent to an optical diffraction pattern of the film, is derived. We processed such images to produce contour diagrams showing the preferred orientation of features having particular spatial frequencies. These diagrams measure the changes which can be detected and described by eye and they document the strength of recurrence of principal spatial frequencies (separation distances) of the main struts in cancellous bone (Figures 10 and 11).

The original image can be reconstructed from the power spectrum in a reverse 2DFFT process, but it is also possible to remove particular frequencies and orientations selectively. This exercise is instructive in demonstrating how obliquely oriented elements are selectively removed in the architectural reorganization occurring in bone aging (Jayasinghe and Boyde, 1990; Jayasinghe, 1991).

F. TBV from Digital Processing of X-Rays of Parallel Slices

High quality digitized radiographs can be further exploited to study variations in the average bone density within local areas. The image is first segmented by repetitively halving the distance between boundaries.

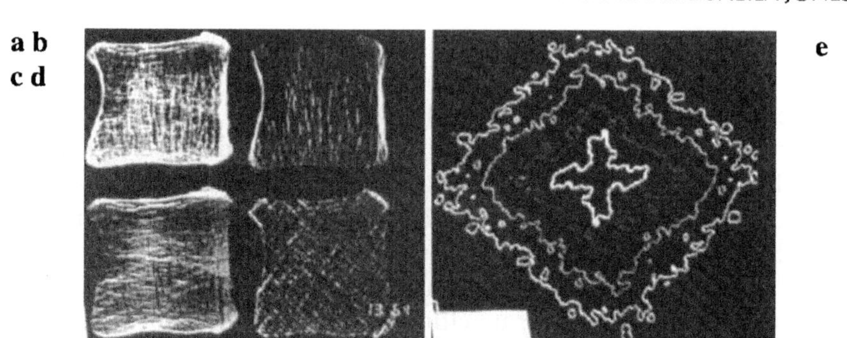

Figure 10. Image analysis and processing in the frequency domain.
a) Digitized x-ray image of 3-mm thick section of L4 body of 35-year-old male
e) 2D FFT: contoured power spectrum of image shown at top left in **a**, with highest intensities, meaning maximum significance, at the center. Highest frequency [least distance in the image] components lie toward the center of the pattern. Vertical trabecular elements are portrayed by the horizontal arm of the central crosslike contour and vice versa, vertical elements by the horizontal.
b) Reverse transformation of power spectrum using binary filter that excludes components in the horizontal direction of the image, leaving the vertical (axial) trabecular features.
c) Reverse transformation to exclude components in the vertical direction of the image, leaving the horizontal (transverse) trabecular features.
d) Reverse transformation to exclude both horizontal and vertical components in the direction of the image, leaving the 45° and 135° oblique trabecular features.

An example to derive 64 roughly equal areas in the x-ray image of a non-osteoporotic 89-year-female L4 body, is shown in Figure 12, with mean values for each of the approximately 3*3*4 mm regions shown in the caption; these indicate the local mean bone (mineral) content on a scale of zero (no bone) to 255. Since the section is plane parallel, the numbers are an index of bone mineral density. If we assume a limited distribution of mineralization densities (Engfeldt and Hjerpe, 1974) and overlook the non-linearities in x-ray photon scattering and absorption and in the use of photographic emulsions for counting, they are also closely related to bone volume fraction or TBV. A very large difference is found between the extremes of the 64 volumes in this example, but each of these is a far larger statistical sample than is ever employed in bone histomorphometry. Thus this method illustrates the extent to which the bone volume fraction (TBV, BMC, BMD) changes globally within the bone, and teaches us how to critically assess values gained from a few sections from small regions.

a b
c d

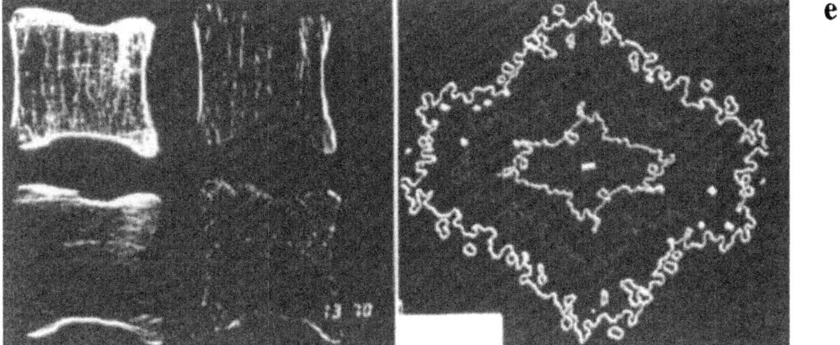

e

Figure 11. **a**) Digitized x-ray image of 3-mm thick section of L4 body of 89-year-old female.
e) 2D FFT: contoured power spectrum of image shown at top left of **c**. Note the dominance of the vertical trabecular elements demonstrated by the horizontal central contours.
b) Reverse transformation of power spectrum using a binary filter that excludes components in the horizontal direction of the image, leaving the vertical (axial) trabecular features.
c) Reverse transformation to exclude components in the vertical direction of the image, leaving the horizontal (transverse) trabecular features.
d) Reverse transformation to exclude both horizontal and vertical components in the direction of the image, leaving the 45° and 135° oblique trabecular features.

G. Giving a Correct Impression of a Complicated 3D Structure From a Single Projection: Advances in SEM Technique

The customary mode for the morphological study of bone in an SEM employs secondary electron (SE) imaging. SE detection is extremely efficient, but the signal intensities do not have a simple physical meaning and there can be severe problems in image interpretation. Because of their low energies, SE are strongly influenced by sub-surface potentials (usually of a few tens or hundreds of volts) induced by the input electron beam, giving rise to a range of undesirable image artefacts. This is minimized by the application of a continuous electrically conductive coating, usually connected to ground potential, but it is a problem to apply good coatings to samples of osteoporotic cancellous bone which are large enough to include a representative survey of tissue undamaged in preparation (our typical sample would be a 2.4 to 4 mm thick plane parallel section of a human lumbar vertebral body). Above all, normal SE images fail to disclose the degree of porosity and relative distance of features.

Leaving the sample unmounted permits coating from both sides, and also allows it to be placed either over a deep black hole from which no SE signal

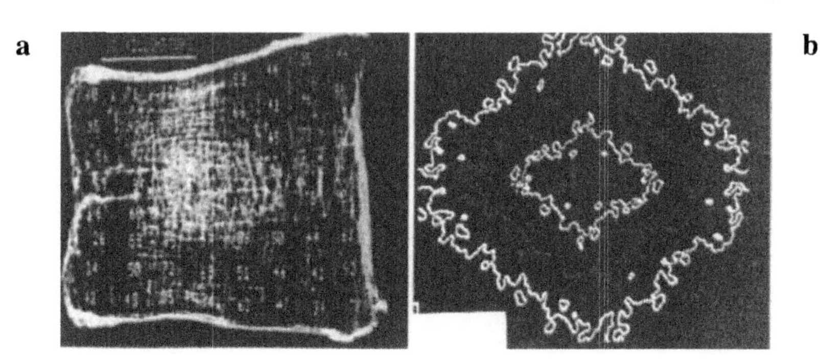

Figure 12. (**a**) Nonosteoporotic 89-year-old female L4 body. Digitized version of high quality x-ray of 3 mm section. The image has been divided into 64 regions (approximately 3*3*4 mm). The local mean fractional bone volume on a scale of zero (no bone) to 255 (solid cortical tissue) is shown in the matrix below. See text for full explanation of procedure. **b**) Contoured power spectrum of same image. The dominance of vertical trabecular elements is shown by the horizontal central contours.

60	71	116	128	53	44	35	46
35	92	131	123	66	41	46	58
56	92	158	143	98	45	73	84
69	84	168	144	122	83	83	82
67	60	150	152	104	92	78	90
26	58	95	97	86	50	44	63
34	50	72	68	51	46	41	52
42	48	85	96	62	47	34	78

is returned, or, conversely over a suitably inclined, high atomic number metal surface at a small distance from the sample from which a high SE signal level is returned where the beam passes directly through the specimen (Figure 8). Both methods can give reasonable discrimination of direct, line of sight pores in digital imaging (Boyde and Jones, 1996; Boyde et al., 1996).

However, if SE are rejected, then so are most charging problems and one gains the advantage of the directionality of signal which is a feature of fast electrons. We have used multiple detectors for forward-scattered (FSE) and transmitted primary (PE) as well as backscattered fast electrons (BSE) in imaging osteoporotic cancellous bone. The signals can be recorded separately, with addition and subtraction off-line by digital image processing, or mixed on-line by using an analog video summing unit (Figures 13 and 14).

Use of a PE detector gives a back lighting effect, making true line of sight pores appear peak white (or black if negated) and measurable (Figure 14). FSE detectors provide sharp images of the edges of trabeculae, making them distinct even where they cross over each other, and excellent topographic detail for steeply sloped surfaces which would normally be beyond imaging (Figure 14). Subtracting opposing BSE signals creates topographic

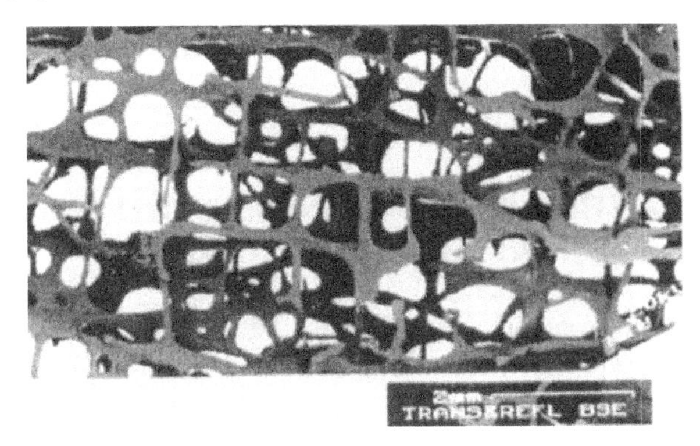

Figure 13. Three mm thick section of human fourth lumbar vertebral body: osteoporotic 89-year-old female, 20 kV BSE and primary transmitted electron image showing line of sight pore space as white.

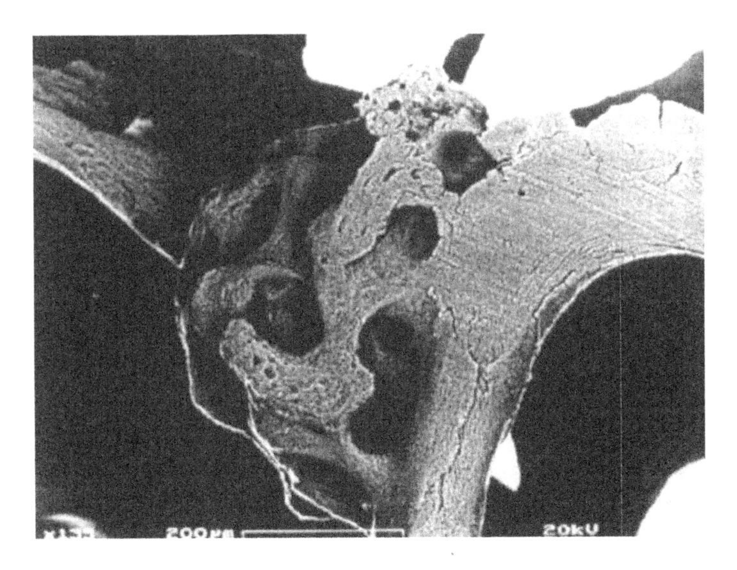

Figure 14. Human fourth lumbar vertebral body: osteoporotic 89-year-old female. Image combines 20 kV BSE, forward scattered (giving high signal for sloped edges) and primary transmitted electrons (showing line of sight pore space as white). Field shows microcallus formation.

shadow contrast. Summing the signal from all the segments of an overhead BSE detector at a longer working distance gives a contrast mechanism where intensity depends mainly upon the distance of the surface facet from the detector, and pseudo-color coding of the intensity gives a reasonable topographic map (Boyde and Jones, 1996). Combining these various possibilities leads to the removal of ambiguities in imaging porotic bone structure.

VII. QUALITATIVE VARIATIONS IN MINERALIZATION PATTERN AND DEGREE

Bone forms a spectrum of tissue types, varying in constitution, organization of matrix and cellularity, with woven bone and lamellar bone at opposite ends of the tissue spectrum but not necessarily confined to one age bracket.

A. Woven Bone

In woven bone, matrix vesicles have been found to act as centers from which mineralization spreads into the surrounding osteoid matrix. These separate centers are seen as minicalcospherites in anorganic and partially anorganic SEM preparations (the early, incomplete osteoid digestion stages show beautifully both fused and early calcospherites). The progression of mineralization is rapid, with invasion of water space in both collagen and extracollagenous matrix domains. The high degree of mineralization achieved reflects the greater water space in the immature bone matrix. There is no maturation phenomenon. The turnover of woven bone may be very rapid; for example, the avian medullary bone associated with the eggshell production cycle is literally here today and gone tomorrow.

Woven bone reaches a higher initial and final level of mineralization than lamellar bone, but lower than either calcified hyaline cartilage or fibrocartilage. However, micro-callus, the type of woven bone deposited upon old lamellar bone trabeculae in crush fractured vertebral bodies, is poorly mineralized in comparison to lamellar bone (Figure 15; Boyde and Jones, 1983a; Reid and Boyde, 1987; Boyde et al., 1992).

B. Lamellar Bone

In lamellar bone, the first phase of mineralization is collagen centered, and only roughly 70% of the final extent of mineralization is achieved during early stages: maturation, or topping up (some) of the remaining water-filled space is a slow process which may continue over months, and at a declining rate over years. It commences as soon as initial mineralization has occurred, but is much slower than the rate of addition of new matrix to complete a packet. Finally, after a new packet is completed, the process continues to

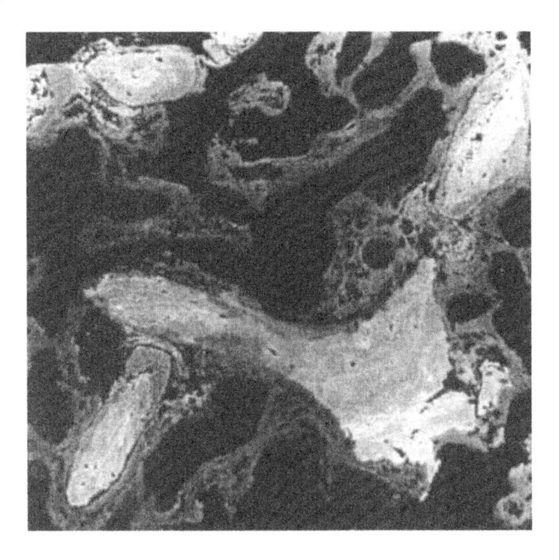

Figure 15. Human second lumbar vertebral body, 74-year-old male, PMMA embedded, micromilled, carbon coated. 20 kV BSE showing repair in crush fracture zone. Densest phase present is old lamellar bone, least dense is micro-callus. Field width = 900 μm.

completion from the last formed surface. Thus, in its longest phase, maturation mineralization in an osteon spreads centripetally from its canal. In bone which is constantly being resorbed and replaced (most bone in man), the tissue is a 3D mosaic of packets of different levels of maturity.

C. Extrinsic Fibers in Bone

The extrinsic or Sharpey fibers may be wholly or partially mineralized, and to a large measure how mineralized they are indicates their rate of incorporation into the matrix (Jones and Boyde, 1974). Those parts of the extrinsic fiber that mineralize are usually highly mineralized.

D. Endochondral Mineralization

Calcified cartilage (CC) is found in bone as a residue of the interstitial growth which permits rapid expansion in physes, and deep to the cartilage which fenders one bone against the next at joints. In hypertrophic growth plate zones, matrix mineralization is mediated via matrix vesicles, giving rise to quite large calcospherites which are a prominent feature in anorganic preparations for the SEM. The pericellular cartilaginous matrix is most highly mineralized, and the centers of the intercolumnar continuum of mineralized matrix in growth plate cartilage may remain unmineralized where the juxta-cellular mineralization is so perfect that it blocks access to the center portions of thicker cartilage. The cartilage remnants within a bone may be incompletely mineralized, but the mineralized parts are always more highly mineralized than bone (Figure 16).

In long bones, secondary centers of endochondral ossification develop within the ends (epiphyses) of the cartilage model, leading to a functional division between articular (slower growth) cartilage and epiphyseal growth plate cartilage (the faster growing physis). Compared to the finer trabeculae in the metaphysis, those in the epiphysis are coarse, but the process is essentially the same, with the same molecular cascade.

Hyaline cartilage remains at the articular surface after epiphyseal closure. The deep tidemark interface between mineralized and nonmineralized tissue is a valuable plane for interpreting age and joint health. Mineralization topography at this junction is the result of three different and merging patterns: creeping collagen-based mineralization along the fibrils, pericellular mineralization, and microcalcospheritic mineralization (Boyde and Jones, 1983b).

Species differences exist in the major events of long bone development, for example, between chick, mouse and man, particularly as to the vascularization and calcification and replacement of the cartilage core, and the development of the bone collar.

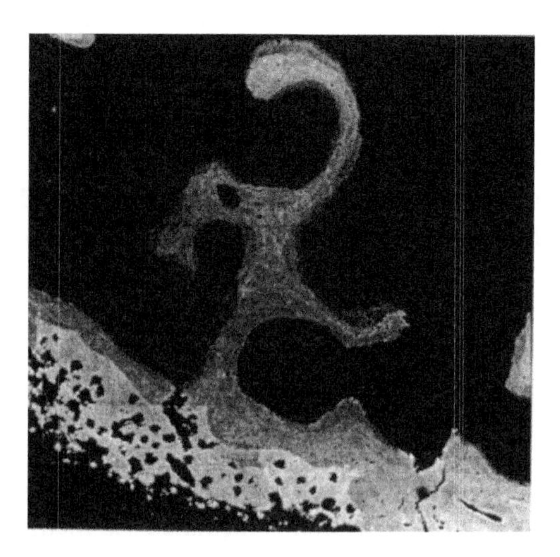

Figure 16. Human second lumbar vertebral body, 74-year-old male, PMMA embedded, micromilled, carbon coated. 20 kV BSE showing end plate zone next to nucleus pulposus. Densest phase present is calcified cartilage. Field width = 1.8 mm.

E. Calcified Fibrocartilage

Calcified fibrocartilage (CFC) is located in regions where tendons or ligaments or the annulus fibrosus of vertebrae meet bone. For various reasons, it may accumulate to become a substantial fraction of the mineralized tissue present in a bone. The progression of the mineralization front reflects the matrix structure (Boyde and Jones, 1983b). CFC is again considerably more densely mineralized than bone (Boyde et al., 1995b).

F. Reversal and Resting Cement Lines

Within all these tissues, resting and reversal lines marking changes in cellular activity at then current surfaces can always be detected because of their increased mineral content compared with the surrounding tissue.

G. Osteocytic Death and Mineralization

The functions of the osteocyte must depend upon its vitality. We speculate that osteocyte death may, indirectly, predispose to microdamage, which may also be the cause of further osteocyte death. Our knowledge of the functions of osteocytes is scant (Aarden et al., 1994). Osteocytes have been held to be involved in the lysis of their encumbrancing matrix (Belanger, 1968), a view which was challenged by SEM findings (Boyde, 1980). However, recent work which shows that mRNA for collagenase is seen in osteocytes adjacent to osteoclasts revives the subject (Fuller and Chambers, 1995). A pivotal role for live osteocytes in the signaling system involved in the control of modeling, where the reactive cell is the osteoblast, and remodeling, where the most reactive cell is the osteoclast, is most interesting (Skerry et al., 1989; Kufahl and Saha, 1990; Harrigan and Hamilton, 1993; Lanyon, 1993; Aarden et al., 1994; Weinbaum et al., 1994). Whatever the role(s) of osteocytes, they can no longer perform their duties if they die (Figure 17).

We propose that a major function of vital osteocytes within bone and osteoblasts or bone lining cells is to keep channels of communication open. Tissue fluid flow over the plasma membrane of the osteocyte is currently a favored candidate for the strain sensing mechanism (Kufahl and Saha, 1990; Harrigan and Hamilton, 1993; Aarden et al., 1994; Weinbaum et al., 1994). Some transport occurs within the cells, their processes, and via their gap junctional complexes, but any remainder must occur primarily in the fluid filled space between the cell process and the wall of the canaliculus, a process which we have studied *in vivo* using confocal fluorescence

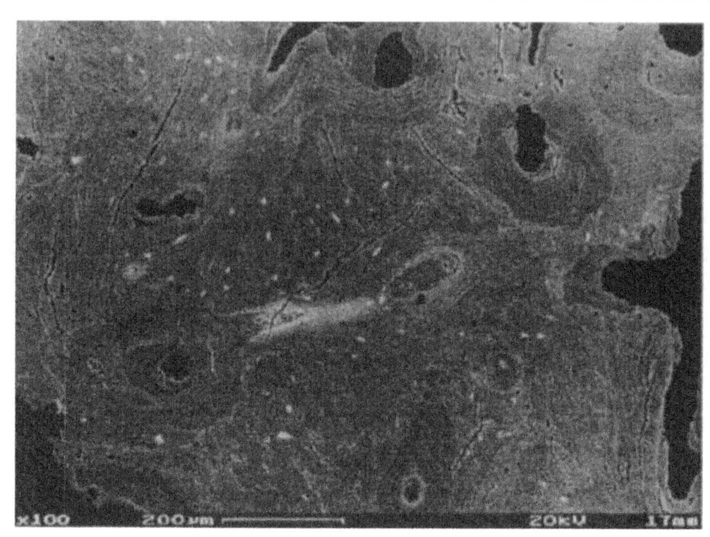

Figure 17. Vertical section of human mandible in mental foramen region, 87-year-old female, PMMA embedded, micromilled, carbon coated. 20 kV BSE. Densest phase is calcified osteocyte lacunae. Courtesy of V.J. Kingsmill.

microscopy after intravenous administration of fluorescent dyes (Boyde et al., 1995c). The residual water-filled space in mineralized bone matrix is probably insignificant for diffusion processes related to cell survival per se, but this space is accessible to ions, and is the compartment within which a super-mineralization or hyper-maturation may occur, an event which we believe is normally inhibited by live osteocytes as well as the covering osteoblasts or lining cells (Fleisch et al., 1966; Steendijk and Boyde, 1973; Boyde et al., 1978).

Dead bone becomes more highly mineralized. Such increased mineralization would lead to excessive stiffness with a heightened risk of microdamage. As little as 0.25 μm of shear could interrupt all canaliculi, eventually leading to cell death. Osteocytes might also die if they are walled off by cement lines, which may occur in normal remodeling (Dunstan et al., 1993), following vascular impedance (Cruess, 1986; Li et al., 1985), or by design as a consequence of apoptosis (programmed cell death, PCD).

A perilacunar, hypermineralized matrix, analogous with peritubular dentine, forms in a proportion of osteocyte lacunae in some mammalian species (Boyde, 1972); the matrix for this phase may possibly be secreted by osteocytes, or it may accumulate via canalicular flow and be analogous to material seen at surfaces as laminae limitantes and resting lines.

VIII. QUANTITATIVE STUDY OF MINERALIZATION DEGREE

A. The Meaning of Bone Density

The density of bone has different meanings to workers in various branches of applied osteology. Many workers think that bone mass is all that matters, and that all variations in bone mechanical properties can be satisfactorily explained by variations in bone density (TBV, BVF, BMC, BMD; Currey, 1984, 1996). We find it difficult to accept such a simplified view on the grounds that chains are as strong as their weakest links: tiny structural defects may be important. It is highly likely that the elastic moduli of bone are nonlinearly dependent on mineral content, and that stiffness and brittleness will change sharply when critical levels of packing and crystal phase continuity are reached. We apply the term mineralization density to refer to the density of bone tissue at a scale of a volume of matrix smaller than that occupied by an osteocyte and excluding the lacunae as space. The distribution of phases within microscopic volumes of bone can be imaged by microradiography and measured by μCT, but not to this resolution. The use of density gradient fractionation leads to the loss of context and histology.

B. Determination of Bone Mineralization Density Using BSE-SEM

A roughly 1,000-fold improvement in volume resolution over microradiography is obtained with digital BSE imaging, and we have established routines for quantification of the distribution of mineral densities within the skeletal mineralized tissues using this approach (Boyde and Jones, 1983a,b; Boyde et al., 1983, 1992, 1995a,b, 1996; Howell and Reid, 1986; Reid and Boyde, 1987). Bone samples are obtained at autopsy or biopsy, fixed in 70% ethanol, dehydrated in ethanol, and embedded in PMMA. Block faces are diamond micromilled with the front face parallel to the back. Conductive tracks are applied around the rim of the milled face by using aqueous colloidal graphite painted to the tops of the sides, prior to coating lightly with carbon by evaporation.

In our equipment, batches of typically 120 fields are located within 80 mm * 80 mm specimen arrays in a digital SEM (Zeiss DSM962) operated at 20 kV and 0.5 nA. Using an annular solid state BSE detector (KE Electronics, Toft, Cambs, UK) and a preset nominal working distance (WD) of 17 mm from the polepiece of the final lens gives a sample-detector distance of 11 mm. The maximum field dimension that we use at this WD is at a nominal magnification of 33x, giving a field width of 2.7 mm. We take this as the lower limit of acceptability for the defocus in the corners of the scanned field. Naturally, this improves at higher magnifications (reduced scan area) and at longer WD; however, if we increase the WD; we reduce the solid angle for collecting BSE. The choice is a compromise.

All focusing is done mechanically to keep the specimen-detector geometry constant. Filament saturation is determined by reference to the BSE signal level. Digital images are recorded under automatic control and after the instrument has been running for more than an hour. The recorded data do not constitute an image (as usually understood), since neighboring pixels in 512*512 scans are separated measuring points in a dense stereological grid.

The relative backscattering of the calcified tissues is determined by comparison with halogenated dimethacrylate standards (Boyde et al., 1995a). Fields containing both $C_{22}H_{25}O_{10}Br$ (mean BSE coefficient calculated by the procedure given by Lloyd (1987) = 0.1159) to $C_{22}H_{25}O_{10}I$ (mean BSE coefficient 0.1519) is recorded at the beginning of each run, after the tenth and every successive twentieth field, and after the last field: the data from these standards documents any temporal variation in instrument function. The image histograms are rescaled such that value 0 is assigned to the mean for the monobromo standard, and 255 to that for the monoiodo compound. This range is then further divided into 8 or 16 equal intervals and we derive a histogram with the corresponding reduced number of bins: for image display purposes, we use a standard pseudo-color LUT which features the same bins.

The bone volume fraction is calculated as the ratio of the area occupied by calcified tissue to the total field of view in which bone could have been measured. The volume fractions of the different density phases are calculated as the ratio of the summed bins in that range to the total amount of bone (and not to the total field of view).

The system has been applied to several problem areas, including analyses of human iliac crest autopsies (Boyde et al., 1995b) and biopsies, lumbar vertebral autopsies, femoral neck autopsies, femoral head and knee replacement implant retrieval autopsies, and biopsies and autopsies of cranial and mandibular and contrasting postcranial bone from the same individuals. We have used it to study changes in mineralization levels in mice with gene knockouts for matrix proteins and enzymes involved in normal bone function (Hayman et al., 1996), and to compare levels of mineralization of rat bone formed *in vivo* and *in vitro* (Jones et al., 1994). It has also been used in studying equine bone, in particular in contrasting low turnover tension cortical bone (with mainly primary osteonal systems) with high turnover compression cortex (with mainly secondary osteons). We now have a large data base of normative values for mammalian bone. A by-product of this research has been to highlight the extent to which tissues other than bone contribute to the total bone mineral content in the bone organ.

Most of the literature on bone density has actually expressed measurements which relate to the total amount of mineral present. Study of the degree of mineralization of the tissue may lead to data which is apparently paradoxical. For example, osteoporotic bone may be more highly mineralized, even though there is less bone within a given volume. There may also be an interesting sex difference, with older women having marginally denser bone than men (Boyde et al., 1995b).

IX. SUMMARY

In summary, it is necessary to think of bone at scales which reflect its functional competence at macroscopic, microscopic and sub-microscopic levels, and to have the imagination to encompass them simultaneously.

ACKNOWLEDGMENTS

The studies which provide the background to this chapter were supported by the Medical Research Council, The Wellcome Trust, and The Veterinary Advisory Committee of the Horserace Betting Levy Board. We are grateful for the technical assistance of R. Radcliffe and M. Arora.

REFERENCES

Aarden, E.M., Burger, E.H., and Nijweide, P.J. (1994). Function of osteocytes in bone. J Cell Biochem. 55, 287-299.

Arnold, J.S. (1970). Focal excessive endosteal resorption an aging and senile osteoporosis. In: Osteoporosis. (Barzel, U.S., Ed.), pp. 80-100. Grune and Stratton, New York.

Belanger, L.F. (1968). Equine Symposium. Osteolysis in pathological material. Cornell Vet 58 (Suppl.), 115-135.

Biewener, A.A., Fazzalari, N.L., Konieczynski, D.D., and Baudinette, R.V. (1996). Adaptive changes in trabecular architecture in relation to functional strain patterns and disuse. Bone 19, 1-8.

Biewener, A.A., Thomason, J., Goodship, A., and Lanyon, L.E. (1983). Bone stress in the horse forelimb during locomotion at different gaits: A comparison of two experimental methods. J. Biomech. 16, 565-576.

Bouligand, Y. (1972). Twisted fibrous arrangements in biological materials and cholesteric mesophases. Tissue Cell 4, 189-217.

Boyde, A. (1972). Scanning electron microscopic studies of bone. In: Biochemistry and Physiology of Bone. (Bourne, G.H., Ed.), Vol. 1. pp. 259-310. Academic Press, New York.

Boyde, A. (1973). Quantitative photogrammetric analysis and qualitative stereoscopic analysis of scanning electron microscope images. J. Microsc. 98, 452-471.

Boyde, A. (1980). Evidence against osteocytic osteolysis. Metab. Bone. Dis. Rel. Res. 2S, 239-255.

Boyde, A. (1984a). Methodology of calcified tissue specimen preparation for scanning electron microscopy. In: Methods of Calcified Tissue Preparation. (Dickson, G.R., Ed.), pp. 251-307. Elsevier, Amsterdam.

Boyde, A. (1984b). Dependence of rate of physical erosion on orientation and density in mineralized tissues. Anat. Embryol. 170, 57-62.

Boyde, A. (1987). Colour-coded stereo images from the tandem scanning reflected light microscope. J. Microsc. 146, 137-142.

Boyde, A., Bianco, P., Portigliatti-Barbos, M., and Ascenzi, A. (1984). Collagen orientation in compact bone: 1. A new method for the determination of the proportion of collagen parallel to the plane of compact bone sections. Metab. Bone Dis. Rel. Res. 5, 299-307.

Boyde, A. Cook, A.D., and Morgan, J.E. (1972). Scanning electron microscope display method and apparatus. U.K. Patent No 1393881.

Boyde, A., Davy, K.W.M., and Jones, S.J. (1995a). Standards for mineral quantitation of human bone by analysis of backscattered electron images. Scanning 17 (Suppl.), V6-V7.

Boyde, A., Haroon, Y., Jones, S.J., and Riggs, C.M. (1996). Highly anisotropic cancellous bone structure in the equine third metacarpal bone: Novel scanning strategies. Scanning 18, 184-185.

Boyde, A., Howell, P.G.T., Bromage, T., Elliott, J.C., Riggs, C.M., Bell, L.S., Kneissel, M., Reid, S.A., Jayasinghe, J.A.P., and Jones, S.J. (1992). Applications of mineral quantitation of bone by histogram analysis of backscattered electron images. In: Chemistry and Biology of Mineralized Tissues, Excerpta Medica Int. Congr. Ser. ICS 1002. (Slavkin, H.C. and Price, P., Eds.), pp. 47-60. Elsevier Science Publishers, Amsterdam.

Boyde, A. and Jones, S.J. (1983a). Backscattered electron imaging of skeletal tissues. Metab. Bone Dis. Rel. Res. 5, 145-150.

Boyde, A., Jones, S.J. (1983b). Scanning electron microscopy of cartilage. In: Cartilage I. (Hall, B.K., Ed.), pp. 105-148. Academic Press, New York.

Boyde, A. and Jones, S.J. (1996). Scanning electron microscopy of bone: Instrument, specimen, and issues. Microsc. Res. and Techn. 33, 93-120.

Boyde, A., Jones, S.J., Aerssens, J., and Dequeker, J. (1995b). Mineral density quantitation of the human cortical iliac crest by backscattered electron image analysis: Variations with age, sex, and degree of osteoarthritis. Bone 16, 619-627.

Boyde, A. and Pawley, J.B. (1975). Transmission electron microscopy of ion erosion thinned hard tissues. Calcif. Tiss. Res. Spec. (Suppl.) (Pors Nielsen, S. and Hjorting-Hansen, E. Eds.), pp. 117-123. FADS Forlag, Copenhagen.

Boyde, A., Radcliffe, R., Watson, T.F., and Jayasinghe, J.A.P. (1990). Continuous motion parallax in the display and analysis of trabecular bone structure. Bone 11, 228. (Abst. p. 29.)

Boyde, A., Reid, S.A., and Howell, P.G.T. (1983). Stereology of bone using both backscattered electron and cathodoluminescence imaging for the SEM. Beitr. elektronenmikroskop. Direktabb. Oberfl. (Münster) 16, 419-430.

Boyde, A., Reith, E.J., and Jones, S.J. (1978). Intercellular attachments between calcified collagenous tissue forming cells in the rat. Cell Tiss. Res. 191, 507-512.

Boyde, A. and Riggs, C.M. (1990). The quantitative study of the orientation of collagen in compact bone slices. Bone 11, 35-39.

Boyde, A., Wolfe, L.A., Maly, M., and Jones, S.J. (1995c). Vital confocal microscopy in bone. Scanning 17, 72-85.

Chambers, T.J. (1988). The regulation of osteoclastic development and function. Ciba Foundation Symposia 136, 92-107.

Cruess, R.L. (1986). Osteonecrosis of bone. Current concepts as to etiology and pathogenesis. Clin. Orthop. 208, 30-39.

Currey, J.D. (1984). Effects of differences in mineralization on the mechanical properties of bone. Phil. Trans. Roy. Soc. Lond. B 304, 509-518.

Currey, J.D. (1996.) Biocomposites: micromechanics of biological hard tissues. Current Opinion in Solid State and Materials Science 1, 440-445.

Dunstan, C.R., Somers, N.M., and Evans, R.A. (1993). Osteocyte death and hip fracture. Calcif. Tissue Int. 53S1, 113-117.

Ejersted, C., Anreassen, T.T., Hauge, E-M., Melsen, F., and Oxlund, H. (1995). Parathyroid hormone (1-34) increases vertebral bone mass, compressive strength, and quality in old rats. Bone 17, 507-511.

Engfeldt, B., Hjerpe, A. (1974). Density gradient fractionation of dentine and bone powder. Calcif. Tiss. Res. 16, 261-275.

Eriksen, E.F., Axelrod, D.W., and Melsen, F. (1994). Bone Histomorphometry. Raven Press, New York.

Fleisch, H., Russell, R.G.G., and Straumann, F. (1966). Effect of pyrophosphate on hydroxyapatite and its implications for calcium homeostasis. Nature 212, 901-903.

Fuller, K., Chambers, T.J. (1995). Localization of mRNA for collagenase in osteocytic, bone surface and chondrocytic cells but not osteoclasts. J. Cell. Sci. 108, 2221-2230.

Gebhardt, W. (1905). Über funktionell wichtige Anordnungsweisen der feineren und gröberen Bauelemente des Wirbeltierknochens. Arch. Entwicklungsmechanik Org. 20, 187-322.

Hahn, M., Vogel, M., Amling, M., Ritzel, H., and Delling, G. (1995). Microcallus formations of the cancellous bone: A quantitative analysis of the human spine. J. Bone Miner. Res. 10, 1410-1416.

Ham, A.W. and Leeson, T.S. (1961). *Histology*, 4th ed. Pitman Medical, London.

Harrigan, T.P. and Hamilton, J.J. (1993). Bone strain sensation via transmembrane potential changes in surface osteoblasts: Loading rate and microstructural implications. J. Biomech. 26, 183-200.

Hayman, A.R., Jones, S.J., Boyde, A., Foster, D., Colledge, W.H., Carlton, M.B., Evans, M.J., and Cox, T.M. (1996). Mice lacking tartrate-resistant acid phosphatase (Acp 5) have disrupted endochondral ossification and mild osteopetrosis. Development 122, 3151-3162.

Howell, P.G.T. and Boyde, A. (1994). Monte Carlo simulations of electron scattering in bone. Bone 15, 285-291.

Howell, P.G.T. and Reid, S.A. (1986). A microcomputer-based system for rapid on-line stereological analysis in the scanning electron microscope. Scanning 8, 139-144.

Jayasinghe, J.A.P. (1991). A study of change in human trabecular bone structure with age and during osteoporosis. PhD thesis, University of London.

Jayasinghe, J.A.P. and Boyde, A. (1990). A preliminary study of normal and osteoporotic trabecular bone using frequency domain analysis. Bone 11, 227 (Abst. p. 26).

Jayasinghe, J.A.P., Jones, S.J., and Boyde, A. (1993). Scanning electron microscopy of human vertebral trabecular bone surfaces. Virchows Archiv. A. Pathol. Anat. 422, 25-34.

Jayasinghe, J.A.P., Jones, S.J., and Boyde, A. (1994). 3D photographic study of cancellous bone in human fourth lumbar vertebral bodies. Anat. Embryol. 189, 259-274.

Jones, S.J. (1973). Morphological and experimental observations on bony tissues using the scanning electron microscope. PhD Thesis, University of London.

Jones, S.J. and Boyde, A. (1970). Experimental studies of the interpretation of bone surfaces studied with the scanning electron microscope. Scanning Electron Microscopy 1970, 193-200.

Jones, S.J. and Boyde, A. (1974). The organization and gross mineralization patterns of the collagen fibers in Sharpey fiber bone. Cell Tiss. Res. 48, 83-96.

Jones, S.J., Boyde, A., and Pawley, J.B. (1975). Osteoblasts and collagen orientation. Calcif. Tissue Res. 159, 73-80.

Jones, S.J., Gray, C., and Boyde, A. (1994). Simulation of bone resorption-repair coupling in vitro. Anat. Embryol. 190, 339-349.

Katz, E.P., Wachtel, E., Yamauchi, M., and Mechanic, G.L. (1989). The structure of mineralized collagen fibrils. Connect Tissue Res. 21, 149-158.

Kneissel, M., Boyde, A., Hahn, M., Teschler-Nicola, M., Kalchhauser, G., and Plenk, H. (1994). Age- and sex-dependent cancellous bone changes in a 4000y BP population. Bone 15, 539-545.

Koch, J.C. (1917). The laws of bone architecture. Am. J. Anat. 21, 177-208.

Kufahl, R.H., Saha, S. (1990). A theoretical model for stress-generated fluid flow in the canaliculi-lacunae network in bone tissue. J. Biomech. 23, 171-180.

Lanyon, L.E. (1973). Analysis of surface bone strain in the calcaneus of sheep during normal locomotion. J. Biomech. 6, 41-49.

Lanyon, L.E. (1993). Osteocytes, strain detection, bone modeling, and remodeling. Calcif. Tiss. Int. 53S1, S102-S107.

Lean, J.M., Jagger, C.J., Chambers, T.J., and Chow, J.W. (1995). Increased insulinlike growth factor I mRNA expression in rat osteocytes in response to mechanical stimulation. Am. J. Physiol. 268, E318-E327.

Li, G.P., Zhang, S.D., Chen, G., Chen, H., and Wang, A.M. (1985). Radiographic and histologic analyses of stress fracture in rabbit tibias. Am. J. Sports Med. 13, 285-294.

Lloyd, G.E. (1987). Atomic number and crystallographic contrast images with the SEM: A review of backscattered electron techniques. Mineralol. Mag. 51, 3-19.

Logue, F.C., Fraser, W.D., O'Reilly, D.S., Cameron, D.A., Kelly, A.J., and Beastall, G.H. (1990). The circadian rhythm of intact parathyroid hormone-(1-84): Temporal correlation with prolactin secretion in normal men. J. Clin. Endocrinol. Metab. 71, 1556-1560.

Marotti, G., Muglia, M.A., and Palumbo, C. (1994). Structure and function of lamellar bone. Clin. Rheumatol. 13 (Suppl. 1), 63-68.

McKee, M.D. and Nanci, A. (1996). Osteopontin at mineralized tissue interfaces in bone, teeth, and osseointegrated implants. Microsc. Res. and Techn. 33, 141-164.

McMahon, J.M., Boyde, A., and Bromage, T.G. (1995). Pattern of collagen fiber orientation in the ovine calcaneal shaft and its relation to locomotor-induced strain. Anat. Rec. 242, 147-158.

Mosekilde, L. (1990). Consequences of the remodelling process for vertebral trabecular bone structure: A scanning electron microscopy study (uncoupling of unloaded structures). Bone and Mineral 10, 13-35.

Mosekilde, L. (1993). Vertebral structure and strength in vivo and in vitro. Calcif. Tiss. Int. 53 (Suppl.), S121-S126.

Oxnard, C.E. (1980), The problem of stress bearing and architecture in bone: Analysis of human vertebrae. J. Am. Osteopath. Assoc. 80, 280-287.

Oxnard, C.E. (1986). The measurement of form: Beyond biometrics. Sausages and stars, dumbbells and doughnuts: Peculiar views of anatomical structures. Cleft Palate J. 23 (Suppl. 1), 110-128.

Reid, S.A. (1986). A study of lamellar organization in juvenile and adult human bone. Anat and Embryol. 174, 329-338.

Reid, S.A. and Boyde, A. (1987). Changes in the mineral density distribution in human bone with age: Image Analysis using backscattered electrons in SEM. J. Bone Min. Res. 2, 13-22.

Riggs, C.M., Lanyon, L.E., Boyde, A. (1993a). Functional associations between collagen fiber orientation and locomotor strain direction in cortical bone of the equine radius. Anat. Embryol. 187, 231-238.

Riggs, C.M., Vaughan, L.C., Evans, G.P., Lanyon, L.E., and Boyde, A. (1993b). Mechanical implication of collagen fiber orientation in cortical bone of the equine radius. Anat. Embryol. 187, 239-248.

Skedros, J.G., Bloebaum, R.D., Mason M.W., and Bramble, D.M. (1994a). Analysis of a tension/compression skeletal system: Possible strain-specific differences in the hierarchical organization of bone. Anat. Rec. 239, 396-404.

Skedros, J.G., Mason, M.W., and Bloebaum, R.D. (1994b). Differences in osteonal micromorphology between tensile and compressive cortices of a bending skeletal system: indications of potential strain-specific differences in bone microstructure. Anat. Rec. 239, 405-413.

Skerry, T.M., Bitensky, L., Chayen, J., and Lanyon, L.E. (1989). Early strain-related changes in enzyme activity in osteocytes following bone loading in vivo. J. Bone Miner. Res. 4, 783-788.

Steendijk, R. and Boyde, A. (1973). Osteocytic control of mineralization—an hypothesis. Calc. Tiss. Res. 11, 249.

Vesterby, A. (1990). Star volume of marrow space and trabeculae in iliac crest: Sampling procedure and correlation to star volume of first lumbar vertebra. Bone 11, 149-155.

Wakamatsu, E. and Sissons, H.A. (1969). The cancellous bone of the iliac crest. Calcif. Tiss. Res. 4, 147-161.

Weinbaum, S., Cowin, S.C., Zeng, Y. (1994). A model for the excitation of osteocytes by mechanical loading-induced bone fluid shear stresses. J. Biomech. 7, 339-360.

Weiner, S. and Traub, W. (1992). Bone structure: From angstroms to microns. FASEB J. 6, 879-885.

Wolff, J. (1892). Das Gesetz der Transformation der Knochen, Hirschwild, Berlin. Translated into English by Maquet, P. and Furlong, R. (1986). The Law of Bone Remodelling. Springer-Verlag, Berlin.

PHYSIOLOGY OF BONE REMODELING

Chantal Chenu and Pierre Dominique Delmas

I. Introduction 46
II. Structure of Bone Tissue 46
A. Organization of Osteons 47
B. Organization of Trabecular Bone 47
III. Osteoclasts and Bone Resorption 48
A. Osteoclast Lineage 49
B. Osteoclast Function 49
C. Osteoclast Regulation 50
IV. Osteoblasts and Bone Formation 50
A. Osteoblastic Lineage 50
B. Mechanisms of Bone Formation and Mineralization 51
V. Bone Remodeling 52
A. Theoretical Basis of Bone Remodeling at the Cellular Level 52
B. Coupling 55
VI. Evaluation of Bone Remodeling 55
A. Bone Histomorphometry 55
B. Biochemical Markers 56
VII. Age and Menopause-Related Bone Loss 57
VIII. Summary 60
Acknowledgments 60

Advances in Organ Biology
Volume 5A, pages 45-64.

ISBN: 0-7623-0390-5

I. INTRODUCTION

Bone remodeling occurs throughout life in the adult, and is the process by which bone is turned over through the removal of bone (bone resorption) and the formation of new bone to replace it (bone formation). In the adult skeleton, at a given time, there is a multitude of remodeling sites, called bone remodeling units (BRUs) or basic multicellular units (BMUs), at different stages. Every remodeling cycle is initiated at a previously quiescent bone surface, by the recruitment of osteoclastic precursors which become multinucleate and then start resorbing bone (Hattner et al., 1965). After bone resorption is terminated, the resorbed area is invaded by preosteoblasts that differentiate into osteoblasts and form a new matrix that will subsequently become mineralized. The sequence of events in each remodeling site is therefore an activation-resorption-formation (A-R-F) sequence (Frost, 1964a).

In order to preserve bone mass and integrity of the skeleton, the breakdown and formation of bone should be balanced, i.e., the amount of bone removed during bone resorption should be equal to the amount of bone laid down during bone formation. Disturbances in the function and arrangement of these bone remodeling units lead to the changes in bone mass and structure that are observed during aging and in metabolic bone diseases.

II. STRUCTURE OF BONE TISSUE

The anatomy of bone is described in the chapter by Boyde (this volume). Here, we recapitulate this briefly, particularly in reference to the bone remodeling sequence. Namely, there are two types of bone. The cortical bone plays a major role in the support function and makes up 80% of skeletal mass. Trabecular or cancellous bone constitutes the remaining 20%, and is metabolically more active and predominates in the vertebrae.

Cortical bone is very compact and is present in ribs or in the external part of the long bone. In the diaphysis of the long bone, it encloses the medullary cavity. Towards the metaphysis and the epiphysis, the cortex becomes progressively thinner and the internal space is filled with trabecular bone. The external surface of cortical bone, the periosteal surface, is important in appositional growth and fracture repair. It displays a lack of balance between bone formation and resorption which results in an increase in the diameter of the long bones. The endosteal surface of cortical bone is in contact with trabecular bone and has a higher level of remodeling activity. On the endosteal surface, resorption tends to exceed formation, leading to expansion of the marrow space in long bones and endocortical thinning.

A. Organization of Osteons

Compact bone is made by the juxtaposition and close association of a large number of osteons which represent the structural units of this type of bone (Jaworski, 1971). The BRU creates advancing tunnels known as "cutting cones." The latter are formed through the cortex by resorbing osteoclasts. New bone is left behind creating the "closing cone." When completed, this so-called Haversian system consists of lamellar bone which is disposed in a cylindric way around the Haversian canal containing blood vessels (Figure 1). This typical lamellar structure is due to an alternating orientation of the collagen fibers when these are laid down. In some pathological conditions of high turnover, the collagen fibers are deposited in a disorganized manner and mechanical properties of such bone, termed woven bone, suffer. The cement line at the periphery of an osteon indicates the borders of the resorption cavity which has occured before the osteon, and the central border is the border of the Haversian canal in an achieved adult osteon.

B. Organization of Trabecular Bone

Trabecular bone consists of trabeculae, i.e., interconnected thin plates or spicules that provide maximal mechanical strength. Cancellous BRUs

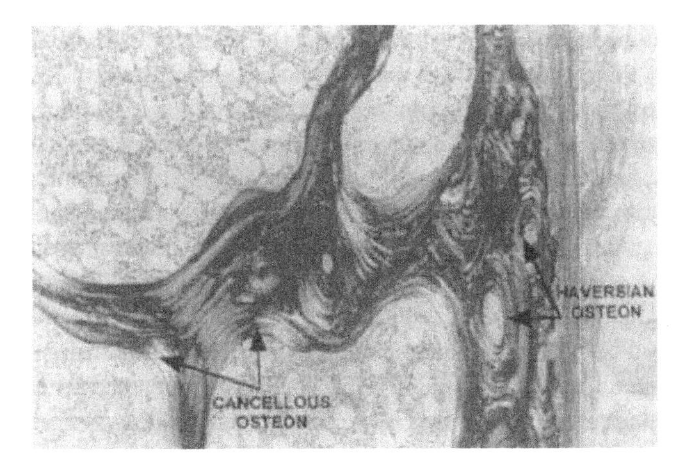

Figure 1. Haversian and cancellous osteons, as seen in bone biopsy sections studied in polarized light to show the lamellar pattern of bone (x 100).

resorb and form bone on the trabecular surface and leave behind units of new lamellar bone which may be observed as semilunar structures (walls or packets of bone), where the collagen fibers are deposited parallel to each other. The cancellous osteon can be represented as an uncoiled Haversian osteon (Figure 1). Resorption takes place at the bone surface adjacent to bone marrow and progresses linearly forming scalloped interruptions of the lamellar system, termed Howship's lacunae, that are refilled with bone by osteoblasts invading the lacunae (Eriksen, 1986).

The bone structural unit (BSU) represents the end result of a remodeling cycle. In cortical bone it constitutes new Haversian systems (cortical osteons), and in cancellous bone it forms new packets of bone (trabecular osteons).

All bone remodeling originates on a bone surface and occurs in four anatomical envelopes defined by Frost (1964a,b): viz. (1) the periosteal envelope, (2) the Haversian envelope, (3) the endosteal envelope, and (4) the trabecular envelope. Since bone turnover is dependent on the surface area and trabecular bone has a larger area involved in remodeling activity compared to cortical bone, it will turnover more rapidly and consequently will be the major site of bone loss (Parfitt, 1983).

III. OSTEOCLASTS AND BONE RESORPTION

Osteoclasts are large, very active and specialized multinucleated cells which resorb mineralized bone and cartilage. They are located on bone surfaces at sites of actively remodeling bone, mainly at the endosteal surface in Haversian systems, but occasionally on their periosteal surfaces.

These multinucleate cells contain an average of 10–20 nuclei, but sometimes many more in pathological conditions. The osteoclast is a cell that is morphologically and functionally polarized with an apical pole facing the bone matrix and a basolateral pole facing the soft tissues which provides mostly regulatory functions (see Baron, 1989, for review). Active osteoclasts have a specific area of their cell membrane in the apical domain, known as the ruffled border, which is comprised of folds and invaginations that allow close contact with the bone surface. This compartment functions as an extracellular lysosome into which acid and enzymes are secreted by the osteoclast. It is surrounded by a clear zone or sealing zone which contains contractile proteins (Marchisio et al., 1984) and which appears to anchor the ruffled border area to the bone surface undergoing resorption. The presence in osteoclasts of multiple vacuolar structures, composed of both primary and secondary lysosomes, as well as numerous mitochondria, provide evidence of an active cell.

A. Osteoclast Lineage

A variety of studies have clearly indicated that the osteoclast is derived from hemopoetic tissue (Walker, 1975; Coccia et al., 1980). However, the specific cell lineage that gives rise to the osteoclast is still not very clear. The monocyte-macrophage lineage is the leading candidate for the osteoclast lineage but differentiation into osteoclasts may occur directly from early hematopoietic precursors or from more differentiated cells in the monocyte-macrophage lineage (for review, see Mundy and Roodman, 1987).

B. Osteoclast Function

Osteoclasts make rapid transitions from a resting phase to an active phase. This activation involves the sealing of the subosteoclastic extracellular resorption zone by the clear zone, and the development of the ruffled border which corresponds to the area of ion transport and enzyme secretion.

The mechanisms of osteoclast attachment to the bone matrix are not completely understood. Osteoclasts express a number of cytoskeletal proteins that are linked to actin filaments and which form podosomes at sites of cell-substratum or cell-cell interaction (for review, see Teti et al., 1991). These cytoskeletal complexes stabilize the interaction of the osteoclast with the bone surface, due to integrins which bind proteins of the bone matrix by their extracellular domains and interact with the cytoskeleton with their cytoplasmic domains (Hynes, 1987). Osteoclasts express several integrins (Nesbitt et al., 1993) among which the vitronectin receptor, $\alpha_v\beta_3$, seems to play a major role in osteoclast attachment to bone and bone resorption (Ross et al., 1993).

The osteoclast resorbs bone by the secretion, at its apical pole, of hydrogen ions and proteolytic enzymes. Hydrogen ions are generated in the cell by a type-II carbonic anhydrase and are secreted into the resorption lacunae by an ATP-dependent proton pump which is localized to the ruffled border. The protons are responsible for the dissolution of the mineral phase, and provide an acidic environment in which proteolytic enzymes can degrade the bone matrix (Baron et al., 1985). The lysosomal cysteine proteinases, which are secreted by the osteoclasts and are able to degrade collagen in an acidic environment, play a major role in matrix degradation (Blair et al., 1993). Among them, cathepsin K is a newly identified cysteine protease, selectively expressed by osteoclasts, than can fully digest collagen and may play a pivotal role in bone resorption (Votta et al., 1997). Matrix metalloproteinases (MMPs) such as collagenase are active at more neutral pH and may also play a role in osteoclastic bone resorption, although the osteocast MMPs seem to be more required for cell migration. A cooperation between the two clases of enzymes during the resorption of bone has been described (Everts et al., 1992).

The basolateral membrane of the osteoclast contains ion transport systems that maintain the electrochemical balance of the osteoclast during bone resorption.

C. Osteoclast Regulation

Many hormones and growth factors have been shown to regulate osteoclast formation and activity. They may act directly on osteoclasts or their precursors or indirectly on members of the osteoblastic lineage. In that case, the osteoblastic cells will release a soluble signal which activates the osteoclasts or will establish a direct contact with the cells of the osteoclastic lineage (Rodan and Martin, 1981; Martin and Ng, 1994).

IV. OSTEOBLASTS AND BONE FORMATION

Osteoblasts are metabolically very active mononucleate cells. They contain an extensive network of rough endoplasmic reticulum with a lot of ribosomes associated with mRNA and an intricate Golgi apparatus, all consistent with their important synthetic and secretory activity. They also characteristically contain large amounts of alkaline phosphatase in their plasma membrane, an enzyme which seems to be involved in bone formation (Whyte, 1994).

A. Osteoblastic Lineage

Osteoblasts originate from mesenchymal cells located in the periosteum or in the bone marrow. These progenitor cells can give rise to a number of other cell types such as cartilaginous, muscular, fibroblastic, or adipocytic cells (Aubin et al., 1992). Cbfa 1 is a recently discovered transcription factor that is essential for osteoblast differentiation and is presumably necessary for determinating the pathway of differentiation of pluripotent mesenchymal cells into the osteoblast lineage (Rodan et Harada, 1997). It has been possible to characterize different stages of osteoblast differentiation beginning with the early cell committed to the osteoblastic lineage to the mature osteoblast.

Preosteoblasts are the early cells committed to the osteoblast lineage and are generally found near the bone surface where active osteoblasts form bone. They still divide and have a low alkaline phosphatase activity. Mature osteoblasts are cuboidal cells that have reached the bone surface and no longer divide. There are metabolically very active, synthesize collagen as well as noncollagenous bone proteins, and ultimately promote mineralization of the bone matrix. At the end of their secretory period they become lining cells or osteocytes.

Lining cells are flattened cells located at the endosteal surfaces and trabeculae. They have a low synthetic activity, but share with osteoblasts the expression of hormone receptors as well as the capacity to produce growth factors and cytokines. They can participate in the transmission of intercellular signals within bone.

Osteocytes are osteoblasts which have become entrapped in the calcified bone within a lacunae. Approximately 10–20% of osteoblasts eventually become osteocytes. These cells are connected to each other and to bone lining cells by gap junctions and canaliculi which give them access to nutrients. Their metabolic activity is very low; nevertheless several functions have been assigned to them (Aarden et al., 1994). The osteocytes are able to synthesize new bone matrix at the surface of the lacunae. They can contribute to the transport of minerals and are well located to sense and adapt to mechanical strain, and to transmit information to the cells of the bone surface.

B. Mechanisms of Bone Formation and Mineralization

Bone formation is characterized by the synthesis of an organic matrix, composed of 90% type I collagen and 10% noncollagenous proteins, followed by its mineralization. The two processes, matrix synthesis and osteoid maturation and mineralization, are described below.

Bone matrix is mainly composed of type I collagen, but also contains small amounts of type V collagen. The molecule of type I collagen is a triple helix, made of three polypeptide chains, two α_1 chains and one α_2 chain. The noncollagenous proteins of bone have been extensively characterized (see Delmas and Malaval, 1993, for review). These proteins have attracted considerable interest as they seem to play a very important role in bone remodeling, regulating bone cell migration and adhesion, cell proliferation and differentiation, as well as mineralization processes. In addition to bone proteins, the osteoblastic cells synthesize and secrete into the bone matrix a number of growth factors, as well as proteinases, and their inhibitors, which are also involved in the regulation of bone metabolism.

The synthesis of the extracellular matrix is followed by a phase of maturation of the osteoid matrix which is necessary for mineralization to occur. The osteoblasts, which modulate bone matrix composition, are the promoters of mineralization, and are involved in the onset and the process of mineralization. Two mechanisms of mineralization have been described, one is predominant in calcified cartilage and woven bone, and the other in lamellar bone. Matrix vesicles, which are membrane-bound bodies exocytosed from the plasma membrane of the skeletal cells, are sites of mineral initiation in calcified cartilage and woven bone. In lamellar bone, the collagen fibrils are well ordered and form gap regions which are sites for hydroxyapatite crystal formation in association with complexes of collagen and phosphoproteins (Anderson and Morris, 1993).

With the exception of calcitonin, all hormones, growth factors, and cytokines which regulate bone remodeling have receptors on osteoblasts or elicit responses from the osteoblasts which have, therefore, a central place in bone cell biology.

V. BONE REMODELING

A. Theoretical Basis of Bone Remodeling at the Cellular Level

Bone cells are involved in the bone replacement mechanisms which occur throughout life in the adult skeleton. This process of bone remodeling was described by Frost over 30 years ago (Frost, 1964a). Frost developed the theory that the skeleton is remodeled in quanta or packets, the BMUs or BRUs, in which the extracellular matrix is sequentially removed and replaced by teams of cells. Bone remodeling occurs in anatomically discrete foci which are active for four to eight months, and is described as an activation-resorption-reversal-formation-quiescence sequence.

The activation of the cycle of remodeling first requires the recruitment of osteoclasts. The bone surface is converted from a quiescent state, characterized by the presence of a thin layer of lining cells, to a state in which precursor cells of hematopoietic lineage are recruited and begin to proliferate and differentiate to form osteoclasts (Tran Van et al., 1982a,b). Although this phase has been identified morphologically, the conditions that precipitate activation are unknown. Different regulatory mechanisms may exist; the activation frequency has been shown to be regulated by systemic hormones and locally by mechanical loading. To be activated, osteoclasts need to gain access to bone and it has been proposed that MMPs, produced by osteoblasts or bone lining cells, remove the nonmineralized osteoid from the bone surface (Vaes, 1988). Osteoclasts may also invade collagen by using MMP activity in the absence of other cells (Sato et al., 1998). It has recently been shown that the interstitial collagenase MMP-13, produced by stromal cells and osteoblasts, may play a role in the initiation of osteoclast bone resorption by generating collagen degradation fragments that activate osteoclasts (Holliday et al., 1997).

Once in contact with bone, the osteoclasts dissolve the mineral and hydrolyze the organic matrix to form a resorption cavity of characteristic shape and dimensions. Eriksen et al. (1984) have proposed that bone resorption occurs in three phases, the first being performed by multinucleated osteoclasts, the second by a mixed osteoclast/mononuclear cell population, and the third by terminal mononuclear cells; the highest rate of resorption takes place in the first phase and the lowest rate in the final phase. This theory has been recently confirmed in cortical bone (Agerbaek et al., 1991). In trabecular bone, osteoclasts erode bone rapidly (approximately seven days), down to a depth of about two-thirds of the final cavity. The remainder is eroded much more slowly by mononuclear cells. When the cavity reaches a mean depth of about 50 μm from the trabecular surface, which takes about 43 days, resorption ceases. In cortical bone, the resorption phase has a mean duration of about 30 days. During that period of time, a tunnel with a diameter of approximately 150 μm and 2.5 mm in length is created by osteoclastic and mononuclear cells.

The resorption process in trabecular and cortical bone includes a final period of a few days during which preosteoblasts are attracted to the resorption cavities and begin to differentiate into osteoblasts. This corresponds to the reversal phase of the remodeling cycle and represents the transition period during which formation is coupled to resorption. After this reversal phase, bone formation occurs in two stages, matrix synthesis and mineralization. The team of new osteoblasts begins to deposit a layer of bone matrix, the osteoid seam. The new matrix begins to mineralize after about 15 to 20 days of maturation (Parfitt, 1990). The cortical osteon is constructed within an interval of approximately 90 days and the trabecular osteon is built over a total period of about 145 days. Whenever a new remodeling site is initiated at the bone surface, the same sequence of events will occur. Figures 2 and 3 illustrate the organization of the BMUs in cortical and trabecular bone, respectively.

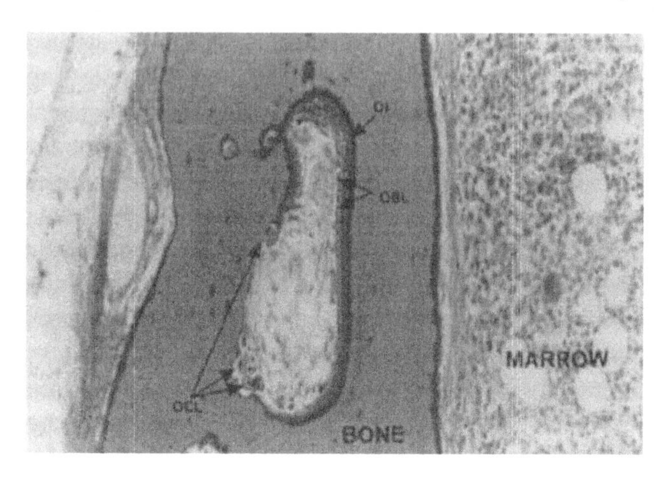

Figure 2. Light micrograph of a basic multicellular unit (BMU) in cortical bone (x 100). The cortical BMU is cylindrical in shape. Osteoclasts (OCL) are resorbing bone while osteoblasts (OBL) are depositing osteoid (OI) on the previously resorbed cavity.

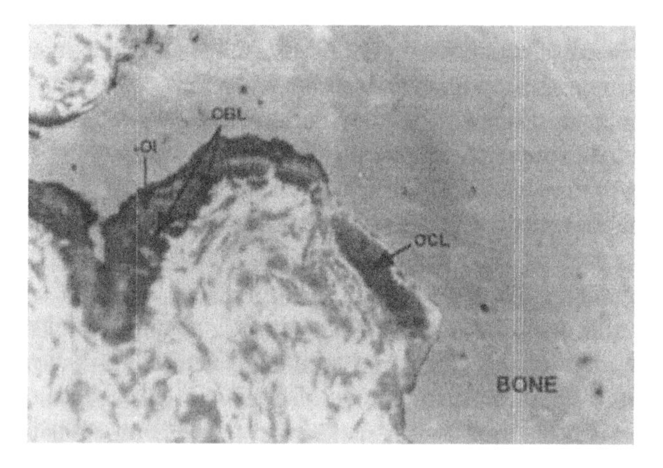

Figure 3. Light micrograph of a basic multicellular unit (BMU) in trabecular bone (x 200). A multinucleated osteoclast (OCL) is resorbing bone, followed by osteoblasts (OBL) which are forming matrix (osteoid, OI).

It has been suggested that bone formation and mineralization may not be a continuous process, but may undergo temporary interruptions prior to their completion (Frost, 1980). Similarly, bone resorption may be also subject to interruptions and/or permanent arrest in the early stages of resorption cavity development, during the process of remodeling (Croucher et al., 1995).

B. Coupling

The term coupling refers to the normal temporal and spatial relationship between bone formation and bone resorption. The rationale for coupling is the conservation of bone mass at each BRU site. The amount of bone removed during bone resorption and the amount of bone laid down during bone formation must be equal. The mechanisms underlying the coupling phenomenon are poorly understood. The agents involved in the coupling process have been restricted to those influencing osteoblast recruitment (Parfitt, 1982). Coupling agents are messengers generated by osteoclastic activity that influence osteoblastic proliferation, migration, differentiation, matrix synthesis, and/or cessation of synthesis. Some growth factors, including insulin-like growth factors and transforming-growth factor-β, as well as noncollagenous proteins might be implicated in that process (Mundy and Roodman, 1987; Dempster, 1992, 1995; Delmas and Malaval, 1993). Coupling has also been used to describe the balance between the resorption depth and the amount of bone refilling the resorption space, which depends on the recruitment, life span, and work efficiency of both osteoclasts and osteoblasts (Jaworski, 1984). The term coupling also could refer to all messages that explain why osteoclasts and osteoblasts generally perform their functions sequentially in the same site, consistent with the fact that many messages to osteoclasts are transmitted via cells of the osteoblastic lineage.

VI. EVAULATION OF BONE REMODELING

Bone remodeling can be studied at different levels of activity, viz. (1) the cellular level, (2) the BMU level, (3) the tissue level, and (4) the organ level. At the tissue level, bone remodeling depends on the individual rates of activity of resorptive and formative cells in each BMU, the consequence being the balance between resorption and formation, and the number of new BMUs initiated at the bone surface per unit of time. The latter is also called the activation frequency. Presently, bone remodeling can be evaluated either through bone histomorphometry or through the measurements of biochemical markers of bone turnover.

A. Bone Histomorphometry

The purpose of bone histomorphometry is to provide *in vivo* information on the cellular and tissue mechanisms of bone remodeling. The events occuring at such levels are neither detectable through the noninvasive methods of bone mineral measurements, nor through the measurement of bone turnover markers. The latter assess whole skeletal turnover. Assessment of bone structure can be achieved by bone histomorphometry, which has developed in recent years a number of new parameters for the quantitative evaluation of the connectivity of the trabecular network (see Meunier, 1995, for review). This method is supposed to give insight in some aspects of bone quality whereas the classical histomorphometry evaluated only quantitative parameters. Those include the direct measurement of static parameters expressing the amount of bone as well as some bone formation and bone resorption parameters. Resorption can be evaluated by measurement of eroded surface, osteoclast number, and more recently depth and surface of resorption cavities, but there is no dynamic measurement available for assessing the rate of bone resorption.

Bone histomorphometry makes possible measurements of static parameters of bone formation, like osteoid surface and width and osteoblast surface. Moreover, it allows the obtaining of dynamic information though the use of fluorescent labels such as tetracycline. When these are administered at timed intervals, they integrate into the mineralization front of forming bone (Frost, 1969). The distances between the labeled regions will then provide information about the mineral apposition rate and the mineralizing surface. This process has allowed the introduction of time into the quantitative analysis, thus providing information on organ, tissue, and cell-level turnover kinetics. The measurement of the mineral apposition rate makes possible the deduction of the bone formation rate, the duration of osteoblast activity, and the activation frequency.

B. Biochemical Markers

Although histomorphometric analyses allow the evaluation of bone metabolism at the tissue level, these are invasive for the patient and time-consuming. They are limited to a small area of the cancellous and of the cortico-endosteal envelope bone of the iliac crest, which may not reflect bone turnover of other sites of the skeleton. The recent development of specific and sensitive biochemical markers, reflecting the overall rate of bone formation and bone resorption, has markedly improved the noninvasive assessment of bone turnover in various metabolic bone diseases (see Garnero and Delmas, 1996, for review). Bone formation can be assessed by serum measurements of total or bone-specific alkaline phosphatase, osteocalcin, and N-terminal, type-I collagen peptide. The most reliable evaluation of bone resorption is the measurement in urine of bone type–I collagen degradation products, such as pyridinium crosslinks and telopeptides containing these crosslinks, and new immunoassays for these markers have been expanded. Recently, an enzyme-linked immunoassay for measuring the isomerized form of type-I collagen degradation product has been developed in serum and it represents a sensitive and specific index of bone resorption, based on the measurements of serum samples from patients with metabolic bone diseases (Bonde et al., 1997). Some of the new immunoassays for pyridinium crosslinks and associated peptides could allow the prediction of the risk of hip fracture in elderly women (Garnero et al., 1996a). The importance of measurements of bone markers levels in combination with bone mass measurements by dual x-ray absorptiometry, to improve the assessment of the osteoporotic risk in elderly women, has been recently demonstrated (Garnero et al., 1998).

VII. AGE AND MENOPAUSE-RELATED BONE LOSS

Imbalance between bone resorption and formation results in the net irreversible loss of bone during each remodeling cycle, as seen during aging. When this imbalance is combined with increased activation frequency, the outcome is an accelerated and irreversible bone loss, with increased risk of disintegration of the trabecular bone structure and loss of its connectivity, as observed in osteoporosis (Riggs and Melton, 1986).

Decrease in cancellous and cortical bone mass is a phenomenom that occurs with age, in both sexes but with significant differences between men and women (Mosekilde, 1989). Some loss of trabeculae occurs in man with aging but not to the same extent as women, for which there is a dramatic acceleration of bone loss after menopause. This postmenopausal cancellous bone loss is rapid and results in the complete removal of trabecular plates and disruption of the trabecular connectivity.

Histomorphometric studies have clearly reported age-related decrease in cancellous bone volume for both sexes (Parfitt et al., 1983b; Aaron et al., 1987; Weinstein and Huston, 1987; Mellish et al., 1989; Eriksen et al., 1990). There are however sex-related differences in magnitude and in the changes in trabecular architecture with age. In women there is an acceleration of bone turnover early after menopause resulting in a higher tendency toward perforation of the trabeculae. They become more widely separated and their connectivity is reduced (Mosekilde, 1989). This fragmentation of the continuous trabecular network is due to an osteoclastic resorption enhanced by more than 50%, which results in an increase of resorption cavity depth (Parfitt, 1983). In older women, there is a decrease of the trabecular thickness due to a reduction of the wall thickness, which is the width of a complete cancellous bone packet (Lips et al., 1978). The cellular mechanism responsible for this process is generally accepted to be an age-related decline in the amount of matrix synthesized by osteoblasts. With increasing age, the osteoblast population becomes progressively less able to reconstitute previously resorbed bone, leading to a bone formation deficit in the individual bone remodeling units (Meunier et al., 1979; Parfitt 1992). Bone histomorphometry has not shown any evidence for an age-related trabecular thinning (Dempster et al., 1995), which could be explained by a preferential perforation of thinner trabeculae (Parfitt et al., 1983).

There is also an accelerated loss of cortical bone after menopause. The decrease in cortical bone is the result of two processes—cortical thinning and an increase in cortical porosity. This change in cortical porosity would be due to progressive osteoblastic incompetence with age, in terms of both reduced preosteoblast recruitment and matrix deposition capacity (Parfitt, 1990). Cortical thinning is the result of a negative bone balance due to enhanced resorption depth on the endosteal surface (Parfitt, 1988; Kimmel et al., 1992), a process that is not compensated for by the slightly positive bone balance on the periosteal surface. Cortical bone loss is less in men, because endocortical resorption is less and periosteal formation is greater.

Biochemical marker measurements confirm that both bone formation and bone resorption rates increase with aging in women, with a bone turnover acceleration after menopause. However, in contrast to the decline with age of the osteoblast function shown at the cellular level, the marked increase of specific markers of bone formation and bone resorption observed in the first years following the menopause is maintained in elderly women for a long time after the menopause and induces a faster bone loss (Garnero et al., 1996b). These data suggest that the age-related bone loss does not result primarily from a decrease in bone formation rate, and is rather consistent with a high bone turnover.

It is now well-established that a low bone mass is the major determinant of all osteoporotic fractures. The concept of two distinct osteoporotic syndromes has been proposed: Type I osteoporosis is characterized by a deficit in spinal trabecular bone and by crush fracture syndrome related to postmenopausal estrogen deficiency; type II osteoporosis is characterized by a dominant loss of cortical bone and by hip fractures related to vitamin-D deficiency and secondary hyperparathyroïdism. Hip-fracture patients have biochemical evidence of increased bone resorption when compared to age-matched controls, which may be a determinant of the low bone mass that characterizes patients with hip fracture (Akesson et al., 1993).

Women classified as "fast bone losers" who have a high bone turnover rate in the first years after the menopause, have an accelerated bone loss that leads to a lower bone mass (Hansen et al., 1991) and a higher risk of fracture than women classified "normal" or "slow losers" (Riis et al., 1995). In a prospective study of risk factors for hip fractures conducted in elderly healthy women, it was shown that an increase in bone resorption markers was associated with a higher risk of hip fracture (Garnero et al., 1996a). Women with both a low bone mass and a fast rate of bone loss, which are two independent risk factors, have a higher risk to sustain fractures.

Hip fractures in men account for one third to one fourth of all hip fractures and their number increases with the increasing life expectancy. Moreover, although the hip fractures are less frequent in men than in women, the hip fracture-related mortality is actually higher in men than in women. The pathophysiology of male osteoporosis is less known and various causes of osteoporosis have been found in men. Reduced gonadal androgen synthesis as well as decreased bioavailable estrogen concentrations (Riggs et al., 1998) may contribute to the continuous bone loss of aging men. Other factors have been implicated as causes of bone loss in aging men, including reduced growth hormone and IGF-I levels, secondary hyperparathyroidism related to vitamin-D deficiency, and life habits (see Orwoll and Klein, 1995, for review). The mechanisms responsible for this bone loss in men are not very well known. Secondary hyperparathyroïdism results in an increased bone resorption but testosterone as well as estrogen deficiency in men may also contribute to a decreased bone formation (Riggs et al., 1998).

The challenge in osteoporosis treatment is to restore bone density, either by increasing bone formation or by decreasing bone resorption. The best rationale for reducing bone loss is to use therapeutic agents that would act on both resorption and formation. They should inhibit osteoclast recruitment and function resulting in a decrease of bone loss. On the other hand, they should activate the recruitment and activity of osteoblasts resulting in the restoration of lost bone. Most therapeutic agents currently used for the prevention of bone loss, such as estrogen and biphosphonates, act mainly as inhibitors of bone resorption. They are effective in the inhibition of bone turnover, as shown by biochemical marker measurements (Harris et al., 1993), resulting in the preservation of bone mass confirmed by bone densitometry and histomorphometry. However, their use may be somewhat limited by their side effects. In this context, an estrogen analogue, Raloxifene, which has a beneficial effect, on the incidence of osteoporosis and cardiovascular disease without stimulating the endometrium (Delmas et al., 1997), may be useful in the prevention of osteoporosis. The question for the future of osteoporosis treatment is the development of agents that stimulate bone formation. In this respect, the agents tested until now have given either negative or positive but preliminary results or side effects precluding their therapeutic use.

VIII. SUMMARY

Bone remodeling is an active process thoughout the skeleton and is mediated through the coupled processes of bone resorption and bone formation which are both finely regulated. The cells that remodel bone act within BMUs, located on bone surfaces. Changes in whole-skeleton resorptive and formative activity can be attributed to changes in the number of new BMUs initiated at the bone surface in unit time and/or changes in the work performed in unit time of the individual bone cells. Any disturbance in the number and the function of these BMUs leads to changes in bone mass observed during aging and menopause, two well-known determinants of osteoporosis. Bone histomorphometry, which provides information on the cellular mechanisms of bone remodeling, is very useful for the evaluation of the mechanisms underlying bone loss in oteoporosis. The recent development of bone marker measurements, which assess whole skeletal turnover, certainly contributes in improving the osteoporotic risk in elderly women.

ACKNOWLEDGMENTS

We would like to thank J.P. Roux for providing the light microscopy pictures of biopsy sections and M. Arlot for helpful discussion. This chapter was submitted in January 96 and briefly updated in August 98.

REFERENCES

Aarden, E.M., Burger, E.H., and Nijweide, P.J. (1994). Function of osteocytes in bone. J. Cell Biochem., 55 287-299.

Aaron, J.E., Makins, N.B., and Sagreiya, K. (1987). The microanatomy of trabecular bone loss in normal aging men and women. Clin. Orthop. Relat. Res. 215, 260-271.

Agerbaek, M.O., Eriksen, E.F., Kragstrup, J., and Mosekilde L. (1991). A reconstruction of the remodeling cycle in normal human cortical iliac bone. Bone Min. 12, 101-112.

Akesson, K., Vergnaud, P., Gineyts, E., Delmas, P.D., and Obrant K. (1993). Impairment of bone turnover in elderly women with hip fractures. Calcif. Tissue Int. 53, 162-169.

Anderson, H.C. and Morris D.C. (1993). Mineralization. In: Physiology and Pharmacology of Bone. (Mundy, G.R. and Martin, T.J., Eds.), pp. 267-298. Springer-Verlag, Berlin.

Aubin, J.E., Bellows, C.G., Turksen, K., Liu, F., and Heersche, J.N. (1992). Analysis of the osteoblast lineage and regulation of differentiation. In: Chemistry and Biology of Mineralized Tissues. (Slavkin, H. and Price, P., Eds.), pp. 267-276. Elsevier Science Publishers, Amsterdam.

Baron, R., Neff, L., Louvard, D., and Courtoy, P.J. (1985). Cell-mediated extracellular acidification and bone resorption: Evidence for a low pH in resorbing lacunae and localization of a 100-kDa lysosomal membrane protein at the osteoclast ruffled border. J. Cell. Biol. 101, 2210-2222.

Baron, R. (1989). Molecular mechanisms of bone resorption by the osteoclast. Anat. Rec. 224, 317-324.

Blair, H.C., Teitelbaum, S.L., Grosso, L.E., Lacey, D.L., Tan, H., McCourt, D.W., and Jeffreys, J.J. (1993). Extracellular-matrix degradation at acid pH. Biochem. J. 290, 873-884.

Bonde, M., Garnero, P., Fledelius, C., Qvist, P., Delmas, P.D., and Christiansen, C. (1997). Measurement of bone degradation products in serum using antibodies reactive with an isomerized form of an 8 amino acid sequence of the C-telepeptide of type I collagen. J. Bone Miner. Res., 12(7), 1028-1034.

Coccia, P.F., Krivit, W., Cervenka, J., Clawson, C., Kersey, J.H., Kim, T.H., Nesbit, M.E., Ramsay, N.K.C., Warkentin, P.I., Teitelbaum, S.L., Kahn, A.J., and Brown, D.M. (1980). Successful bone marrow transplantation for infantile malignant osteopetrosis. N. Engl. J. Med. 302, 701-708.

Croucher, P.I., Gilks, W.R., and Compston, J.E. (1995). Evidence for interrupted bone resorption in human iliac cancellous bone. J. Bone Min. Res. 10 (10), 1537-1543.

Delmas, P.D., Bjarnason, N.H., Mitlak, B.H., Ravoux, A.C., Shah, A., Huster, W.J., Draper, M., and Christiansen, C. (1997). Effects of raloxifene on bone mineral density, serum cholesterol concentrations, and uterine endometrium in postmenopausal women. N. Engl. J. Med., 337(23), 1641-1647.

Delmas, P.D. and Malaval, L. (1993). The proteins of bone. In: Physiology and Pharmacology of Bone (Mundy, G.R. and Martin, T.J.), pp. 673-724. Springer-Verlag, Berlin.

Dempster, D.W. (1992). Bone remodeling. In: Disorders of Bone and Mineral Metabolism (Coe, F.L. and Favus, M.J., Eds.), pp 355-380. Raven Press, New York.

Dempster, D.W. (1995). Bone Remodeling. In: Osteoporosis (Riggs, L.W. and Melton, L.J., Eds.), pp. 67-91. Lippincott-Raven Publishers, Philadelphia.

Eriksen, E.F., Gundersen, H.J.G., Melsen, R., and Mosekilde, L. (1984). Reconstruction of the formative site in trabecular bone in 20 normal individuals employing a kinetic model for matrix and mineral apposition. Metab. Bone Dis. Relat. Res. 5, 243-252.

Eriksen, E.F. (1986). Normal and pathological remodeling of human trabecular bone: Three dimensional reconstruction of the remodeling sequence in normals and in metabolic bone disease. Endoc. Rev. 7 (4), 379-408.

Eriksen, E.F., Hodgson, S.F., Eastell, R., Cedel, S.L., O'Fallon, W.M., and Riggs, B.L. (1990). Cancellous bone remodeling in type I (postmenopausal) osteoporosis: Quantitative assessment of rates of formation, resorption, and bone loss at tissue and cellular levels. J. Bone Min. Res. 5 (4), 311-319.

Everts, V., Delaisse, J-M., Korper, W., Niehof, A., Vaes, G., and Beertsen, W. (1992). Degradation of collagen in the bone-resorbing compartment underlying the osteoclast involves both cysteine-proteinases and matrix metalloproteinases. J. Cell. Physiol. 150, 221-231.

Frost, H.M. (1964a). Dynamics of bone remodeling. In: Bone Biodynamics (Frost, H.M., Ed.), pp. 315-333. Little Brown and Co., Boston.

Frost, H.M. (1964b). Mathematical Elements of Lamellar Bone Remodeling. Charles C. Thomas, Springfield, IL.

Frost, H.M. (1969). Tetracycline-based histological analysis of bone remodeling. Calcif. Tiss. Res. 3, 211-237.

Frost, H.M. (1980). Resting seams: "On" and "Off" in lamellar bone-forming centers. Metab. Bone Dis. Rel. Res. 2S, 167-170.

Garnero, P., Dargent-Molina, P., Hans, D., Schott, A.M., Breart, G., Meunier, P.J., and Delmas, P.D. (1998). Do markers of bone resorption add to bone mineral density and ultrasonographic heel measurement for the prediction of hip fracture in elderly women? The EPIDOS prospective study. Osteoporis Int. (In Press.)

Garnero, P. and Delmas, P.D. (1996). Measurement of biochemical markers: Methods and limitation. In: Principles of Bone Biology (Bilezikian, J.P., Raisz L.G., and Rodan, G.A., Eds.), Academic Press Inc., San Diego, 1277-1292.

Garnero, P., Hauser, E., Chapuy, M.C., Marcelli, C., Gandjean, H., Muller, C., Cormier, C., Breard, G., Meunier, P.J., and Delmas, P.D. (1996a). Markers of bone resorption predict hip fracture in elderly women: The EPIDOS prospective study. J. Bone Miner. Res. 11(10), 1531-1538.

Garnero, P., Sornay-Rendu, E., Chapuy, M.C., and Delmas, P.D. (1996b). Increased bone turnover in late postmenopausal women is a major determinant of osteoporosis. J. Bone Min. Res. 11(3): 337-349.

Hansen, M.A., Kirsten, O., Riss, B.J., and Christiansen, C. (1991). Role of peak bone mass and bone loss in postmenopausal osteoporosis: 12-year study. Brit. Med. J. 303, 961-964.

Harris, E.T., Gertz, B.J., Genant, H.K., Eyre, D.R., Survill, T.T., Ventura, J.N., DeBrock, J., Ricerca, E., and Chesnut, C.H. (1993). The effect of short-term treatment with alendronate on vertebral density and biochemical markers of bone remodeling in early postmenopausal women. J. Clin. Endocrinol. and Metab. 76, 1399-1403.

Hattner, R., Epker, B.N. , Frost, H.M. (1965). Suggested sequential mode of control of changes in cell behaviour in adult bone remodeling. Nature (London) 206, 489-490.

Holliday, L.S. Welgus, H.G., Fliszar, C.J., Veith, M. Jeffrey, J.J., and Gluck, S.L. (1997). Initiation of osteoclast bone resorption by interstitial collagenase. J. Biol. Chem. 272(35), 22053-22058.

Hynes, R.O. (1987). Integrins: A family of cell surface receptors. Cell 48, 549-554.

Jaworski, Z.F. (1971). Some morphologic and dynamic aspects of remodelling on the endosteal-cortical and trabecular surfaces. In: Calcified Tissue, Structural, Functional, and Metabolic Aspects (Menczel, J. and Harell, A.), pp. 159-160. Academic Press, New York.

Jaworski, Z.F. (1984). Coupling of bone formation to bone resorption: A broader view. Calcif. Tissue Int. 36, 531-535.

Kimmel, D.B., Recker, R.R., and Lappe, J.M. (1992). Histomorphometry of normal pre- and postmenopausal women. Bone 13, A18 (Abstract).

Lips, P., Courpron, P., and Meunier, P.J. (1978). Mean wall thickness of trabecular bone packets in the human iliac crest: Changes with age. Calcif. Tissue Int. 26, 13-17.

Marchisio, P.C., Naldini, L., Cirillo, D., Primavera, M.V., Teti, A., and Zambonin-Zallone, A. (1984). Cell-substratum interactions of cultured avian osteoclasts is mediated by specific adhesion structures. J. Cell. Biol. 99, 1696-1705.

Martin, T.J. and Ng, K.W. (1994). Mechanisms by which cells of the osteoblast lineage control osteoclast formation and activity. J. Cell. Biochem. 56, 357-366.

Mellish, R.W.E., Garrahan, N.J., and Compston, J.E. (1989). Age-related changes in trabecular width and spacing in human iliac crest biopsies. Bone Min. 6, 331-338.

Meunier, P.J., Courpron, P., Edouard, C., Alexandre, C., Bressot, C., Lips, P., and Boyce, B.F. (1979). Bone histomorphometry in osteoporotic states. In: Osteoporosis II (Barzel, U.S., Ed.), pp. 27-47. Grune and Stratton, New York.

Meunier, P.J. (1995). Bone Histomorphometry. In: Osteoporosis: Etiology, Diagnosis and Management (Riggs, B.L. and Melton, L.J., Eds.), pp 299-318. Lippincott-Raven Publishers, Philadelphia.

Mosekilde, L. (1989). Sex differences in age-related loss of vertebral trabecular bone mass and structure: Biochemical consequences. Bone, 10 425-432.

Mundy, G.R. and Roodman, G.D. (1987). Osteoclast ontogeny and function. In: Bone and Mineral Research, (Peck, W.A., Ed.) Vol. 5, pp. 209-280. Elsevier, New York.

Nesbitt, S., Nesbitt, A., Helfrich, M., Horton, M. (1993). Biochemical characterization of human osteoclast integrins. J. Biol. Chem. 268, 16737-16745.

Orwoll, E.S. and Klein, R.F. (1995). Osteoporosis in men. Endocrine Reviews 16 (1), 298-327.

Parfitt, A.M. (1982). The coupling of bone formation to bone resorption: A critical analysis of the concept and its relevance to the pathogenesis of osteoporosis. Metab. Bone. Dis. Relat. Res. 4, 1-6.

Parfitt, A.M. (1983). The physiological and clinical significance of bone histomorphometry data. In: Bone Histomorphometry, Techniques and Interpretation. (Recker, R.R., Ed.), pp. 143-223. CRC Press Inc, Boca Raton, FL.

Parfitt, A.M., Mathews, C.H.E., Villanueva, A.R., Kleerekoper, M., Frame, B., Rao, D.S. (1983). Relationships between surface volume and thickness of iliac trabecular bone in ageing and in osteoporosis: Implications for the microanatomic and cellular mechanisms of bone. J. Clin. Invest. 72, 1396-1409.

Parfitt, A.M. (1988). Bone remodeling: relationship to the amount and structure of bone, and the pathogenesis and prevention of fractures. In: Osteoporosis: Etiology, diagnosis, and management (Riggs, B.L. and Melton L.J., Eds.), pp. 45-93, Raven Press, New York.

Parfitt, A.M. (1990). Bone-forming cells in clinical conditions. In: Bone, Vol. 1, The osteoblast and osteocyte. (Hall, B.K., Ed.), pp. 351 429. Telford Press, Caldwell, N.J.

Parfitt, A.M. (1992). The physiologic and pathogenetic significance of bone histomorphometric data. In: Disorders of bone and mineral metabolism. (Coe, F.L. and Favus, M.J., Eds.), pp. 475-489. Raven Press, New York.

Riggs, B.L., Khosla, S., and Melton, L.J. (1998). A unitary model for involutional osteoporosis: Estrogen deficiency causes both type I and type II osteoporosis in postmenopausal women and contributes to bone loss in aging men. J. Bone Min. Res. 13(5), 763-773.

Riggs, B.L. and Melton L.J. (1986). Involutional osteoporosis. N. Engl. J. Med. 314, 1676-1686.

Riis, S.B.J., Hansen, A.M., Jensen, K., Overgaard, K., and Christiansen, C. (1995). Low bone mass and fast rate of bone loss at menopause-equal risk factors for future fracture. A 15-year follow-up study. J. Bone Min. Res. 10, S1: S146 (Abstract).

Rodan, G.A., and Harada, S. (1997). The missing bone. Cell 89, 677-680.

Rodan, G.A. and Martin, T.J. (1981). The role of osteoblasts in hormonal control of bone resorption. Calcif. Tissue Int. 33, 349-351.

Ross, F.P., Chappel, J., Alvarez, J.I., Sander, D., Butler, W.T., Farach-Carson, M.C., Mintz, K.A., Gehron Robey, P., Teitelbaum, S.L., and Cheresh, D.A. (1993). Interactions between the bone matrix proteins osteopontin and bone sialoprotein and the osteoclast integrin $\alpha_v\beta_3$ potentiate bone resorption. J. Biol. Chem. 268 (13), 9901-9907.

Sato, T. Foged, N.T., and Delaisse, J.M. (1998). The mirgration of purified osteoclasts through collagen is inhibited by matrix metalloproteinase inhibitors. J. Bone Min. Res. 13(1), 59-66.

Teti, A., Marchisio, P.C. and Zallone, A.W. (1991). Clear zone in osteoclast function: Role of podosomes in regulation of bone resorbing activity. Am. J. Physiol. 261, C1-C7.

Tran Van, P., Vignery, A. and Baron, R. (1982a). Cellular kinetics of the bone remodeling sequence in the rat. Anat. Rec. 202, 441-451.

Tran Van, P., Vignery, A. and Baron, R. (1982b). An electron microscopic study of the bone remodeling sequence in the rat. Cell Tissue Res. 225, 283-292.

Vaes, G. (1988). Cellular biology and biochemical mechanism of bone resorption. Clin. Orthop. 231, 239-271.

Votta, B.J., Levy, M.A., Badger, A., Dodds, R.A., James, I.E., Thompson, S., Bossard, M.J., Carr, T., Connor, J.R., Tomaszek, T.A., Szewczuk, L., Drake, F.H., Verber, D.F., and Gowen, M. (1997). Peptide aldehyde inhibitors of cathepsin K inhibit bone resorption both in vitro and in vivo. J. Bone Miner. Res. 12, 1396-1406.

Walker, D.G. (1975). Control of bone resorption by hematopoietic tissue. The induction and reversal of congenital osteopetrosis in mice through use of bone marrow and splenic transplants. J. Exp. Med. 142, 651-663.

Weinstein, R.S. and Huston, M.S. (1987). Decreased trabecular width and increased trabecular spacing contribute to bone loss with aging. Bone 8, 137-142.

Whyte, M.P. (1994). Hypophosphatasia and the role of alkaline phosphatase in skeletal mineralization. Endocrine Reviews 15 (4), 439-461.

HORMONAL REGULATION OF BONE REMODELING

Kong Wah Ng and T. John Martin

I. Introduction 65
II. Role of Osteoblasts, Hormones, and Cytokines in Osteoclast Formation 66
A. Cytokines 68
B. Common Pathway for Osteoclast Formation? 76
III. Plasminogen Activator-Inhibitor System and the Coupling of Bone Resorption to Bone Formation 78
A. Plasminogen Activator-Inhibitor System 78
B. Transforming Growth Factor β 79
C. Bone Morphogenetic Proteins 81
D. Insulinlike Growth Factors 82
IV. Summary 83

I. INTRODUCTION

Bone remodeling refers to the renewal process whereby small pockets of old bone, dispersed throughout the skeleton and separated from others geo-

Advances in Organ Biology
Volume 5A, pages 65-100.

ISBN: 0-7623-0390-5

graphically as well as chronologically, are replaced by new bone throughout adult life. Remodeling is essential for the balance of bone formation and resorption, maintenance of normal bone structure, and calcium homeostasis. This sequence of resorption and formation has been referred to as a basic multicellular unit (BMU) of bone turnover (Frost, 1984) (see Chapter 2, this volume). A remodeling site is initiated by the appearance of osteoclasts (and precursors) following any of several humoral or local stimuli to resorption. Osteoclasts proceed to resorb an amount of bone to produce a small resorption pit, following which the cells move to resorb at another site. This resorptive phase is followed by an active reversal phase when the cement line is deposited (Baron et al., 1980). During the subsequent formative phase, actively synthesizing cuboidal osteoblasts appear and begin to deposit uncalcified matrix (osteoid) which is later mineralized. As the lacuna fills, the osteoblasts become less cuboidal and eventually become flattened lining cells (Vaughan, 1981), while the osteoid seam narrows and eventually disappears. Resorption and formation always occur successively in the same location and always in the same order (Parfitt, 1983, 1993); this process of bone resorption followed by an equal amount of formation has been termed coupling (Frost, 1964). One of the intriguing issues of bone cell biology is to determine how osteoclast precursors are recruited and induced to differentiate into mature osteoclasts and, in turn, how osteoblasts are instructed to replace just exactly that amount of bone which has been resorbed. Bone remodeling is controlled by several circulating hormones and locally produced factors, and intercellular communication among the different bone cells is an integral part of these mechanisms. This chapter examines the processes of communication between osteoblasts and osteoclasts, with particular attention to the interactive roles of osteotropic hormones and cytokines.

II. ROLE OF OSTEOBLASTS, HORMONES, AND CYTOKINES IN OSTEOCLAST FORMATION

There is increasing evidence for the functional and developmental interdependence of osteoclasts and osteoblasts. The concept that cells of the osteoblast lineage control the formation and activity of osteoclasts through the actions of a number of cytokines and growth factors generated locally stems from the observation that isolated osteoblasts, but not osteoclasts, respond to the bone resorbing hormones and possess receptors for them (Rodan and Martin, 1981).

There is little doubt that the formation of new osteoclasts from precursors is an important part of the response to bone-resorbing hormones (Suda et al., 1992, 1995; Martin and Ng, 1994). Several *in vitro* systems have provided strong evidence that accessory cells are necessary for the generation of osteoclasts from hemopoietic precursors (Burger et al., 1982; Scheven et al., 1986; Hagenaars et al., 1990). The more recently developed system of mouse bone marrow culture has allowed reproducible assays of osteoclast-forming capability and added greatly to the understanding of the processes leading to osteoclast development (Takahashi et al., 1988a,b,c; Shinar and Rodan, 1990; Suda et al., 1992). The latter experiments have clearly shown that recruitment of osteoclasts from precursors is an indirect effect mediated by viable cells of the osteoblast lineage and other cells of the bone marrow stroma. Coculture experiments use either bone marrow or spleen cells as a source of osteoclast precursors that are grown in the presence of 1α,25-dihydroxy vitamin D_3 (1α,25$(OH)_2D_3$). In this system, the requirement for osteoblasts or stromal cells to be cultured on the same surface strongly implies that cell-cell contact is necessary for the promotion of osteoclast formation (Akatsu et al., 1991). Such contact might be necessary to allow the action of a molecule expressed on the cell membrane of osteoblast/stromal cells capable of promoting osteoclast formation.

A factor termed 'osteoclast differentiation factor' (ODF) has recently been cloned and fulfils the functions of such a putative membrane-associated peptide. ODF encodes a 316-amino-acid, type-II transmembrane protein and is a member of the TNF ligand family (Yasuda et al., 1998). Recombinant protein corresponding to the extracellular domain of ODF stimulates the formation of active, bone-resorbing osteoclasts from hemopoietic cells within the spleen even in the absence of stromal cells. A peptide identical to ODF has also been cloned from T cells and given the terms tumor necrosis factor-related activation-induced cytokine (TRANCE) or receptor activator of NF-κB ligand (RANKL) (Wong et al., 1997). When released by T cells following activation of the T cell receptor, it mediates the interaction of T and dendritic cells resulting in stimulation and increased survival of the naive T cells. RANK, another member of the TNF-receptor family, has been identified on dendritic cells and acts as the receptor for ODF/TRANCE/RANKL (Anderson et al., 1997). The action of ODF is antagonized by Osteoprotegerin (OPG), a soluble factor secreted by osteoblastic stromal cells. Overexpression of OPG in transgenic mice resulted in severe osteopetrosis with a loss of marrow cavities and profound depletion of osteoclasts. The same effects were observed upon administration of OPG in normal mice. Furthermore,

OPG blocked ovariectomy-associated bone loss in rat. OPG mRNA transcripts have been identified within bone and cartilage, vascular structures, midgut and kidney, and several osteoblast cell lines. Current data suggests that OPG blocks the terminal stages of osteoclast differentiation but not the formation of mononuclear osteoclast precursors (Simonet et al., 1997, Tsuda et al., 1997).

A. Cytokines

The systemic hormones, parathyroid hormone (PTH) (and parathyroid hormone-related protein, PTHrP) and $1\alpha,25(OH)_2D_3$ are well-recognized stimulators of osteoclast formation in bone marrow cultures and in co-cultures of osteoblast/stromal cells with hemopoietic cells (Suda et al., 1992). In addition to systemic factors, the bone marrow microenvironment plays an essential role in bone remodeling as a source of cytokines. Cytokines are soluble peptides that regulate cell growth and differentiation. They exert their effects by interacting with specific cell surface receptors, and a characteristic feature of cytokines is their functional pleiotropy and redundancy. Unlike classical endocrine hormones, they are produced locally from diverse sources, acting mainly as paracrine or autocrine regulators. In the discussion to follow, many examples will be provided to show that cytokine action is typified by intricate interactive networks that serve to amplify the responses triggered by the initiating event.

Cytokines relevant to bone cell function are the interleukins (ILs) 1, 3, 4, 6, 11, 13, and 18; tumor necrosis factors (TNFs) α and β; leukemia inhibitory factor (LIF); colony stimulating factors M-CSF and GM-CSF. Production of many of these is under the control of circulating hormones such as PTH and $1\alpha,25(OH)_2D_3$ (reviewed in Suda et al., 1992, 1995; Mundy, 1993; Martin and Ng, 1994; Horwood et al., 1998).

The functional pleiotropy and redundancy of cytokines can be explained by the molecular biology of the cytokine receptor system. Most cytokine receptors consist of a low affinity ligand-binding receptor (α chain) and a class-specific, non-ligand binding, signal-transducing component (β chain) (Miyajima et al., 1992; Taga and Kishimoto, 1992). In the case of IL-6, IL-11, oncostatin-M (OSM), LIF, ciliary neurotrophic factor (CNTF) and the recently described cytokine, cardiotrophin-1 (CT-1), the common signal transducing β chain is a 130 kDa glycoprotein (gp 130) (Yin et al., 1992, 1993; Hilton et al., 1994; Kishimoto et al., 1994, 1995; Pennica et al., 1995). Signaling occurs after the ligand-bound receptor dimerizes with gp 130 to form a high affinity receptor complex, resulting in activation of downstream molecules which include members of the JAK family of nonreceptor kinases and a latent transcription factor, signal transducer and activator of transcription factor 3 (STAT3) (Lutticken et al., 1994; Narazaki et al., 1994; Yin et al., 1994; for reviews, see Ihle, 1995; Kishimoto et al., 1995). Sharing a signal transducer allows these cytokines to mediate similar functions on various tissues (Taga et al., 1989, 1992; Gearing et al., 1992; Ip et al., 1992; Yin et al., 1992; Zhang et al., 1994). Since gp 130 is ubiquitously expressed (Saito et al., 1992), the specificity of action of each cytokine may be determined by the differential expression of the ligand-specific receptors on target cells. A common β chain is used by receptors for IL-3, IL-5, and GM-CSF (Miyajima et al., 1992) while the γ chain of the IL-2 receptor system is shared by receptors for IL-2, IL-4, and IL-7 (Kondo et al., 1993; Noguchi et al., 1993; Russell et al., 1993; Zurawski et al., 1993; Kishimoto et al., 1994). Conceptually, the action of cytokines may be influenced by the relative amounts of any of the cytokines, their specific receptors, or their signal transducers.

Colony-Stimulating Factors

A mutation in the coding region of the M-CSF gene in the mouse impairs the ability to form multinucleate osteoclasts, resulting in one variant of murine osteopetrosis, the op/op mouse (Felix et al., 1990; Yoshida et al., 1990). Marrow hemopoietic cells pretreated with M-CSF, GM-CSF, and IL-3 before coculture with osteoblast/stromal cells in the presence of $1\alpha,25(OH)_2D_3$ show increased formation of osteoclasts, with M-CSF the most effective. These results suggest that CSFs, especially M-CSF, stimulate the growth of osteoclast progenitors, which then differentiate into osteoclasts in response to $1\alpha,25(OH)_2D_3$ in the presence of osteoblasts (Takahashi et al., 1991).

Interleukin 1 and Tumor Necrosis Factors

IL-1α and IL-1β are potent *in vitro* and *in vivo* bone resorbing factors (Gowen et al., 1983; Lorenzo et al., 1987; Sabatini et al., 1988; Boyce et al., 1989) produced by macrophages, monocytes, and other cells of bone marrow. IL-1 has two receptors, an 80 kDa high affinity receptor mediating the effects of IL-1 and present in T lymphocytes and fibroblasts, and a 40 kDa receptor present on pre-B cells, bone marrow granulocytes, and macrophages. The latter is a naturally occurring IL-1 receptor antagonist, competing with the binding of IL-1 to the 80 kDa receptor (Mundy, 1993).

IL-1 promotes the proliferation and differentiation of osteoclast precursors (Pfeilschifter et al., 1989) through several mechanisms which include the secretion of prostaglandins (Akatsu et al., 1991) and the induction of IL-6 (Feyen et al., 1989; Löwik et al., 1989), M-CSF (Felix et al., 1989), as well as IL-11 secretion (Romas et al., 1995). IL-1 also upregulates mRNA for gp 130 (Romas et al., 1996). Its bone-resorbing activity is enhanced by synergizing with cytokines like TNFα (Sabatini et al., 1987) and IL-6 (Black et al., 1990) or with systemic hormones, such as PTH or PTHrP (Sato et al., 1989).

The action of IL-1 on bone resorption may be blocked by the naturally occurring IL-1 receptor antagonist which blocks the 80 kDa IL-1 receptor. IL-1-induced osteoclastogenesis is blocked by a monoclonal antibody against gp 130 (Romas et al., 1996) and at least partially by a monoclonal antibody against IL-11 (Girasole et al., 1994), but not by an antibody against the IL-6 receptor (IL6Rα) (Tamura et al., 1993). This would suggest that the osteoclastogenic actions of IL-1 are more likely to be mediated by IL-11 rather than IL-6. Although there is much evidence to support the concept that IL-1 stimulates osteoclast activity indirectly via a primary effect on osteoblasts (Thomson et al., 1986), the recent demonstration of mRNA expression for IL-1 receptors in human osteoclasts raises the possibility that osteoclasts and their precursors may also be capable of responding directly to this cytokine (Sunyer et al., 1995).

TNFα is a multifunctional cytokine produced chiefly by infiltrating macrophages and monocytes at sites of inflammation to regulate cell function locally. The actions of TNFα are mediated by specific cell surface receptors present on virtually all cells that have been examined. Binding of TNFα results in the activation of multiple signal transduction pathways, transcription factors, and regulation of transcription of a wide array of genes (Fiers, 1991; Jaattela, 1991; Vilcek and Lee, 1991; Aggarwal and Vilcek, 1992; Chaturvedi et al., 1994). Many of its functions are synergistic with those of IL-1. Together, they have pivotal roles in immune and inflammatory responses. Its predominant effect in bone is the stimulation of osteoclastic bone resorption (Bertolini et al., 1986). Chinese hamster ovary (CHO) cells transfected with the human TNF gene and inoculated into nude mice resulted in a marked increase on osteoclastic bone resorption and hypercalcemia (Johnson et al., 1989). TNFα also shows important interactions with other cytokines. It can induce IL-1 production (Roodman et al., 1987) and, together with IL-1, enhance IL-6 secretion by osteoblasts and stromal cells (Akira et al., 1990a). TNFα is also a potent inducer of IL-11 (Romas et al., 1995), M-CSF, and prostaglandin production by osteoblasts

(Sato et al., 1987). Conversely, the bone-resorbing actions of TNFα can be inhibited by neutralizing antibodies to IL-1 (Mundy, 1993) and *in vitro* osteoclast formation induced by TNFα is partially inhibited by a monoclonal antibody against IL-11 (Girasole et al., 1994).

Interest in the role of IL-1 in postmenopausal bone loss was prompted by the observation that IL-1 activity was increased in peripheral monocytes of postmenopausal women and returns to premenopausal levels with estrogen/progesterone treatment (Pacifici et al., 1989). Similar effects were also observed for TNF and GM-CSF (Pacifici et al., 1990). Estrogen decreases secretion and mRNA expression of both IL-1 and TNF in monocytes (Pacifici, 1992). A current hypothesis suggests that estrogen suppresses the expression of IL-1, IL-6, and TNFα. With the onset of oestrogen deficiency, IL-1, TNFα, and GM-CSF secretion by peripheral monocytes is increased. IL-1 and TNFα then induces IL-6 production by osteoblasts and marrow stromal cells. This leads to an increase in colony-forming units for granulocytes and macrophages, an increased number of osteoclasts, and enhanced bone resorption (Horowitz, 1993).

Interleukin-6

IL-6 is a multifunctional cytokine that regulates pleiotropic functions in many types of cells (Kishimoto et al., 1992, 1995). There is substantial *in vitro* and *in vivo* evidence to support an important role of IL-6 in bone cell physiology, particularly in osteoclast formation and function (Roodman, 1992). IL-6 is produced by osteoblasts in response to PTH, IL-1, TNFα, and lipopolysaccharides but not $1\alpha,25(OH)_2D_3$ (Ishimi et al., 1990; Littlewood et al., 1991; Greenfield et al., 1993). IL-6 induces the release of IL-1 which, in turn, stimulates multinucleate osteoclast formation in long-term human marrow cultures (Kurihara et al., 1990). Some evidence has been obtained that PTH and PTHrP-stimulated bone resorption in mouse bone organ culture is mediated by IL-6 (Löwik et al., 1989). Hypercalcemia is induced in nude mice transplanted with human tumor (MH-85) cells secreting IL-6 (Yoneda et al., 1993) or CHO cells transfected with the IL-6 gene (Black et al., 1991). More recently, IL-6 has been implicated in the bone loss caused by estrogen deficiency. Ovariectomy in the mouse results in increased osteoclast formation in marrow cultures (Kalu, 1990). The latter is prevented by anti-IL-6 antibody indicating a role for IL-6 in the promotion of osteoclast formation in the estrogen-depleted state (Jilka et al., 1992; Manolagas, 1992; Horowitz, 1993). Conversely, sex steroids inhibit IL-6 production by osteoblasts *in vitro* (Girasole et al., 1992; Jilka et al., 1992) and downregulate expression of gp 130 (Manolagas, 1995). Further support for the role of IL-6 in the mediation of bone loss associated with estrogen deficiency was observed with genetically engineered IL-6 deficient mice (Poli et al., 1994). Lack of IL-6 had no effect on the survival, fertility, or maintenance of trabecular bone mass in mice with intact ovarian function although there was increased cortical bone turnover. Following ovariectomy, IL-6 deficient mice maintained their bone mass, whereas wild-type mice lost 50% of trabecular bone volume as a result of a three- to fivefold increase in bone resorption.

Much is known about the regulation of IL-6 expression, the IL-6 receptor complex, and signal transduction initiated by the binding of IL-6 to its receptor. This will be summarized briefly as an example of cytokine action and its regulation.

IL-6 gene expression is enhanced in response to a variety of stimuli such as bacterial endotoxin, viral infections, cytokines such as TNFα or IL-1, platelet-derived growth factor and epidermal growth factor (Ray et al., 1988; Sehgal, 1992). IL-6 expression is inhibited by IL-4, IL-10 (de Waal Malefyt et al., 1991), IL-13 (Minty et al., 1993), glucocorticoids (Ray et al., 1990), and products of the tumor suppressor genes, retinoblastoma and p53 (Santhanam et al., 1991). Expression of IL-6 is mediated by major signal transduction pathways involving a variety of protein kinases and second messenger agonists, which include protein kinase C, cyclic adenosine monophosphate (cAMP), and intracellular calcium. Several regulatory elements have been identified in the 5' flanking region of the IL-6 gene, such as the multiple response element (MRE), AP-1, NF-IL-6 and NF-κB binding sites, glucocorticoid response elements, and a potential recognition sequence for members of the *ets* family of transcripion factors (Dendorfer et al., 1994). The κB-like sequence functions as a major *cis*-acting element for IL-6 induction by TNFα or IL-1 (Zhang et al., 1990) while NF-IL-6 is a nuclear factor that specifically binds to an IL-1 responsive element in the IL-6 gene (Akira et al., 1990b). At least four regulatory elements (MRE, AP-1, NF-IL-6, and NF-κB) cooperate to activate IL-6 gene transcription in response to cAMP and prostaglandins (Dendorfer et al., 1994).

The IL-6/IL-6 receptor complex interacts with gp 130 (Taga et al., 1989) and induces homodimerization and covalent linkage of gp 130 via disulfide bonds (Murakami et al., 1993). Soluble IL-6R (sIL-6R) which lacks the transmembrane and cytoplasmic regions of IL-6R is also capable of mediating the IL-6 signal through gp 130 (Yamasaki et al., 1988; Taga et al., 1989; Saito et al., 1991). In the mouse coculture system, osteoclast formation was strikingly induced by simultaneous treatment with IL-6 and sIL-6R, but not by IL-6 or sIL-6R alone (Tamura et al., 1993). In this system, osteoblastic cells express a very low level of IL-6R mRNA, unlike osteoclast progenitors which constitutively express relatively high levels of IL-6R mRNA. Treatment of osteoblastic cells with dexamethasone induced a marked increase in the expression of IL-6R mRNA to the extent that IL-6 alone could stimulate osteoclast formation without the need for addition of soluble IL-6R (Udagawa et al., 1995). This data emphasizes the importance of IL-6R regulation in IL-6 action. The constitutive expression of IL-6R mRNA in osteoclast progenitors could imply a direct action of IL-6 on osteoclast progenitors to form multinucleate osteoclasts. However, studies with transgenic mice constitutively expressing human IL-6R showed that osteoblasts from these mice could support osteoclast development in cocultures with normal spleen cells, in the presence of human IL-6 alone. In sharp contrast, osteoclast progenitors overexpressing human IL-6R were not able to differentiate into osteoclasts in cocultures with normal osteoblasts, in response to IL-6. This is an important result because it clearly indicates that the ability of IL-6 to induce osteoclast differentiation depends on signal transduction mediated by IL-6R expressed on osteoblasts, but not on osteoclast progenitors (Udagawa et al., 1995). However, direct action of IL-6 on multinucleate osteoclasts remains a possibility since mRNA for IL-6 and IL-6R have been demonstrated in multinucleate giant cells derived from human giant cell tumors (Ohsaki et al., 1992; Sakamuri et al., 1994).

Interleukin-11

IL-11 was isolated from a bone marrow derived stromal cell line based on its ability to stimulate the proliferation of IL-6 dependent cells (Paul et al., 1990). It is also expressed by other mesenchymal cells such as lung fibroblasts, bone marrow and placental stromal cells, articular chondrocytes, osteosarcoma cells, and synoviocytes (Maier et al., 1993; Elias et al., 1994, 1995; Rubin et al., 1995). The IL-11 receptor consists of a unique ligand-binding 150 kDa glycoprotein chain (IL-11Rα) and gp 130. Both components are necessary for high affinity binding and signal transduction (Yin et al., 1992, 1993; Hilton et al., 1994). Similar to the action of the IL-6/IL-6 receptor complex, binding of IL-11 to its receptor induces homodimerization of gp 130 (Kishimoto et al., 1995).

Several lines of evidence suggest that IL-11 is an important osteotropic factor. IL-11 receptor transcripts are present in chondroblastic and osteoblastic progenitor cells during mouse embryogenesis (Neuhas et al., 1994). Girasole et al. (1994) showed that IL-11 dose-dependently stimulated osteoclast-like multinucleate cell formation in co-cultures of mouse osteoblasts and bone marrow cells. In this system, monoclonal anti-IL-11 antibody inhibited osteoclast formation induced by $1\alpha,25(OH)_2D_3$ and PTH.

Work carried out in our laboratory showed that IL-1, TNFα, Prostaglnadin E_2 (PGE_2), PTH, and $1\alpha,25(OH)_2D_3$, but not IL-6, IL-4, or TGFβ, induced production of IL-11 by osteoblasts. PTH, IL-1, and $1\alpha,25(OH)_2D_3$ upregulated expression of mRNA for gp 130 (Bellido et al., 1995; Romas et al., 1995) but not IL-11Rα. An upregulation of gp 130 expression by systemic hormones could modulate the sensitivity of osteoblasts to cytokines such as IL-11 and IL-6/sIL-6Rα (Yang and Yang 1994; Bellido et al., 1995). In cocultures of mouse bone marrow cells with primary osteoblasts, the formation of multinucleate osteoclasts in response to IL-11 or IL-6 together with its soluble IL-6 receptor was dose-dependently inhibited by monoclonal anti-mouse gp 130 antibody which also inhibited osteoclast formation induced by IL-1, PTH, PGE_2, and $1\alpha,25(OH)_2D_3$ (Romas et al., 1996). IL-11Rα mRNA expression was demonstrated by *in situ* hybridization not only in osteoblasts, but also in osteoclast precursors and multinucleate osteoclasts. The presence of IL-11Rα mRNA in mature osteoclasts suggests another important biological function of IL-11 in osteoclasts, perhaps distinct from its role in osteoclast formation. Thus both osteoblasts and osteoclasts may be targets for IL-11 action (Romas et al., 1995, 1996).

Leukemia Inhibitory Factor

LIF is a cytokine that was characterized on the basis of its ability to induce differentiation of murine myeloid leukemia cells (Metcalf et al., 1988; Moreau et al., 1988; Abe et al., 1989; for reviews, see Gearing, 1990; Metcalf, 1991). LIF also stimulates other hemopoietic cells, including bone marrow blasts and megakaryocytes (Metcalf et al., 1991; Verfaillie and McGlave, 1991). Experiments with LIF-deficient mice derived by gene targeting techniques showed that LIF is required for the survival of the normal pool of hemopoietic stem cells, but not for their terminal differentiation (Escary et al., 1993). Other effects of LIF include inhibition of differentiation of embryonic stem cells (Smith et al., 1988), generation of sensory (Murphy et al., 1991) and cholinergic neurons (Yamamori et al., 1989), and stimulation of hepatic acute phase proteins similar to those induced by IL-6 (Baumann and Wong, 1989).

Production of LIF by osteoblasts and osteoblast-like cells suggests an important role as a paracrine or autocrine modulator in bone. Treatment of osteoblasts with LIF resulted in induction of alkaline phosphatase expression as well as a rapid and dose-dependent induction of mRNA for plasminogen activator inhibitor-1 (PAI-1) which correlated with inhibition of plasminogen activator (PA) activity (Allan et al., 1990; Noda et al., 1990; Rodan et al., 1990). The anabolic effects of this cytokine on bone formation *in vivo* could be related to the inhibition of protease activity (Metcalf and Gearing, 1989; Cornish et al., 1993).

The biological activities of LIF and CNTF are mediated through the heterodimers formed between the low-affinity LIF receptor (LIFR) with gp 130 (Gearing et al., 1994; Kishimoto et al., 1994, 1995). Therefore the absence of membrane-bound LIFR by gene knock-out would eliminate binding of LIF as well as signaling induced by ligand-bound CNTFRα. Murine LIFR mutants show disruption of normal placentation leading to poor intrauterine nutrition but still allowing fetuses to continue to term. Fetal bone volume in these LIFR mutants is reduced greater than threefold and the number of osteoclasts is increased sixfold, resulting in severe osteopenia of perinatal bone (Ware et al., 1995).

Interleukin-13 and interleukin-4

The actions of IL-1, TNFα, and IL-6 are antagonized by IL-4 and IL-13. IL-13 is produced by activated T lymphocytes. There is little known about other types of cells which secrete IL-13, the target cells of IL-13 action, or the signaling pathway involved. Mapping of the IL-13 gene shows that it is closely linked to the IL-4 gene on chromosome 5q-23-31 (Morgan et al., 1992). IL-13 exhibits limited homology to IL-4, particularly in the first and last α-helical regions of IL-4 which are critical for its activity. IL-13 strongly inhibits the secretion of IL-6 as well as mRNA expression of IL-1β and TNF induced by bacterial lipopolysaccharide in peripheral blood mononuclear cells (Minty et al., 1993). This action of IL-13 in blocking inflammatory monokine synthesis is shared with IL-4 (de Waal Malefyt et al., 1991). It has recently been proposed that there are two types of IL-4 receptor but only one IL-13 receptor. The binding of IL-4 to IL-4Rα could lead either to the recruitment of the IL-2Rγ to share a signaling pathway which involves activation of the cytoplasmic protein-tyrosine kinase JAK2 (Tanaka et al., 1994), or IL-13Rα. Conversely, binding of IL-13 to IL-13Rα would lead to recruitment of IL-4Rα to form a functional receptor (Lin et al., 1995). Signal transduction resulting from interaction between IL-4Rα and IL-13Rα could explain the similarity in actions between IL-4 and IL-13.

In a murine coculture system, where bone marrow macrophages were cultured with a stromal cell line, IL-4 or IL-13 inhibited the formation of tartrate-resistant acid phosphatase- (TRAP) positive cells (McHugh et al., 1995). In a different *in vitro* system, IL-13 and IL-4 suppressed the release of labeled calcium from prelabeled fetal mouse long bones stimulated by IL-1α, but not by PTH or 1α,25$(OH)_2D_3$. Inhibition of bone resorption by IL-13 and IL-4 was achieved by suppressing IL-1α-induced cyclooxygenase-2 mRNA expression and prostaglandin production in osteoblasts (Akatsu et al., 1991; Miyaura et al., 1995). It is possible that IL-4 and IL-13 may be equally effective in inhibiting the action of IL-11 since the effects of IL-11 on osteoclastogenesis can be abolished by indomethacin (Girasole et al., 1994).

Interleukin-18

Interleukin 18 (IL-18) was originally isolated from activated murine macrophages. In humans, the production of IFN-γ and GM-CSF in peripheral blood mononuclear cells are enhanced by IL-18 while in human T cells, it stimulates the production of T helper-type I cytokines, IL-2, GM-CSF, and IFN-γ. Osteoblastic stromal cells also produce IL-18 and it has inhibited the formation of osteoclasts in a coculture system of mouse hemopoietic and primary osteoblastic stromal cells. While both IFN-γ and GM-CSF were capable of inhibiting osteoclast formation in vitro, the action of IL-18 was shown to be mediated by GM-CSF and not IFN-γ (Udagawa et al., 1997). Recent work highlighted the involvement of T cells in IL-18 action, providing evidence for a new inhibitory pathway whereby IL-18 inhibited osteoclast formation by acting upon T cells to promote the release of GM-CSF (Horwood et al., 1998).

B. Common Pathway for Osteoclast Formation?

Osteoclast formation is induced by at least three different mechanisms (Suda et al., 1992, 1995). The first mechanism is the PTH–IL-1–PGE_2 axis which is mediated by cAMP. The second mechanism is 1α,25$(OH)_2D_3$-induced osteoclast formation. The gp 130 signal, activated by cytokines such as IL-11, IL-6/sIL-6Rα, LIF, OSM, and CNTF, is clearly an additional pathway of osteoclast formation. Although the actions of IL-6 and IL-11 on osteoclast formation are similar, there is evidence to support distinct roles for these two cytokines. Osteoclast formation by IL-6/sIL-6Rα, but not by IL-1 or 1α,25$(OH)_2D_3$, is inhibited by anti-IL-6Rα antibody, indicating that IL-6 is not implicated in osteoclastogenesis stimulated by IL-1 or 1α,25$(OH)_2D_3$ (Tamura et al., 1993). Girasole et al. (1994) showed that anti-IL-6 neutralizing antibody was completely ineffective in inhibiting osteoclast formation induced by IL-1, TNF, or IL-11, whereas a monoclonal anti-IL-11 antibody inhibited PTH, $1\alpha,25(OH)_2D_3$, IL-1, or TNF-mediated osteoclast formation. To take matters a step further, anti-gp 130 antibody completely abolished osteoclastogenesis induced by IL-11, IL-6/sIL-6Rα, IL-1, and significantly inhibited the effects of PGE_2, PTH, and $1\alpha,25(OH)_2D_3$ (Romas et al., 1996). Taken together, these results suggest that IL-11 ultimately mediates IL-1-induced osteoclast formation, consistent with the ability of IL-1 to induce both PGE_2 and IL-11 production by osteoblasts. It also implies that IL-11 contributes substantially to osteoclast formation induced by TNF, PGE_2, PTH, and $1\alpha,25(OH)_2D_3$. These results indicate a pivotal role of IL-11 and gp 130 in osteoclast development. IL-6, on the other hand, may play a significant role in osteoclast formation in the estrogen deficient state.

It has been clearly demonstrated that osteoblastic stromal cells are needed for osteoclast formation to proceed from hemopoietic precursors. A cell membrane stromal factor capable of programming the final stages of osteoclast differentiation may be the final common pathway for these different mechanisms for osteoclast formation (Figure 1; Martin and Ng, 1994;

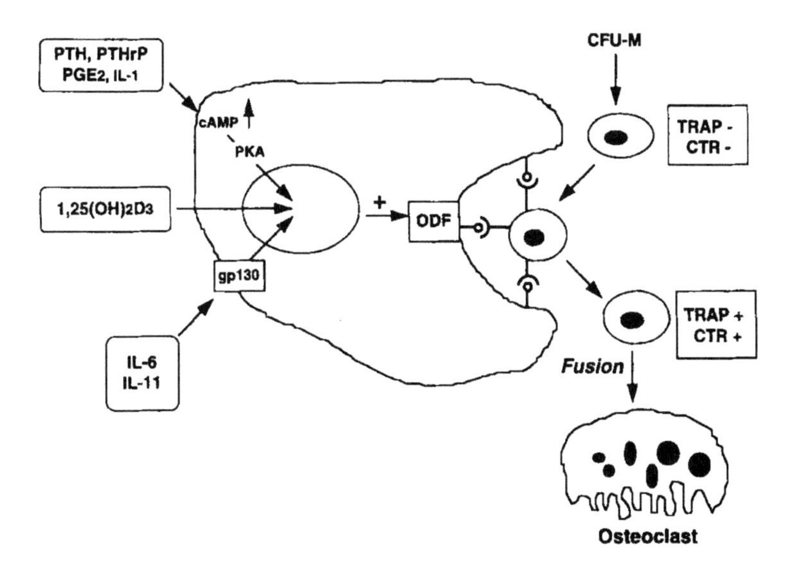

Figure 1. Central role of the stromal osteoblast in enhancing osteoclast formation from hemopoietic precursors. This model proposes the promotion of osteoclast formation by osteoclast differentiation factor (ODF), a factor associated with the stromal cell membrane. It is the common mediator of the effects of the three main classes of stimulators of osteoclast formation, one group acting through AMP and protein kinase A, one through a steroid hormone response pathway, and the cytokine stimulators with varied signaling pathways after interaction with specific cell-surface receptors. TRAP (tartrate-resistant acid phosphatase) and CTR (calcitonin receptor) are markers of the osteoclast phenotype.

Suda et al., 1995). There can be little doubt of the importance of locally generated factors in the control of bone cell formation and function. In the preceding discussion, the cytokines and their receptors have been considered in some detail, and reference made to their regulation by circulating hormones. Increasingly, it is becoming apparent that local events in bone are critical to the remodeling process, and that circulating hormones contribute to an important extent to the formation of cytokines and their receptors. Ultimately however, local concentrations of cytokines and active receptor components are likely to be determined by local, rather than humoral, factors. These include regulation of cytokine formation by other cytokines and by prostanoids.

III. PLASMINOGEN ACTIVATOR-INHIBITOR SYSTEM AND THE COUPLING OF BONE RESORPTION TO BONE FORMATION

As indicated earlier, an essential feature of the bone remodeling process is that when a given amount of bone is resorbed by osteoclasts, the same amount is replaced by bone-forming osteoblasts. It was considered for some time that this tight coupling might be achieved by a locally produced "coupling factor," and some evidence was produced for the existence of such a factor (Howard et al., 1981). Current evidence would suggest that the processes leading to new bone formation in bone remodeling units are orchestrated by growth factors such as transforming growth factor β (TGFβ), insulinlike growth factors, IGF-I, IGF-II, and bone morphogenetic proteins released locally as a result of bone resorption. These growth factors, secreted by osteoblasts and stored in bone matrix, are released and become activated as a result of the proteolytic action of the plasminogen activator system and products of osteoclastic activity. Osteoblast precursors are attracted to resorption pits, stimulated to proliferate and differentiate into mature osteoblasts before laying down new organic matrix to be mineralized.

A. Plasminogen Activator-Inhibitor System

The production of specific proteinases, such as collagenase and plasmin, by osteoblasts and their regulation by hormones and paracrine factors play an important role in the coupling of bone resorption to bone formation and possibly in osteoclast motility. Collagenase in bone is produced solely by osteoblasts (Sakamoto and Sakamoto, 1984), and is secreted in an inactive form which must be activated by proteolysis. The release of latent collagenase and collagenase inhibitor from osteoblast-like cells is promoted by PTH (Partridge et al., 1987). The production by osteoblasts of the serine protease, tissue plasminogen activator, and its inhibitors is very tightly regulated by a number of hormones such as PTH and cytokines (Allan et al., 1990, 1991; Fukumoto et al., 1992; Martin et al., 1993a,b). PA converts plasminogen to plasmin, which activates latent collagenase (Eeckhout and Vaes, 1977), thereby contributing to matrix breakdown or removal of the thin layer of collagen separating osteoclasts from minerals (Chambers and Fuller, 1985). This system is in turn regulated by specific inhibitors of both collagenase and PA (Cawston et al., 1981; Otsuka et al., 1984; Allan et al., 1991). The PAI system may also contribute to the motility of osteoclasts. Inactive single chain urokinase (sc uPA) produced by osteoblasts (Fukumoto et al., 1992) could bind to uPA receptors on osteoclasts and become activated as a result. This generates plasmin at those sites and the resulting pericellular proteolysis would contribute to cell motility which would cease when the appropriate part of the cell reaches PAI-1 present in the matrix. Plasminogen-dependent movement of the osteoclast has been demonstrated *in vitro* (Grills et al., 1990). At the conclusion of osteoclastic activity, the PA system continues to play an important role in coupling resorption to bone formation. TGFβ, IGF-I, and IGF-II, secreted by osteoblasts and stored in a latent form in bone matrix, are proteolytically cleaved from their binding proteins enabling them to modulate osteoblast function in an autocrine/paracrine fashion.

B. Transforming Growth Factor β

TGFβ plays an important role in the control of bone formation and remodeling (Bonewald and Mundy, 1990). This cytokine is abundant in bone and is a product of osteoblasts (Hauschka et al., 1986; Robey et al., 1987; for review, see Centrella et al., 1995b). The *in vitro* effects of TGFβ on osteoblasts have been widely studied and one of the major actions of TGFβ in osteoblasts is the regulation of extracellular matrix protein synthesis (Wrana et al., 1988; Sporn and Roberts, 1989). *In vitro* effects of TGFβ in osteoblasts include stimulation of proliferation in freshly isolated osteoblastlike cells (Centrella et al., 1987; Lomri and Marie 1990), stimulation of collagen synthesis (Centrella et al., 1986, 1987; Hock et al., 1990), regulation of gene expression of pro-α1(I) collagen, osteonectin, osteopontin, fibronectin and osteocalcin (Noda and Rodan, 1987; Noda et al., 1988; Noda, 1989), and chemotaxis (Pfeischifter et al., 1990; Hughs et al., 1992). Extracellular matrix accumulation is also enhanced indirectly through the action of TGFβ in inhibiting expression of matrix-degrading proteases while stimulating expression of protease inhibitors (Kubota et al., 1991; Tomooka et al., 1992). *In vivo* injections of TGFβ into the subperiosteal region resulted in localized intramembranous bone formation in rodent calvaria (Noda and Camilliere, 1989; Joyce et al., 1990; Mackie and Trechsel, 1990) and endochondral bone formation in rat femur (Noda and Camilliere, 1989). Increased intramembranous bone formation was also observed when TGFβ was infused into titanium chambers implanted into tibiae of baboons (Aufdemorte et al., 1992) or rabbits (Zhou et al., 1995) to study *in vivo* bone formation. Short-term systemic administration of recombinant TGFβ substantially increased cancellous bone formation in juvenile and adult rats (Rosen et al., 1994).

A member of a large family of structurally homologous proteins, TGFβ is secreted by virtually all cell types, including osteoblasts, in a latent biologically inactive form (Lawrence et al., 1985; Wakefield et al., 1988; Dallas et al., 1995) that is prevented from binding to the widely distributed TGFβ receptors (Cheifetz et al., 1987; Wakefield et al., 1987). TGFβ may be released from latent complexes at appropriate sites in bone by plasmin generated locally through the action of PA in a manner which is controlled temporally and spatially. In turn, TGFβ has a regulatory action on the PA system in that it decreases PA activity in a dose-dependent manner by increasing the synthesis of PAI-1 in osteoblasts (Laiho et al., 1987; Allan et al., 1991; Martin 1993a,b). Latent TGFβ stored in the bone matrix may also be released and activated by the protease, cathepsin-B (Oursler et al., 1993), or the acid microenvironment within the resorbing zones of active osteoclasts (Baron et al., 1985; Oreffo et al., 1989). TGFβ synthesis is increased by $1\alpha,25(OH)_2D_3$ in intact rodent bone explants or osteoblasts in culture (Petkovich et al., 1988; Finkelman et al., 1991), sex steroids in rat and human osteoblasts (Oursler et al., 1991; Finkelman et al., 1992; Westerlind et al., 1994), PTH in cultured human bone cells (Oursler et al., 1991), and bone morphogenetic protein-2 (BMP-2) in human osteoblasts (Zheng et al., 1994).

Modulation of TGFβ action may also be achieved by altering the ratio of binding to the different TGFβ receptors. TGFβ binds with high affinity to type I and type II receptors (Derynck, 1993; Miyazono et al., 1994) which are transmembrane serine/threonine kinases. Signal transduction occurs with the formation of a heteromeric complex between TGFβ and the two types of receptors. TGFβ is thought to bind directly to the type II receptor which is a constitutively active kinase. Type I receptor is then recruited into the complex to be phosphorylated by the type II receptor. Phosphorylation of type I receptor initiates the first step of a TGFβ signaling pathway (Wrana et al., 1992, 1994; Chen et al., 1993; Ebner et al., 1993). TGFβ binds with lower affinity to betaglycans which has been postulated to increase TGFβ binding to type II receptors to form stable complexes (Lopez-Casillas et al., 1993). These receptors are present on osteoblasts (Robey et al., 1987; Centrella et al., 1991a, 1995b). TGFβ binding at all binding sites in osteoblast-enriched bone cell cultures is enhanced by PTH (Centrella et al., 1988), while glucocorticoid treatment causes a redistribution of TGFβ binding from the signal-transducing type I receptor to nonsignal transducing betaglycans, resulting in a diminished TGFβ response (Centrella et al., 1991b). In contrast, TGFβ effects on fetal rat osteoblast differentiation are enhanced by treatment with BMP-2 which increases TGFβ binding to type I receptors, at the same time reducing its binding to type II receptors and betaglycans (Centrella et al., 1995a).

C. Bone Morphogenetic Proteins

Ectopic bone formation is elicited at intramuscular sites by implantation of bone inducing factors contained in demineralized bone matrix (Urist, 1965; Reddi and Huggins, 1972; Urist et al., 1983). Proteins identified as the active component in bone extract were given the name bone morphogenetic proteins (BMPs), a group of proteins which act to induce the differentiation of mesenchymal-type cells into chondrocytes and osteoblasts before initiating bone formation (Urist 1989).

Several members of this protein family have been isolated, cloned, and expressed as recombinant proteins (Wang et al., 1988, 1990; Wozney et al., 1988; Celeste et al., 1990; Sampath et al., 1990; Ozkaynak et al., 1992). BMP-1 is a novel protein but the other BMPs are all related molecules that share some common characteristics with the TGFβ superfamily (Ozkaynak et al., 1992; Wozney and Rosen, 1993). Implantation of recombinant human BMP-2 forms bone tissue *in vivo* (Wang et al., 1990). Recent *in vitro* evidence demonstrated that BMP-2 is capable of committing undifferentiated mesenchymal cells as well as myoblasts into osteoblast progenitors. BMP-2 inhibits myogenic differentiation in rat osteoblastic progenitor, C26, cells which retain the potential to differentiate into both myotubes and adipocytes (Yamaguchi and Kahn, 1991). In contrast, BMP-2 induces the expression of alkaline phosphatase mRNA, cAMP responses to PTH, and the synthesis of osteocalcin in the presence of $1\alpha,25(OH)_2D_3$ (Yamaguchi et al., 1991). Treatment of rat myoblastic C2C12 cells with human recombinant BMP-2 reversibly inhibits myotube formation by suppression of myogenin mRNA and induction of Id-1 mRNA expression. The cells are also induced to express alkaline phosphastase activity, cAMP responsiveness to PTH, and to produce osteocalcin (Katagiri et al., 1994).

The action of BMPs are not confined to osteogenic progenitors because, in more mature osteoblasts, BMPs stimulate cell proliferation, alkaline phosphatase activity, synthesis of type I collagen (Chen et al., 1991), PTH-specific cAMP responsiveness, and osteocalcin synthesis (Yamaguchi et al., 1991). BMP-2 induces TGFβ synthesis in human osteoblasts (Zheng et al., 1994) and increases TGFβ binding to the signal-transducing type I receptor (Centrella et al., 1995a). Since BMPs are synthesized by osteoblasts and stored in bone matrix, their release from storage in the course of bone resorption should enable them to exert their influence on the differentiation of osteoblast precursors migrating to the resorption pits.

D. Insulinlike Growth Factors

The insulinlike growth factors, IGF-I and IGF-II, are synthesized by osteoblasts and stored in bone matrix and act as paracrine or autocrine regulators of bone formation (McCarthy et al., 1990a; Canalis et al., 1993). In skeletal tissue, IGF-I and IGF-II are among the most prevalent growth factors (Canalis et al., 1988, 1993). IGFs stimulate the proliferation and/or differentiation of preosteoblasts, osteoblasts, fibroblasts, osteoclast progenitors and marrow stromal cells (Canalis et al., 1989a; Mohan and Baylink 1991; Zhang et al., 1991; Andress and Birnbaum, 1992; Mochizuki et al., 1992). IGF-1 secretion by osteoblasts is increased by PTH, PTHrP, $1\alpha,25(OH)_2D_3$ (McCarthy et al., 1989; Canalis et al., 1990; Scharla et al., 1991), PGE_2 (Canalis et al., 1993), and is inhibited by cortisol (McCarthy et al., 1990b). Less is known about the regulation of IGF-II since the factors that regulate IGF-I synthesis have no effect on IGF-II production (McCarthy et al., 1990a).

The IGFs exist primarily as large complexes bound to a family of six specific, high-affinity, binding proteins (IGFBPs) which regulate their bioavailability (Clemmons, 1992; Drop et al., 1992). IGFBP-1 and 6 have not been shown to exert any influence on bone cell function (Bach et al., 1994; Kachra et al., 1994), while IGFBP-2, 3, and 4 inhibit various parameters of bone formation (LaTour et al., 1990; Feyen et al., 1991; Schmid et al., 1991). In contrast, IGFBP-5 increases osteoblast growth and enhances the actions of IGF-I (Andress and Birnbaum, 1992). In turn, IGF-I enhances the synthesis and stability of IGFBP-5 (McCarthy et al., 1994). It was recently shown that the skeletal growth factors, TGFβ, platelet growth factor BB, and basic fibroblast growth factor, decrease the synthesis of IGFBP-5 (Canalis and Gabbitas, 1995); this is likely to be relevant in the control of IGF actions in bone through the regulation of one of its binding proteins.

IGF-I is a potent mitogen of proliferating preosteoblasts (Birnbaum et al., 1995), and plays an important role in collagen synthesis (Canalis et al., 1993). The synthesis and secretion of IGF-I and II from osteoblasts is stimulated by PTH (Canalis et al., 1989b; Linkhart and Mohan, 1989; McCarthy et al., 1989). Furthermore, the stimulatory effect of PTH on type I collagen synthesis is abolished by neutralizing antibodies against IGF-I, demonstrating that IGF-I can mediate some anabolic effects of PTH in bone (Canalis et al., 1989b). Regulation by hormones and cytokines of IGF-1 synthesis, and that of its binding proteins, may contribute to coupling by influencing availability of IGF-1 at appropriate sites in bone. The finding that plasmin can liberate IGF-1 from association with its inhibitory binding protein (Campbell et al., 1992) provides a further level of local control by the regulated PA system.

IV. SUMMARY

Recent advances in knowledge about cell biology of bone have led to a greater appreciation of the heterogeneity of the marrow microenvironment. Interactions between the osteoblast and osteoclast lineages during differentiation highlight the importance of locally generated growth factors in the renewal of the osteoblast and osteoclast populations and, hence, to the process of bone remodeling. The formation, actions, and interactions among cytokines are complex and are under the influence of circulating hormones. PTH and $1\alpha,25(OH)_2D_3$ are not only potent bone resorbing hormones, but also play key roles in modulating osteoblast differentiation and function in the bone formation phase of bone remodeling. The anabolic actions of PTH are mediated through its effects on osteoblast proliferation, regulation of synthesis of growth factors such as IGFs and TGFβ as well as the synthesis of $1\alpha,25(OH)_2D_3$ (Dempster et al., 1995) whose actions on the maturation of both osteoblastic and osteoclastic lineage cells are well documented (Lawson and Muir, 1991; Walters, 1995). Bone formation and resorption cannot be viewed in isolation. Both processes are linked and when balanced, the result is normal healthy bone. These interactions are summarized in Figure 2. There is much to learn about the nature and mechanisms of intercellular communication between osteoblasts and osteoclasts. A better understanding of the regulatory control of cytokine production and action

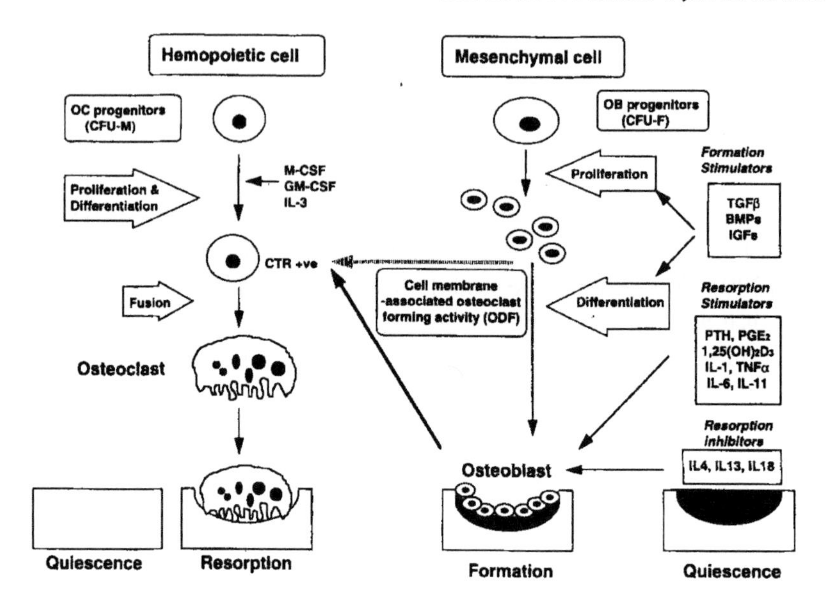

Figure 2. Interactions between the osteoblast and osteoclast lineages during differentiation. The importance of locally generated factors in bone remodeling is emphasized. Hemopoietic growth factors, products of members of the osteoblast lineage, stimulate proliferation of hemopoietic precursors of osteoclasts. Cells at various stages of osteoblast differentiation are capable of promoting osteoclast differentiation, mediating the actions of bone resorption stimulators. These are antagonized by cytokines such as IL-4, IL-13, and IL-18. The growth factor stimulators of bone formation are also locally produced.

may ultimately lead to the development of new strategies in the treatment of metabolic bone diseases.

REFERENCES

Abe, T., Murakami, M., Sato, T., Kajiki, M., Ohno, M., and Kodaira, R. (1989). Macrophage differentiation inducing factor from human monocytic cells is equivalent to murine leukaemia inhibitory factor. J. Biol. Chem. 264, 8941-8945.

Aggarwal, B.B. and Vilcek, J. (1992). In: Tumor necrosis factors: Structure, function, and mechanism of Action. Marcel Dekker, Inc., New York.

Akatsu, T., Takahashi, N., Udagawa, N., Imamura, K., Yamaguchi, A., Sata, K., Nagata, N., and Suda, T. (1991). Role of prostaglandins in interleukin-1 induced bone resorption in mice in vitro. J. Bone Miner. Res. 6, 183-190.

Akira, S., Hirano, T., Taga, T., and Kishimoto, T. (1990a). Biology of multifunctional cytokines: IL 6 and related molecules (IL 1 and TNF). FASEB. 4, 2860-2867.

Akira, S., Isshiki, H., Sugita, T., Tanabe, O., Kinoshita, S. H., Nishio, Y., Jakajima, T., Hirano, T., and Kishimoto, T. (1990b). A nuclear factor for IL-6 expression (NF-IL6) is a member of a C/EBP family. Embo. J. 9, 1897-1906.

Allan, E.H., Hilton, D.J., Brown, M.A., Evely, R.S., Yumita, S., Metcalf, D., Gough, N.M., Ng, K.W., Nicola, N.A., and Martin, T.J. (1990). Osteoblasts display receptors for and responses to leukemia inhibitory factor. J. Cell. Physiol. 145, 110-119.

Allan, E.H., Zeheb, R., Gelehrter, T.D., Heaton, J.H., Fukumoto, S., Yee, J.A., and Martin, T.J. Transforming growth factor beta inhibits plasminogen activator (PA) activity and stimulates production of urokinase-type PA, PA inhibitor-1 mRNA, and protein in rat osteoblastlike cells. (1991). J. Cell. Physiol. 149, 34-43.

Anderson, D.M., Maraskovsky, E., Billingsley, W.L., Dougall, W.C., Tometsko, M.E., Roux, E.R., Teepe, M.C., DuBose, R.F., Cosman, D., and Gaiibert, L. (1997). A homologue of the TNF receptor and its ligand enhance T-cell growth and dendritic-cell function. Nature 390, 175-179.

Andress, D.L. and Birnbaum, R.S. (1992). Human osteoblast-derived insulinlike growth factor (IGF) binding protein-5 stimulates osteoblast mitogenesis and potentiates IGF action. J. Biol. Chem. 26, 22467-22472.

Aufdemorte, T.B., Fox, W.C., Holt, G.R., McGuff, H.W., Ammann, A.J., and Beck, L.S. (1992). An intraosseous device for studies of bone-healing. J. Bone Joint Surg. 74-A, 1153-1161.

Bach, L.A., Hsieh, S., Brown, A.L., and Rechler, M.M. (1994). Recombinant human insulinlike growth factor (IGF)-binding protein-6 inhibits IGF-II induced differentiation of L6A1 myoblasts. Endocrinology 135, 2168-2176.

Baron, R., Neff, L., Louvard, I., and Courtoy, P.J. (1985). Cell-mediated extracellular acidification and bone resorption: Evidence for a low pH in resorbing lacunae and localization of a 100-kDa lysosomal membrane protein at the osteoclast ruffled border. J. Cell Biol. 101, 2210-2222.

Baron, R., Vignery, A., and Tran Van, P. (1980). The significance of lacunar erosion without osteoclasts: Studies on the reversal phase of the remodeling sequence. Metab. Bone Dis. Rel. Res. 2S, 35-40.

Baumann, H. and Wong, G.G. (1989). Hepatocyte-stimulating factor III shares structural and functional identity with leukemia-inhibitory factor. J. Immunol. 143, 1163-1167.

Bellido, T., Stahl, N., Jilka, R.L., Epstein, G., Yancopoulos, G., and Manolagas, S.C. (1995). Bone active hormones regulate the gp 130 signal transduction pathway. Bone. 16 (Suppl.), 142S.

Bertolini, D.R., Nedwin, G.E., Bringman, T.S., Smith, D.D., and Mundy, G.R. (1986). Stimulation of bone resorption and inhibition of bone formation in vitro by human tumour necrosis factors. Nature (London) 319, 516-518.

Birnbaum, R.S., Bowsher, R.R., and Wiren, K.M. (1995). Changes in IGF-1 and -II expression and secretion during the proliferation and differentiation of normal rat osteoblasts. J. Endocrinol. 144, 251-259.

Black, K., Garrett, R., and Mundy, G.R. (1991). Chinese hamster ovarian cells transfected with the murine interleukin-6 gene cause hypercalcemia as well as cachexia, leukocytosis, and thrombocytosis in tumor-bearing nude mice. Endocrinology 128, 2657-2659.

Black, K., Mundy, G.R., and Garrett, I.R. (1990). Interleukin-6 causes hypercalcemia in vivo, and enhances the bone resorbing potency of interleukin-1 and tumor necrosis factor by two orders of magnitude in vitro. J. Bone Miner. Res. 5 (Suppl. 2), S271.

Bonewald, L.F. and Mundy, G.R. (1990). Role of transforming growth factor-β in bone remodeling. Clin. Orthop. Rel. Res. 250, 261-276.

Boyce, B.F., Aufdemorte, T.B., Garrett, I.R., Yates, A.J.P., and Mundy, G.R. (1989). Effects of interleukin-1 on bone turnover in normal mice. Endocrinology 125, 1142-1150.

Burger, E.H., van der meer, J.W.M., van de Gevel, J.S., Gribnau, J.C., Thesingh, C.W., and van Furth, R. (1982). In vitro formation of osteoclasts from long-term cultures of bone marrow mononuclear phagocytes. J. Exp. Med. 156, 1604-1610.

Campbell, P.G., Novak, J.F., Yanosick, T.B., and McMaster, J.H. (1992). Involvement of the plasmin system in dissociation of the insulinlike growth factor binding protein. Endocrinology 130, 1410-1412.

Canalis, E., Centrella, M., Burch, W., and McCarthy, T. (1989b). Insulinlike growth factor mediates selective anabolic effects of parathyroid hormone in bone cultures. J. Clin. Invest. 83, 60-65.

Canalis, E. and Gabbitas, B. (1995). Skeletal growth factors regulate the synthesis of insulinlike growth factor binding protein-5 in bone cell cultures. J. Biol. Chem. 270, 10771-10776.

Canalis, E., McCarthy, T. and Centrella, M. (1988). Isolation and characterization of insulinlike growth factor-I (Somatomedin C) from cultures of fetal rat calvariae. Endocrinology 122, 22-27.

Canalis, E., McCarthy, T.L., and Centrella, M. (1989a). The role of growth factors in skeletal remodeling. Endocr. Metab. Clin. North Am. 18, 903-918.

Canalis, E., McCarthy, T.L., and Centrella, M. (1990). Differential effects of continous and transient treatment with parathyroid hormone related peptide (PTHrP) on bone collagen synthesis. Endocrinology 126, 1806-1812.

Canalis, E., McCarthy, T.L., and Centrella, M. (1993). Factors that regulate bone formation. In: Handbook of Experimental Pharmacology. (Mundy, G.R. and Martin, T.J., Eds). 107, 249-266.

Cawston, T.E., Galloway, W.A., Mercer, E., Murphy, G., and Reynolds, J.J. (1981). Purification of rabbit bone inhibitor of collagenase. Biochem. J. 195, 159-165.

Celeste, A.J., Iannazzi, J.A., Taylor, R.C., Hewick, R.M., Rosen, V., Wang, E.A., and Wozney, J.M. (1990). Identification of transforming growth factor β family members present in bone-inductive protein purified from bovine bone. Proc. Natl. Acad. Sci. U.S.A. 87, 9843-9847.

Centrella, M., Casinghino, S., Kim, J., Pham, T., Rosen, V., Wozney, J., and McCarthy, T.L. (1995a). Independent changes in type I and type II receptors for tranforming growth factor β induced by bone morphogenetic protein 2 parallel expression of the osteoblast phenotype. Mol. Cell. Biol. 15, 3273-3281.

Centrella, M., Massague, J., and Canalis E. (1986). Human platelet-derived transforming growth factor-β stimulates parameters of bone growth in fetal rat calvariae. Endocrinology 119, 2306-2312.

Centrella, M., McCarthy, T.L., and Canalis, E. (1991a). Current concepts review: Transforming growth factor β and remodeling of bone. J. Bone Joint Surg. 73-A, 1418-1428.

Centrella, M., McCarthy, T.L., and Canalis, E. (1991b). Glucocorticoid regulation of transforming growth factor β1 (TGF-β1) activity and binding in osteoblast-enriched cultures from fetal rat bone. Mol. Cell. Biol. 11, 4490-4496.

Centrella, M., McCarthy, T.L., and Canalis, E. (1988). Parathyroid hormone modulates transforming growth factor β activity and binding in osteoblastic cells. Proc. Natl. Acad. Sci. U.S.A. 8, 5889-5893.

Centrella, M., McCarthy, T.L., and Canalis, E. (1987). Transforming growth factor-β is a bifunctional regulator of replication and collagen synthesis in osteoblast-enriched cell cultures from fetal rat bone. J. Biol. Chem. 262, 2869-2874.

Centrella, M., Rosen, V., Horowitz, M.C., Wozney, J.M., and McCarthy, T.L. (1995b). Transforming growth factor-β gene family members, their receptors, and bone cell function. In: Endocr. Rev. Monographs. (Bikle, D.D. and Negro-Vilar, A., Eds), Vol. 4, pp. 211-226. Endocrine Soc., Bethesda, MD.

Chambers, T.J. and Fuller, K. (1985). Bone cells predispose endosteal surfaces to resorption by exposure of bone mineral to osteoclastic contact. J. Cell Sci. 76, 155-165.

Chaturvedi, M.M., LaPushin, R., and Aggarwal, B.B. (1994). Tumor necrosis factor and lymphotoxin. J. Biol. Chem. 269, 14575-14583.

Cheifetz, S., Weatherbee, J.A., Tsang, M.L-S., Anderson, J.K., Mole, J.E., Lucas, R., Massague, J. (1987). The transforming growth factor-β system. A complex pattern of cross-reactive ligands and receptors. Cell 48, 409-415.

Chen, R-H., Ebner, R., and Derynck, R. (1993). Inactivation of the type II receptor reveals two receptor pathways for the diverse TGF-β activities. Science 260, 1335-1338.

Chen, T.L., Bates, R.L., Dudley, A., Hammonds, Jr., R.G., and Amento, E.P. (1991). Bone morphogenetic protein 2b stimulation of growth and osteogenic phenotypes in rat osteoblastlike cells: Comparison with TGF-β. J. Bone Miner. Res. 6, 1387-1393.

Clemmons, D.R. (1992). IGF Binding proteins: Regulation of cellular actions. Growth Regul. 2, 80-87.

Cornish, J., Callon, K., King, A., Edgar, S., and Reid, I.R. (1993). The effect of leukemia inhibitory factor on bone in vivo. Endocrinology 132, 1359-1366.

Dallas, S.L., Miyazono, K., Skerry, T.M., Mundy, G.R., and Bonewald, L.F. (1995). Dual role for the latent transforming growth factor-β binding protein in the extracellular matrix and as a structural matrix protein. J. Cell Biol. 131, 539-550.

de Waal Malefyt, R., Abrams, J., Bennett, B., Figdor, C.G., and de Vries, J.E. (1991). Interleukin 10 (IL-10) inhibits cytokine synthesis by human monocytes: An autoregulatory role of IL-10 produced by monocytes. J. Exp. Med. 174, 1209-1220.

Dempster, D.W., Cosman, F., Parisien, M., Shen, V., and Lindsay, R. (1995). Anabolic actions of parathyroid hormone on bone. In: Endocr. Rev. Monographs (Bikle, D.D., and Negro-vilar, A., Eds). Vol. 4, pp.227-246. Endocrine Soc., Bethesda.

Dendorfer, U., Oettgen, P., and Libermann, T. (1994). Multiple regulatory elements in the interleukin-6 gene mediate induction by prostaglandins, cyclic AMP, and lipopolysaccharide. Mol. Cell. Biol. 14, 4443-4454.

Derynck, R. (1993). TGF-β-receptor-mediated signaling. Trends Biochem. Sci. 19, 548-553.

Drop, S.L.S., Schuller, A.G.P., Lindenbergh-Korteive, D.J., Groffen, C., Brinkman, A., and Zwarthoff, E.C. (1992). Structural aspects of the IGF BP family. Growth Regul. 2, 69-79.

Ebner, R., Chen, R-H., Shum, L., Lawler, S., Zioncheck, T.F., Lee, A., Lopez, A.R., and Derynck, R. (1993). Cloning of a type I TGF-b receptor and its effect on TGF-β binding to the type II receptor. Science 260, 1344-1348.

Eeckhout, Y. and Vaes, G. (1977). Further studies on the activation of procollagenase, the latent precursor of bone collagenase. Effects of lysosomal cathepsin B: Plasma and kallikrein, and spontaneous activation. Biochem. J. 166, 21-31.

Elias, J.A., Tang, W., and Horowitz, M. (1995). Cytokine and hormonal stimulation of human osteosarcoma interleukin-11 production. Endocrinology 136, 489-498.

Elias, J.A., Zheng, T., Whiting, N.L., Trow, T.K., Merrill, W.W., Zitnik, R., Ray, P., and Alderman, E.M. (1994). Interleukin-1 and transforming growth factor-β regulation of fibroblast derived interleukin-11. J. Immunol. 152, 2421-2429.

Escary, J-L., Perreau, J., Duménil, D., Ezine, S., and Brûlet, P. (1993). Leukaemia inhibitory factor is necessary for maintenance of haematopoietic stem cells and thymocyte stimulation. Nature (London) 363, 361-364.

Felix, R., Cecchini, M.C., and Fleisch, H. (1990). Macrophage colony-stimulating factor restores in vivo bone resorption in the op/op osteopetrotic mouse. Endocrinology 127, 2592-2594.

Felix, R., Fleish, H., and Elford, P.R. (1989). Bone-resorbing cytokines enhance release of macrophage colony-stimulating activity by osteoblastic cell MC3T3-E1. Calcif. Tissue Int. 44, 356-360.

Feyen, J.H.M., Elford, P., Dipadova, F.E., and Trechsel, U. (1989). Interleukin-6 is produced by bone and modulated by parathyroid hormone. J. Bone Miner. Res. 4, 633-638.

Feyen, J.H., Evans, D.B., Binkert, C., Heinrich, G.F., Geisse, S., and Kocher, H.P. (1991). Recombinant human [cys^{281}]-insulinlike growth factor-binding protein 2 inhibits both basal and insulin-like growth factor I-stimulated proliferation and collagen synthesis in fetal rat calvariae. J. Biol. Chem. 266, 19469-49474.

Fiers, W. (1991). Tumor necrosis factor. Characterization at the molecular, cellular and in vivo level. FEBS Lett. 285, 199-212.

Finkelman, R.D., Bell, N.H., Strong, D.D., Demers, L.M., and Baylink, D.J. (1992). Ovariectomy selectively reduces the concentration of transforming growth factor β in rat bone: Implications for estrogen deficiency-associated bone loss. Proc. Natl. Acad. Sci. U.S.A. 89, 12190-12193.

Finkelman, R.D., Linkhart, T.A., Mohan, S., Lau, K.H., Baylink, D.J., and Bell, N.H. (1991). Vitamin D deficiency causes a selective reduction in deposition of transforming growth factor β in rat bone: Possible mechanism for impaired osteoinduction. Proc. Natl. Acad. Sci. U.S.A. 88, 3657-3660.

Frost, H.M. (1964). Dynamics of bone remodeling. In: Bone biodynamics (Frost, H.M., Ed.), pp. 315-333. Little Brown, Boston.

Frost, H.M. (1984). The origin and nature of transients in human bone remodeling dynamics. In: Clinical Aspects of Metabolic Bone Disease. (Frame, B., Parfitt, A.M., and Duncan, H., Eds), p. 124. Excerpta Medica, Amsterdam.

Fukumoto, S., Allan, E.H., Yee, J.A., Gelehrter, T.D., and Martin, T.J. (1992). Plasminogen activator regulation in osteoblasts: Parathyroid hormone inhibition of type I plasminogen activator inhibitor and mRNA. J. Cell. Physiol. 152, 346-355.

Gearing, D.P., Comeau, M.R., Friend, D.J., Gimpel, S.D., Thut, C.J., McGourty, J., Brasher, K.K., King, J.A., Gillies, S., Mosley, B., Ziegler, S.F., and Cosman, D. (1992). The IL-6 signal transducer, gp130: An oncostatin M receptor and affinity converter for the LIF receptor. Science 255, 1434-1437.

Gearing, D.P., Ziegler, S.F., Comeau, M.R., Friend, D., Thoma, B., Cosman, D., Park, L., and Mosley, B. (1994). Proliferative responses and binding properties of hematopoietic cells transfected with low-affinity receptors for leukemia inhibitory factor, oncostatin M, and ciliary neurotrophic factor. Proc. Natl. Acad. Sci. U.S.A. 91, 1119-1123.

Gearing, D.P. (1990). Leukemia inhibitory factor: Does the cap fit? Ann. N. Y. Acad. Sci. 628, 9-18.

Girasole, G., Jilka, R.L., Passeri, G., Boswell, S., Boder, G., Williams, D.C., and Manolagas, S.C. (1992). 17β-estradiol inhibits interleukin-6 production by bone marrow-derived stromal cells and osteoblasts in vitro: A potential mechanism for the antiosteoporotic effect of estrogens. J. Clin. Invest. 89, 883-891.

Girasole, G., Passeri, G., Jilka, R. L., and Manolagas, S.C. (1994). Interleukin-11: A new cytokine critical for osteoclast development. J. Clin. Invest. 93, 1516-1524.

Gowen, M., Wood, D.D., Ihrie, E.J., McGuire, M.K.B., and Russell, R.G.G. (1983). An interleukin-1-like factor stimulates bone resorption in vitro. Nature (London) 306, 378-380.

Greenfield, E.M., Gornik, S.A., Horowitz, M.C., Donahue, H.J., and Shaw, S.M. (1993). Regulation of cytokine expression in osteoblasts by parathyroid hormone: Rapid stimulation of interleukin-6 and leukemia inhibitory factor mRNA. J. Bone Miner. Res. 8, 1163-1171.

Grills, B., Gallagher, J.A., Allan, E.H., Yumita, S., and Martin, T.J. (1990). Identification of plasminogen activator in osteoclasts. J. Bone Miner. Res. 5, 499-506.

Hagenaars, C. E., van der Kraan, A.A.M., Kawilarang-deHaas, E.W.M., Spooncer, E., Dexter, T.M., and Nijweide, P.J. (1990). Interleukin-3 dependent hemopoietic stem cell lines are capable of osteoclast formation in vitro: A model system for the study on osteoclast formation. In: Calcium Regulation and Bone Metabolism (Cohn, D.V., Glorieux, E.H., and Martin, T.J., Eds.), p. 280. Elsevier Science Publishers, New York.

Hauschka, P.V., Maurakos, A.E., Iafrati, M.D., Doleman, S.E., and Klagsburn, M. (1986). Growth factors in bone matrix. J. Biol. Chem. 261, 12665-12674.

Hilton, D.J., Hilton, A.A., Raicevic, A., Rakar, S., Harrison-Smith, M., Gough, N.M., Begley, G., Metcalf, D., Nicola, N., and Willson, T. (1994). Cloning of a murine IL-11 receptor α-chain: Requirement for gp130 for high affinity binding and signal transduction. EMBO. J. 13, 4765-4775.

Hock, J.M., Canalis, E., and Centrella, M. (1990). Transforming growth factor-β stimulates bone matrix apposition and bone cell replication in cultured fetal rat calvariae. Endocrinology 126, 421-426.

Horowitz, M.C. (1993). Cytokines and estrogen in bone: Anti-osteoporotic effects. Science 260, 626-627.

Horwood, N.J., Udagawa, N., Elliott, J., Grail, D., Okamura, H., Kurimoto, M., Dunn, A.R., Martin, T., and Gillespie, M.T. (1998). Interleukin 18 inhibits obsteoclast formation via T cell production of granulocyte macrophage colony-stimulating factor. J. Clin. Invest. 101, 595-603.

Howard, G.A., Bottemieller, B.L., Turner, R.T., Rader, J.I., and Baylink, D.J. (1981). Parathyroid hormone stimulates bone formation and resorption in organ culture: Evidence for a coupling mechanism. Proc. Natl. Acad. Sci. U.S.A. 78, 3204-3207.

Hughs, F.J., Aubin, J.E., and Heersche, J.N.M. (1992). Differential chemotactic responses of different populations of fetal rat calvarial cells to PDGF and TGF-β. Bone Miner. 19, 63-74.

Ihle, J.N. (1995). Cytokine receptor signalling. Nature (London) 377, 591-594.

Ip, N.Y., Nye, S.H., Boulton, T.G., Davis, D., Taga, T., Li, Y., Birren, S.J., Yashkawa, K., Kishimoto, T., Anderson, D.J., Stahl, N., and Yancopoulos, G.D. (1992). CNTF and LIF act on neuronal cells via shared signaling pathways that involve the IL-6 signal transducing receptor component gp130. Cell 69, 1121-1132.

Ishimi, Y., Miyaura, C., Jin, C.H., Akatsu, T., Abe, E., Nakamura, Y., Yamaguchi, A., Yoshiki, S., Matsuda, T., Hirano, T., Kishimoto, T., and Suda, T. (1990). IL-6 is produced by osteoblasts and induces bone resorption. J. Immunol. 145, 3297-3303.

Jaattela, M. (1991). Biologic activities and mechanisms of action of tumor necrosis factor-α/cachectin. Lab. Invest. 64, 724-742.

Jilka, R.L., Hangoc, G., Girasole, G., Passeri, G., Williams, D.C., Abrams, J.S., Boyce, B., Broxmeyer, H., and Manolagas, S.C. (1992). Increased osteoclast development after estrogen loss: Mediation by interleukin-6. Science 257, 88-91.

Johnson, R.A., Boyce, B.F., Mundy, G.R., and Roodman, G.D. (1989). Tumors producing human TNF induce hypercalcemia and osteoclastic bone resorption in nude mice. Endocrinology 124, 1424-1427.

Joyce, M.E., Roberts, A.B., Sporn, M.B., and Bolander, M.E. (1990). Transforming growth factor-β and the initiation of chondrogenesis and osteogenesis in the rat femur. J. Cell Biol. 110, 2195-2207.

Kachra, Z., Chang-Ren, Y., Murphy, L.J., and Posner, B.I. (1994). The regulation of insulinlike growth factor-binding protein I messenger ribonucleic acid in cultured hepatocytes: The roles of glucagon and growth hormone. Endocrinology 135, 1722-1728.

Kalu, D.N. (1990). Proliferation of tartrate-resistant, acid phosphatase positive, multinucleate cells in ovariectomized animals. Proc. Soc. Exp. Biol. Med. 195, 70-74.

Katagiri, T., Yamaguchi, A., Komaki, M., Abe, E., Takahashi, N., Ikeda, T., Rosen, V., Wozney, J.M., Fujisawa-Sehara, A., and Suda, T. (1994). Bone morphogenetic protein-2 converts the differentiation pathway of C2C12 myoblasts into the osteoblast lineage. J. Cell Biol. 127, 1755-1766.

Kishimoto, T., Akira, S., Narazaki, N., and Taga, T. (1995). Interleukin-6 family of cytokines and gp130. Blood 86, 1243-1254.

Kishimoto, T., Akira, S., and Taga, T. (1992). Interleukin-6 and its receptor: A paradigm for cytokines. Science 25, 593-597.

Kishimoto, T., Taga, T., and Akira, S. (1994). Cytokine signal transduction. Cell 76, 253-262.

Kondo, K., Takeshita, T., Ishii, N., Nakamura, M., Watanabe, S., Arai, K-I., and Sugamura, K. (1993). Sharing of the interleukin-2 (IL-2) receptor γ chain between receptors for IL-2 and IL-4. Science 262, 1874-1877.

Kubota, S., Fridman, R., and Yamada, Y. (1991). Transforming growth factor-β suppresses the invasiveness of human fibrosarcoma cells in vitro by increasing expression of tissue inhibitor of metalloprotease. Biochem. Biophys. Res. Commun. 197, 129-136.

Kurihara, N., Bertolini, D., Suda, T., Akiyama, Y., and Roodman, G.D. (1990). Interleukin-6 stimulates osteoclastlike multinucleated cell formation in long term human marrow cultures by inducing IL-1 release. J. Immunol. 144, 426-430.

Laiho, M., Saksela, O., and Keski-Oja, J. (1987). Transforming growth factor-β induction of type-1 plasminogen activator inhibitor. J. Biol. Chem. 262, 17467-17474.

LaTour, D., Mohan, S., Linkhart, T.A., Baylink, D.J., and Strong, D.D. (1990). Inhibitory insulinlike growth factor binding protein: Cloning, complete sequence, and physiological regulation. Mol. Endocrinol. 4, 1806-1814.

Lawrence, D.A., Pircher, R., and Jullien, P. (1985). Conversion of a high molecular weight latent β-TGF from chicken embryo fibroblasts into a low molecular weight active β-TGF under acidic conditions. Biochem. Biophys. Res. Commun. 133, 1026-1034.

Lawson, D.E.M. and Muir, E. (1991). Molecular biology and vitamin D function. Proc. Nutr. Soc. 50, 131-137.

Lin, J.-X., Migone, T.S., Tsang, M., Friedmann, M., Weatherbee, J.A., Zhou, L., Yamauchi, A., Bloom, E.T., Mietz, J., John, S., and Leonard, W.J. (1995). The role of shared receptor motifs and common STAT proteins in the generation of cytokine pleiotropy and redundancy by IL-2, IL-4, IL-7, IL-13, and IL-15. Immunity 2, 331-339.

Linkhart, T.A. and Mohan, S. (1989). Parathyroid hormone stimulates release of insulinlike growth factor-1 (IGF-I), IGF-II from neonatal mouse calvaria in organ culture. Endocrinology 125, 1484-1491.

Littlewood, A.J., Russell, J., Harvey, G.R., Hughes, D.E., Russell, R.G.G., and Gowen, M. (1991). The modulation of the expresson of IL-6 and its receptor in human osteoblasts in vitro. Endocrinology 129, 1513-1520.

Lomri, A. and Marie, P.J. (1990). Bone cell responsiveness to TGF-β, parathyroid hormone, and prostaglandin E_2 in normal and osteoporotic women. J. Bone Miner. Res. 5, 1149-1155.

Lopez-Casillas, F., Wrana, J.L., and Massague, J. (1993). Betaglycan presents ligand to the TGF-β signaling receptor. Cell 73, 1435-1444.

Lorenzo, J.A., Sousa, S.L., Alander, C., Raisz, L.G., and Dinarello, C.A. (1987). Comparison of the bone-resorbing activity in the supernatants from phytohemaglutinin-stimulated human peripheral blood mononuclear cells with that of cytokines through the use of an antiserum to interleukin 1. Endocrinology 121, 1164-1170.

Löwik, C.W.G.M., Vanderpluijm, G., Bloys, H., Hoekman, K., Bijvoet, O.L.M., Aarden, L.A., and Papapoulos, S.E. (1989). Parathyroid hormone (PTH) and PTH-like protein (PLP) stimulate interleukin-6 production by osteogenic cells—a possible role of interleukin-6 in osteoclastogenesis. Biochem. Biophys. Res. Comun. 162, 1546-1552.

Lutticken, C., Wegenka, U. M., Yuan, J., Buschmann, J., Schindler, C., Ziemiecki, A., Harpur, A. G., Wilks, A. F., Yasukawa, K., Taga, T., Kishimoto, T., Barbieri, G., Pellegrini, S., Sendtner, M., Heinrich, P. C., and Horn, F. (1994). Association of transcription factor APRF and protein kinase JAK 1 with the interleukin 6 signal transducer gp130. Science 26, 89-92.

Mackie, E.J. and Trechsel, U. (1990). Stimulation of bone formation in vivo by transforming growth factor-β: Remodeling of woven bone and lack of inhibition by indomethacin. Bone 11, 295-300.

Maier, R, Ganu, V., and Lotz, M. (1993). Interleukin 11, an inducible cytokine in human articular chondrocytes and synoviocytes, stimulates the production of the tissue inhibitor of metalloproteinases. J. Biol. Chem. 268, 21527-21532.

Manolagas, S.C. (1992). Cytokines, hematopoiesis, osteoclastogenesis, and estrogens. Editorial. Calcif. Tissue Int. 50, 199-202.

Manolagas, S.C. (1995). Role of cytokines in bone resorption. Bone 16, (Suppl.) 91S.

Martin, T.J., Allan, E.H., and Fukumoto, S. (1993a). The plasminogen activator and inhibitor system in bone remodeling. Growth Regul. 3, 209-214.

Martin, T.J., Findlay, D.M., Heath, J.K., and Ng, K. (1993b). Osteoblasts: Differentiation and function. In: Handbook of Experimental Pharmacology (Mundy, G.R., and Martin, T.J., Eds), Vol. 107, pp. 149-183. Springer-Verlag, Heidelberg, Germany.

Martin, T.J. and Ng, K.W. (1994). Mechanisms by which cells of the osteoblast lineage control osteoclast formation and activity. J. Cell. Biochem. 58, 1-10.

McCarthy, T.L., Casinghino, S., Centrella, M., and Canalis, E. (1994). Complex pattern of insulinlike growth factor binding protein expression in primary rat osteoblast enriched cultures: Regulation by prostaglandin E_2, growth hormone, and the insulinlike growth factors. J. Cell. Physiol. 160, 163-175.

McCarthy, T.L., Centrella, M., and Canalis, E. (1989). Parathyroid hormone enhances the transcript and polypeptide levels of insulinlike growth factor-I in osteoblast enriched cultures from fetal rat bone. Endocrinology 124, 1247-1253.

McCarthy, T.L., Centrella, M., and Canalis, E. (1990a). Cyclic AMP induces insulinlike growth factor-I synthesis in osteoblast-enriched cultures. J. Biol. Chem. 265, 15353-15356.

McCarthy, T.L., Centrella, M., and Canalis, E. (1990b). Cortisol inhibits the synthesis of insulinlike growth factor-I in skeletal cells. Endocrinology 126, 1569-1575.

McHugh, K.P., Teitelbaum, S.L., and Ross, F.P. (1995). Interleukin-13, like interleukin-4, modulates murine osteoclastogenesis and expression of the integrin $\alpha_v\beta_3$ on osteoclast precursors. J. Bone Miner. Res. 10 (Suppl.) S487.

Metcalf, D. and Gearing, D.P. (1989). Fatal syndrome in mice engrafted with cells producing high levels of the leukemia inhibitory factor. Proc. Natl. Acad. Sci. U.S.A. 86, 5948-5952.

Metcalf, D., Hilton, D., and Nicola, N.A. (1991). Leukemia inhibitory factor can potentiate murine megakaryocyte production in vitro. Blood 77, 2150-2153.

Metcalf, D., Hilton, D.J., and Nicola, N.A. (1988). Clonal analysis of the actions of the murine leukemia inhibitory factor on leukaemic and normal murine haematopoietic cells. Leukemia 2, 216-221.

Metcalf, D. (1991). The leukemia inhibitory factor (LIF). Int. J. Cell Cloning. 9, 95-108.

Minty, A., Chalon, P., Derocq, J.-M., Dumont, X., Guillemot, J.-C., Kaghad, M., Labit, C., Leplatois, P., Liauzun, P., Miloux, B., Minty, C., Casellas, P., Loison, G., Lupker, J., Shire, D., Ferrara, P., and Caput, D. (1993). Interleukin-13 is a new human lymphokine-regulating inflammatory and immune responses. Nature (London) 362, 248-250.

Miyajima, A., Kitamura, T., Harada, N., Yokota, T., and Arai, K-I. (1992). Cytokine receptors and signal transduction. Annu. Rev. Immunol. 10, 295-331.

Miyaura, C., Onoe, Y., Ohta, H., Nozawa, S., Nagai, Y., Kaminakayashiki, T., Kudo, I., and Suda, T. (1995). Interleukin-13 inhibits bone resorption by suppressing cyclooxygenase-2 (COX2) mRNA expression and prostaglandin production in osteoblasts. J. Bone Miner. Res. 10, (Suppl.) S158.

Miyazono, K., ten Dijke, P., Ichijo, H., and Heldin, C-H. (1994). Receptors for transforming growth factor-β. Advances Immunol. 55, 181-220.

Mochizuki, H., Hakeda, Y., Wakatsuki, N., Usui, N., Akashi, S., Sato, T., Tanaka, K., and Kumegawa, M. (1992). Insulinlike growth factor-1 supports formation and activation of osteoclasts. Endocrinology 131, 1075-1080.

Mohan, S. and Baylink, D. J. (1991). Bone growth factors. Clin. Orthop. Rel. Res. 263, 30-48.

Moreau, J.F., Donaldson, D.D., Bennet, F., Witek-Giannotti, J., Clark, S.C., and Wong, G.G. (1988). Leukaemia inhibitory factor is identical to the myeloid growth factor human interleukin for DA cells. Nature (London) 336, 690-692.

Morgan, J.G., Dolganov, G.M., Robbins, S.E., Hinton, L.M., and Lovett, M. (1992). The selective isolation of novel cDNAs encoded by the regions surrounding the human interleukin-4 and 5 genes. Nucleic Acids Res. 20, 5173-5179.

Mundy, G.R. (1993). Cytokines of bone. In: *Handbook of Experimental Pharmacology.* (Mundy, G.R. and Martin, T. J., Eds), Vol. 107, pp. 185-207. Springer-Verlag, Heidelberg, Germany.

Murakami, M., Hibi, M., Nakagawa, N., Nakagawa, T., Yasukawa, K., Yamanishi, K., Taga, T., and Kishimoto, T. (1993). IL-6-induced homodimerization of gp-130 and associated activation of a tyrosine kinase. Science 260, 1808-1810.

Murphy, M., Reid, K., Hilton, D.J., and Bartlett, P.F. (1991). Generation of sensory neurons is stimulated by leukemia inhibitory factor. Proc. Natl. Acad. Sci. U.S.A. 88, 3498-3501.

Narazaki, M., Witthuhn, B.A., Ihle, J.N., Kishimoto, T., and Taga, T. (1994). Activation of JAK2 kinase mediated by the IL-6 signal transducer, gp130. Proc. Natl. Acad. Sci. U.S.A. 91, 2285-2289.

Neuhaus, H., Bettenhausen, B., Bilinski, P., Simon-Chazottes, D., Guenet, J., and Gossler, A. (1994). Etl2, a novel putative type-1 cytokine receptor expressed during mouse embryogenesis at high levels in skin and cells with skeletogenic potential. Dev. Biol. 166, 531-542.

Noda, M. and Camilliere, J.J. (1989). In vivo stimulation of bone formation by transforming growth factor-β. Endocrinology 124, 2291-2994.

Noda, M. and Rodan, G.A. (1987). Type β transforming growth factor (TGFβ) regulation of alkaline phosphatase expression and other phenotype-related mRNAs in osteoblastic rat osteosarcoma cells. J. Cell. Physiol. 133, 426-437.

Noda, M., Vogel, R. L., Hasson, D. M., and Rodan, G. A. (1990). Leukemia inhibitory factor suppresses proliferation, alkaline phosphatase activity, and type I collagen messenger ribonucleic acid level and enhances osteopontin mRNA level in murine osteoblastlike (MC3T3E1) cells. Endocrinology 127, 185-190.

Noda, M., Yoon, K., Prince, C.W., Butler, W.T., and Rodan, G.A. (1988). Transcriptional regulation of osteopontin production in rat osteosarcoma cells by type ? transforming growth factor. J. Biol. Chem. 263, 13916-13921.

Noda, M. (1989). Transcriptional regulation of osteocalcin production by transforming growth factor-β in rat osteoblast-like cells. Endocrinology 124, 612-617.

Noguchi, M., Nakamura, Y., Russel, S.M., Ziegler, S.F., Tsang, M., Cao, X., and Leonard, W. J. (1993). Interleukin-2 receptor γ chain: A functional component of the interleukin-7 receptors. Science 262, 1877-1880.

Ohsaki, Y., Takahashi, S., Scarcez, T., Demulder, A., Nishihara, T., Williams, R., and Roodman, G.D. (1992). Evidence for an autocrine-paracine role for interleukin-6 in bone resorption by giant cells from giant cell tumors of bone. Endocrinology 131, 2229-2234.

Oreffo, R.O., Mundy, G.R., Seyedin, S.M., and Bonewald, L.F. (1989). Activation of the latent bonederived TGF β complex by isolated osteoclasts. Biochem. Biophys. Res. Commun. 158, 817-823.

Otsuka, K., Sodek, J., and Limiback, H. (1984). Synthesis of collagenase and collagenase inhibitors by osteoblastlike cells in culture. Eur. J. Biochem. 145, 123-129.

Oursler, M.J., Cortese, C., Keeting, P., Anderson, M.A., Bonde, S.K., Riggs, B.L., and Spelsberg, T.C. (1991). Modulation of transforming growth factorβ production in normal human osteoblast-like cells by 17β-estradiol and parathyroid hormone. Endocrinology 129, 3313-3320.

Oursler, M.J., Riggs, B.L., and Spelsberg, T.C. (1993). Glucocorticoid induced activation of latent transforming growth factorβ by normal human osteoblast-like cells. Endocrinology 133, 2187-2196.

Ozkaynak, E., Schnegelsberg, P.N.J., Jin, D.F., Clifford, G.M., Warren, F.D., Drier, E.A., and Oppermann, H. (1992). Osteogenic protein 2: A new member of the transforming growth factorβ superfamily expressed early in embryogenesis. J. Biol. Chem. 2676, 25220-25227.

Pacifici, R., Rifas, L., McCracken, R., Vered, I., McMurtry, C., Avioli, L.V., and Peck, W.A. (1989). Ovarian steroid treatment blocks a postmenopausal increase in blood monocyte interleukin 1 release. Proc. Natl. Acad. Sci. U.S.A. 86, 2398-2402.

Pacifici, R. (1992). Is there a causal role for IL-1 in postmenopausal bone loss? Editorial. Calcif. Tissue Int. 50, 295-299.

Pacifici, R., Brown, C., Rifas, L., and Avioli, L.V. (1990). TNFα and GM-CSF secretion from human blood monocytes: Effect of menopause and estrogen replacement. J. Bone Miner. Res. 5, (Suppl. 2), S110.

Parfitt, A.M. (1983). The physiological and clinical significance of bone histomorphometric data. In: Bone histomorphometry: Techniques and Interpretation (Recker, R.R., Ed.), pp. 143-223. CRC Press, Boca Raton, Florida.

Parfitt, A.M. (1993). Calcium homeostasis. In: Handbook of Experimental Pharmacology (Mundy, G.R. and Martin, T.J., Eds) 107, 1-65.

Partridge, N.C., Jeffrey, J.J., Ehlick, L.S., Teitelbaum, S.L., Fliszar, C., Welgus, H.G., and Khan, A.J. (1987). Hormonal regulation of the production of collagenase and a collagenase inhibitor activity by rat osteogenic sarcoma cells. Endocrinology 120, 1956-1962.

Paul, S.R., Bennett, F., Calvetti, J.A., Kelleher, K., Wood, C.R., O'Hara, Jr., R.M, Leary, A.C., Sibley, B., Clark, S.C., Williams, D.A., and Yang, Y.C. (1990). Molecular cloning of a cDNA encoding interleukin 11, a stromal cell-derived lymphopoietic cytokine. Proc. Natl. Acad. Sci. U.S.A. 87, 7512-7516.

Pennica, D., Shaw, K., Swanson, T., Moore, M., Shelton, D., Zioncheck, K., Rosenthal, A., Taga, T., Paoni, N., and Wood, W. (1995). Cardiotrophin-1. J. Biol. Chem. 270, 10915-10922.

Petkovich, P.M., Wrana, J.L., Grigoriadis, A.E., Heersche, J.N.M., and Sodek, J. (1988). 1,25-Dihydroxyvitamin D_3 increases epidermal growth factor receptors and transforming growth factor β-like activity in a bone-derived cell line. J. Biol. Chem. 2, 13424-13428.

Pfeilschifter, J., Wolf, O., Nautmann, A., Minne, H.W., Mundy, G.R., and Ziegler, R. (1990). Chemotactic response of osteoblastlike cells to TGF-β. J. Bone Miner. Res. 5, 825-830.

Pfeilschifter, J., Chenu, C., Bird, A., Mundy, G.R., and Roodman, G.D. (1989). Interleukin-1 and tumor necrosis factor stimulate the formation of human osteoclastlike cells in vitro. J. Bone Miner. Res. 4, 113-118.

Poli, V., Balena, R., Fattori, E., Markatos, A., Yamamoto, M., Tanaka, H., Ciliberto, G., Rodan, G.A., and Costantini, F. (1994). Interleukin-6-deficient mice are protected from bone loss caused by estrogen depletion. EMBO. J. 13, 1189-1196.

Ray, A., LaForge, K.S., and Sehgal, P.B. (1990). On the mechanism for efficient repression of the interleukin-6 promoter by glucocorticoids: Enhancer, TATA box and RNA start site (Inr motif) occlusion. Mol. Cell. Biol. 10, 5736-5746.

Ray, A., Tatter, S. B., May, L.T., and Sehgal, P.B. (1988) Activation of the human "B2-interferon/hepatocyte-stimulating factor/interleukin 6" promoter by cytokines, viruses, and second messenger agonists. Proc. Natl. Acad. Sci. U.S.A. 85, 6701-6705.

Reddi, A.H. and Huggins, C. (1972). Biochemical sequences in the transformation of normal fibroblasts in adolescent rats. Proc. Natl. Acad. Sci. U.S.A. 6, 1601-1605.

Robey, P.G., Young, M.F., Flanders, K.C., Roche, N.S., Kondiah, P., Reddi, A.H., Termine, J.D., Sporn, M.B., and Roberts, A.B. (1987). Osteoblasts synthesize and respond to transforming growth factor-type β (TGF-β) in vitro. J. Cell Biol. 105, 457-463.

Rodan, G., and Martin, T.J. (1981). Role of osteoblasts in hormonal control of bone resorption. Calcif. Tissue Int. 33, 349-351.

Rodan, S.B., Wesolowski, G., Hilton, D.J., Nicola, N.A., and Rodan, G.A. (1990). Leukemia inhibitory factor binds with high affinity to preosteoblastic RCT-1 cells and potentiates the retinoic acid induction of alkaline phosphatase. Endocrinology 127, 1602-1608.

Romas, E., Udagawa, N., Hilton, D.J., Suda, T., Ng, K.W., and Martin, T.J. (1995). Osteotropic factors regulate interleukin-11 production by osteoblasts. J. Bone Miner. Res. 10 (Suppl.), S142.

Romas, E., Udagawa, N., Zhou, H., Tamura, T., Saito, M., Taga, T., Hilton, D.J., Suda, T., Ng, K.W., and Martin, T.J. (1998). The role of gp130-mediated signals in osteoclast development: Regulation of interleukin 11 production by osteoblasts and distribution of its receptor in bone marrow cultures. J. Exp. Med. 183, 2581-2591.

Roodman, G.D., Takahashi, N., Bird, A., and Mundy, G.R. (1987). Tumor necrosis factor α (TNF) stimulates formation of osteoclastlike cell (OCL) in long-term human marrow cultures by stimulating production of interleukin-1 (IL-1). Clin. Res. 35, 515A.

Roodman, G.D. (1992). Interleukin-6: An osteotropic factor ? J. Bone Miner. Res. 7, 475-478.

Rosen, D., Miller, S.C., DeLeon, E., Thompson, A.Y., Bentz, H., Mathews, M., and Adams, S. (1994). Systemic administration of recombinant transforming growth factor β 2 (rTGF-β2) stimulates parameters of cancellous bone formation in juvenile and adult rats. Bone 15, 355-359.

Rubin, L.P., Tsai, S.W., Ireland, R.C., McGinnis, R., Iovene, C., and Paul, S.R. (1995). Parathyroid hormone related protein can regulate human trophoblast differentiation via induction of interleukin 11 in placental stromal cells. J. Bone Miner. Res. 10 (Suppl.), S342.

Russell, S.M., Keegan, A.D., Narada, N., Nakamura, Y., Noguchi, M., Leland, P., Friedman, M. C., Miyajima, A., Puri, R.K., Paul, W.E., and Leonard, W.J. (1993). Interleukin-2 receptor γ chain: A functional component of the interleukin-4 receptor. Science 262, 1880-1883.

Sabatini, M., Boyce, B., Aufdemorte, T., Bonewald, L., and Mundy, G.R. (1988). Infusions of recombinant human interleukin 1 α and β cause hypercalcemia in normal mice. Proc. Natl. Acad. Sci. U.S.A. 85, 5235-5239.

Sabatini, M., Garrett, I.R., and Mundy, G.R. (1987). TNF potentiates the effects of interleukin-1 on bone resorption in vitro. J. Bone Miner. Res. 2 (Suppl. 1), 34 (Abstract).

Saito, M., Yoshida, K., Hibi, M., Taga, T., and Kishimoto, T. (1992). Molecular cloning of a murine IL-6 receptor-associated signal transducer, gp130, and its regulated expression in vivo. J. Immunol. 148, 4066-4071.

Saito, T., Yasukawa, K., Suzuki, H., Futatsugi, K., Fukunaga, T., Yokomizo, C., Koishihara, Y., Fukui, H., Ohsugi, Y., Yawata, H., Kobayashi, I., Hirano, T., Taga, T., and Kishimoto, T. (1991). Preparation of soluble murine IL-6 receptor and anti-murine IL-6 receptor antibodies. J. Immunol. 147, 168-173.

Sakamoto, M. and Sakamoto, S. (1984). Immunocytochemical localization of collagenase in isolated mouse bone cells. Biomed. Res. 5, 29-38.

Sakamuri, V.R., Takahashi, S., Dallas, M., Williams, R.E., Neckers, L., and Roodman, G.D. (1994). Interleukin-6 antisense deoxyoligonucleotides inhibit bone resorption by giant cells from human giant cell tumors of bone. J. Bone Miner. Res. 9, 753-757.

Sampath, T.K., Coughlin, J.E., Whetsone, R.M., Banach, D., Corbett, C., Ridge, R.J., Ozkaynak, E., Oppermann, H., and Rueger, D.C. (1990). Bovine osteogenic protein is composed of dimers of OP-1 and BMP-2A, two members of the transforming growth factor-β superfamily. J. Biol. Chem. 265, 13198-13205.

Santhanam, U., Ray, A., and Sehgal, P.B. (1991). Repression of the interleukin-6 gene promoter by *p53* and the retinoblastoma susceptibility gene product. Proc. Natl. Acad. Sci. U.S.A. 88, 7605-7609.

Sato, K., Fujii, Y., Kasono, K., Ozawa, M., Imamura, H., Kanaji, Y., Kurasawa, H., Tsushima, T., and Shizume, K. (1989). Parathyroid hormone-related protein and interleukin-1α synergistically stimulate bone resorption in vitro and increase the serum calcium concentrationin mice in vivo. Endocrinology 124, 2172-2178.

Sato, K., Kasono, K., Fujii, Y., Kawakami, M., Tsushima, T., and Shizume, K. (1987). Tumor necrosis factor type α (cachectin) stimulates mouse osteoblastlike cells (MC3T3-E1) to produce macrophage-colony stimulating activity and prostaglandin E_2. Biochem. Biophy. Res. Commun. 145, 323-329.

Scharla, S.H., Strong, D.D., Mohan, S., Baylink, D.J., and Linkhart, T.A. (1991). 1,25-Dihydroxyvitamin D_3 differentially regulates the production of insulinlike growth factor-I (IGF-1) and IGF-binding protein-4 in mouse osteoblasts. Endocrinology 129, 3139-3146.

Scheven, B.A.A., Visser, J.W.M., and Nijweide, P.J. (1986). In vitro osteoclast generation from different bone marrow fractions, including a highly enriched haematopoietic stem cell population. Nature (London) 321, 79.

Schmid, C., Rutishauser, J., Schlapfer, I., Froesch, E.R., and Zapf, J. (1991). Intact but not truncated insulinlike growth factor binding protein-3 (IGFBP-3) blocks IGF I-induced stimulation of osteoblasts: Control of IGF signalling to bone cells by IGFBP-3-specific proteolysis? Biochem. Biophys. Res. Commun. 179, 579-585.

Sehgal, P. B. (1992). Regulation of IL6 expression. Res. Immunol. 143, 724-734.

Shinar, D.M., and Rodan, G.A. (1990). Biphasic effects of transforming growth factor-β on the production of osteoclastlike cells in mouse bone marrow cultures: The role of prostaglandins in the generation of these cells. Endocrinology 126, 3153-3158.

Simonet, W.S., Lacey, D.L., Dunstan, C.R., Kelly, M., Chang, M.S., Luthy, R., Nguyen, H.Q., Wooden, S., Bennett, K., Boone, T., Shimamoto, G., DeRose, M., Elliott, R., Colombero, A., Tan, H.L., Trail, G., Sullivan, J., Davy, E., Bucay, N., Renshaw-Gegg, L., Hughes, T.M., Hill, D., Pattison, W., Campbell, P., Sander, S., Van, G., Tarpley, J., Derby, P., Lee, R., Amgen EST program, and Boyle, W.J.

(1997). Osteoprotegerin: A novel secreted protein involved in the regulation of bone density. Cell. 89, 309-319.

Smith, A.G., Heath, J.K., Donaldson, D.D., Wong, G.G., Moreau, J., Stahl, M., and Rogers, D. (1988). Inhibition of pluripotential embryonic stem cell differentiation by purified polypeptides. Nature (London) 336, 688-690.

Sporn, M.B. and Roberts, A.B. (1989). Transforming growth factor-β. Multiple actions and potential clinical applications. JAMA 262, 938-941.

Suda, T., Takahashi, N., and Martin, T.J. (1992). Modulation of osteoclast differentiation. Endocr. Rev. 13, 66-80.

Suda, T., Takahashi, N., and Martin, T.J. (1995). Modulation of osteoclast differentiation: Update. In: Endocr. Rev. Monographs. (Bikle, D.D. and Negro-Vilar, A. Eds.), Vol. 4, pp. 266-270. Endocrine Soc., Bethesda, MD.

Sunyer, T., Lewis, J., Jiana, X., and Osdoby, P. (1995). Human osteoclastlike cells express mRNA for IL-1 receptors type I and type II. J. Bone Miner. Res. 10 (Suppl.), S428.

Taga, T., Hibi, M., Hirata, Y., Yamasaki, K., Yasukawa, K., Matsuda, T., Hirano, T., and Kishimoto, T. (1989). Interleukin-6 triggers the association of its receptor with a possible signal transducer, gp130. Cell 58, 573-581.

Taga, T., and Kishimoto, T. (1992). Cytokine receptors and signal transduction. FASEB J. 6, 3387-3396.

Taga, T., Narazaki, M., Yasukawa, K., Saito, T., Miki, D., Hamaguchi, M., Davis, S., Shoyab, M., Yancopoulos, G. D., and Kishimoto, T. (1992). Functional inhibition of hematopoietic and neurotrophic cytokines by blocking the interleukin-6 signal transducer gp130. Proc. Natl. Acad. Sci. U.S.A. 89, 10998-11001.

Takahashi, N., Akatsu, T., Sasaki, T., Nicholson, G. C., Moseley, J. M., Martin, T. J., and Suda, T. (1988a). Induction of calcitonin receptors by 1-25 dihydroxyvitamin D_3 in osteoclastlike multinucleated cells formed from mouse bone marrow cells. Endocrinology 123, 1504-1510.

Takahashi, N., Akatsu, T., Udagawa, N., Sasaki, T., Yamaguchi, A., Moseley, J.M., Martin, T. J., and Suda, T. (1988b). Osteoblastic cells are involved in osteoclast formation. Endocrinology 123, 2600-2602.

Takahashi, N., Udagawa, N., Akatsu, T., Tanaka, H., Shionome, M., and Suda, T. (1991). Role of colony-stimulating factors in osteoclast development. J. Bone Miner. Res. 6, 977-985.

Takahashi, N., Yamana, H., Yoshiki, S., Roodman, D.G., Mundy, G.R., Jones, S.J., Boyde, A., and Suda, T. (1988c). Osteoclastlike cell formation and its regulation by osteotropic hormones in mouse bone marrow cultures. Endocrinology 122, 1373-1380.

Tamura, T., Udagawa, N., Takahashi, N., Miyaura, C., Tanaka, S., Yamada, Y., Koishihara, Y., Ohsugi, Y., Kumaki, K., Taga, T., Kishimoto, T. and Suda, T. (1993). Soluble IL-6 receptor triggers osteoclast formation by interleukin 6. Proc. Natl. Acad. Sci. USA 90, 11924-11928.

Tanaka, N., Asao, H., Ohbo, K., Ishii, N., Takeshita, T., Nakamura, M., Sasaki, H., and Sugamura, K. (1994). Physical association of JAK1 and JAK2 tyrosine kinases with the interleukin 2 receptor β and γ chains. Proc. Natl. Acad. Sci. USA 91, 7271-7275.

Thomson, B.M., Saklatvala, J., and Chambers, T.J. (1986). Osteoblasts mediate interleukin 1 stimulation of bone resorption by rat osteoclasts. J. Exp. Med. 164, 104-112.

Tomooka, S., Border, W.A., Marshall, B.O., and Noble, W.A. (1992). Glomerular matrix accumulation is linked to inhibition of the plasmin protease system. Kidney Int. 42, 1462-1469.

Tsuda, E., Goto, M., Mnochizuki, S., Yano, K., Kobayashi, F., Morinaga, T., and Higashio, K. (1997). Isolation of a novel cytokine from human fibroblasts that specifically inhibits osteoclastogenesis. Biochem. Biophys. Res. Commun. 234, 137-142.

Udagawa, N., Horwood, N.J., Elliott, J., Mackay, A., Owen, J., Okamura, H., Kurimoto, M., Chambers, T.J., Martin, T.J., and Gillespie, M.T. (1997). Interleukin 18 (interferon-γ-inducing factor) is produced by osteoblasts and acts via granulocyte-macrophage colony-stimulating factor and not via interferon-γ to inhibit osteoclast formation. J. Exp. Med. 185, 1005-1012.

Udagawa, N., Takahashi, N., Katagiri, T., Tamura, T., Wada, S., Findlay, D.M., Martin, T.J., Hirota, H., Taga, T., Kishimoto, T., and Suda, T. (1995). Interleukin (IL)-6 induction of osteoclast differentiation depends on IL-6 receptors expressed on osteoblastic cells but not on osteoclast progenitors. J. Exp. Med. 182, 1461-1468.

Urist, M.R., DeLange, R.J., and Finerman, G.A.M. (1983). Bone cell differentiation and growth factors. Science 220, 680-686.

Urist, M.R. (1989). Bone morphogenetic protein, bone regeneration, heterotopic ossification and the bone-bone marrow consortium. In: Bone and Mineral Research. (Peck, W.A., Ed), Vol. 6, pp. 57-113, Elsevier, Amsterdam.

Urist, M.R. (1965). Bone: Formation by autoinduction. Science 15, 893-899.

Vaughan, J. (1981). The Physiology of Bone, 3rd ed. Clarendon Press, Oxford, United Kingdom.

Verfaillie, C. and McGlave, P. (1991). Leukemia inhibitory factor/human interleukin for DA cells: A growth factor that stimulates the in vitro development of multipotential human hematopoietic progenitors. Blood 77, 263-270.

Vilcek, J. and Lee, T.H. (1991). Tumor necrosis factor. New insights into the molecular mechanisms of its multiple actions. J. Biol. Chem. 266, 7313-7316.

Wakefield, L.M., Smith, D.M., Flanders, K.C., and Sporn, M.B. (1988). Latent transforming growth factor-β from human platelets. J. Biol. Chem. 263, 7646-7654.

Wakefield, L.M., Smith, D.M., Masui, T., Harris, C.C., and Sporn, M.B. (1987). Distribution and modulation of the cellular receptor for transforming growth factor-b. J. Cell Biol. 105, 965-975.

Walters, M.R. (1995). Newly identified actions of the vitamin D endocrine system. In: Endocr. Rev. Monographs (Bikle, D.D. and Negro-Vilar, A., Eds.), Vol. 4, pp.1-56, Endocrine Soc., Bethesda, MD. Endocrine Reviews Monographs..

Wang, E.A., Rosen, V., D'Alessandro, J.S., Bauduy, M., Cordes, P., Harada, T., Israel, D., Hewick, R.M., Kerns, K., LaPan, P., Luxenberg, D.P., McQuaid, D., Moutsatsos, I., Nove, J., and Wozney, J.M. (1990). Recombinant human bone morphogenetic protein induces bone formation. Proc. Natl. Acad. Sci. U.S.A. 87, 2220-2224.

Wang, E.A., Rosen, V., Cordes, P., Hewick, R.M., Kriz, M.J., Luxenberg, D.P., Sibley, B.S., and Wozney, J.M. (1988). Purification and characterization of other distinct bone-inducing factors. Proc. Natl. Acad. Sci. USA 85, 9484-9488.

Ware, C.B., Horowitz, M.C., Renshaw, B.R., Hunt, J.S., Liggitt, D., Koblar, S.A., Gliniak, B.C., McKenna, H.J., Papayannopoulou, T., Thoma, B., Cheng, L., Donovan, P.J., Peschon, J.J., Bartlett, P.F., Willis, C.R., Wright, B.D., Carpenter, M.K., Davison, B.L., and Gearing, D.P. (1995). Targeted disrupted of the low-affinity leukemia inhibitory factor receptor gene causes placental, skeletal, neural, and metabolic defects and results in perinatal death. Development 121, 1283-1299.

Westerlind, K., Wronski, T.J., Evans, G.L., and Turner, R.T. (1994). The effect of long-term ovarian hormone deficiency on transforming growth factor-β and bone matrix protein mRNA expression in rat femora. Biochem. Biophys. Res. Commun. 200, 283-289.

Wong, B.R., Rho, J., Arron, J., Robinson, E., Orlinick, J., Chao, M., Kalachikov, S., Cayani, E., Bartlett, R.S., III, Franke, W.N., Lee, S.Y., and Choi, Y. (1997). TRANCE is a novel ligand of the tumor necrosis factor family that activates c-Jun N-terminal kinase in T cells. J. Biol. Chem. 272, 25190-25194.

Wozney, J.M., Rosen, V., Celeste, A.J., Mitsock, L.M., Whitters, M.J., Kriz, R.W., Hewick, R. M., and Wang, E.A. (1988). Novel regulators of bone formation: Molecular clones and activities. Science 242, 1528-1534.

Wozney, J.M. and Rosen, V. (1993). Bone morphogenetic proteins. In: Handbook of Experimental Pharmacology. (Mundy, G.R. and Martin, T.J., Eds.) Vol. 107, pp. 723-748. Springer-Verlag Heidelberg, Germany.

Wrana, J.L., Attisano, L., Carcamo, J., Zentella, A., Doody, J., Laiho, M., Wang, X-F., and Massague, J. (1992). TGFβ signals through a heteromeric protein kinase receptor complex. Cell 71, 1003-1014.

Wrana, J.L., Attisano, L., Wieser, R., Ventura, F., and Massague, J. (1994). Mechanism of activation of the TGF-β receptor. Nature (London) 370, 341-347.

Wrana, J. L., Maeno, M., Hawrylyshyn, B., Yao, K.L., Domenicucci, C., and Sodek, J. (1988). Differential effects of transforming growth factor-β on the synthesis of extracellular matrix proteins by normal fetal rat calvarial bone cell populations. J. Cell Biol. 106, 915-924.

Yamaguchi, A. and Kahn, A.J. (1991). Osteogenic cell lines express myogenic and adipocytic developmental potential. Calcif. Tissue Int. 49, 221-225.

Yamaguchi, A., Katagiri, T., Ikeda, T., Wozney, J. M., Rosen, V., Wang, E.A., Kahn, A.J., Suda, T., and Yoshiki, S. (1991). Recombinant human bone morphogenetic protein-2 stimulates osteoblastic maturation and inhibits myogenic differentiation in vitro. J. Cell Biol. 113, 681-687.

Yamamori, T., Fukada, K., Aebersold, R., Korsching, S., Fann, M. J., and Patterson, P.H. (1989). The cholinergic neuronal differentiation factor from heart cells is identical to leukemia inhibitory factor. Science 246, 1412-1416.

Yamasaki, K., Taga, T., Hirata, Y., Yawata, H., Kawanishi, Y., Seed, B., Taniguchi, T., Hirano, T., and Kishimoto, T. (1988). Cloning and expression of the human interleukin-6 (BSF-2/IFNb2) receptor. Science 241, 825-828.

Yang, L. and Yang, Y. (1994). Regulation of interleukin (IL)-11 gene expression in IL-1 induced primate bone marrow cells. J. Biol. Chem. 269, 32732-32739.

Yasuda, H., Shima, N., Nakagawa, N., Yamaguchi, K., Kinosaki, M., Mochizuki, S., Tomoyasu, A., Yano, K., Goto, M., Murakami, A., Tsuda, E., Morinaga, T., Higashio, K., Udagawa, N., Takahashi, N., and Suda, T. (1988). Osteoclast differentiation factor is a ligand for osteoprotegerin/osteoclastogenesis-inhibitory factor and is identical to TRANCE/RANKL. Proc. Natl. Acad,. Sci. U.S.A. 95, 3597-3602.

Yin, T., Miyazsawa, K., and Yang, Y. (1992). Characterization of interleukin-11 receptor and protein tyrosine kinase phosphorylation induced by interleukin-11 in mouse 3T3-L1 cells. J. Biol. Chem. 267, 8347-8351.

Yin, T., Taga, T., Tsang, L., Yasukawa, K., Kishimoto, T., and Yang, Y. (1993). Involvement of IL-6 signal transducer gp130 in IL-11 mediated signal transduction. J. Immunol. 151, 2555-2561.

Yin, T., Yasukawa, K., Taga, T., Kishimoto, T., and Yang, Y.C. (1994). Identification of 130-kDa tyrosine phosphorylated protein induced by interleukin-11 as JAK 2 tyrosine kinase, which associates with gp130 signal transducer. Exp. Hematol. 22, 467-472.

Yoneda, T., Nakai, M., Moriyama, K., Scott, L., Ida, N., Kunitomo, T., and Mundy, G.R. (1993). Neutralizing antibodies to human interleukin-6 reverse hypercalcaemia associated with a human squamous carcinoma. Cancer Res. 53, 737-740.

Yoshida, H., Hayashi, S-I., Kunisada, T., Ogawa, M., Nishikawa, S., Okamura, T., Sudo, T., Shultz, L.D., and Nishikawa, S-I. (1990). The murine mutation osteopetrosis is in the coding region of the macrophage colony stimulating factor gene. Nature (London). 345, 442-444.

Zhang, R.W., Simmons, D.J., Crowther, R.S., Mohan, S., and Baylink, D.J. (1991). Contribution of marrow stromal cells to the regulation of osteoblast proliferation in rats: Evidence for the involvement of insulinlike growth factors. Bone and Mineral 13, 201-215.

Zhang, X-G., Gu, J-J., Lu, Z-Y., Yasukawa, K., Yancopoulos, G.D., Turner, K., Shoyab, M., Taga, T., Kishimoto, T., Bataille, R., and Klein, B. (1994). Ciliary neurotropic factor, interleukin 11, leukemia inhibitory factor, and oncostatin M are growth factors for human myeloma cell lines using the interleukin-6 signal transducer gp130. J. Exp. Med. 177, 1337-1342.

Zhang, Y., Lin, J-X., and Vilcek, J. (1990). Interleukin-6 induction by tumor necrosis factor and interleukin-1 in human fibroblasts involves activation of a nuclear factor binding to a kB-like sequence. Mol. Cell. Biol. 10, 3818-3823.

Zheng, M.H., Wood, D.J., Wysocki, S., Papadimitriou, J.M., and Wang, E.A. (1994). Recombinant human bone morphogenetic protein-2 enhances expression of interleukin-6 and transforming growth factor-β1 genes in normal human osteoblast-like cells. J. Cell. Physiol. 159, 76-82.

Zhou, H., Choong, P.C., Chou, S.T., Kartsogiannis, V., Martin, T.J., and Ng, K.W. (1995). Transforming growth factor β1 stimulates bone formation and resorption in an in-vivo model in rabbits. Bone. Vol. 17, No. 4 (Suppl.), 443S-448S.

Zurawski, S.M., Vega, Jr., F., Huyghe, B., and Zurawski, G. (1993). Receptors for interleukin-13 and interleukin-4 are complex and share a novel component that functions in signal transduction. EMBO. J. 12, 2663-2670.

COUPLING OF BONE FORMATION AND BONE RESORPTION: A MODEL

James T. Ryaby, Robert J. Fitzsimmons, Subburaman Mohan, and David J. Baylink

I. Introduction 101
II. A Model for Coupling of Bone Formation to Bone Resorption 103
A. Local Source of Growth Factors and Cytokines that Affect Resorption Cavity Fill-in 103
B. Systemic Source of Growth Factors and Hormones that Affect Resorption Cavity Fill-in 111
III. Clinical Correlates of Impaired Formation-Resorption Coupling 113
A. Secondary Hyperparathyroidism in the Elderly 113
B. Pathogenesis of Glucocorticoid-Induced Osteoporosis 115
IV. Summary 116

I. INTRODUCTION

Bone is a remarkably versatile tissue which functions both in mechanical support and as a calcium reservoir. These two functions are accomplished

Advances in Organ Biology
Volume 5A, pages 101-122.

ISBN: 0-7623-0390-5

by remodeling, or the restructuring of bone through the processes of bone resorption and formation in response to mechanical demands and requirements for calcium stored in the skeleton. Remodeling occurs throughout adult life at a rate of approximately 5 to 10% per year (Frost, 1961). The remodeling enables bone to be restructured following any occupational change in mechanical strain and the replacement of bone which has suffered from microdamage with new bone. In addition, bone remodeling provides a mechanism by which the skeleton provides calcium to the body fluids whenever the loss of calcium from the organism exceeds its intake, such as during a calcium deficient diet or during lactation.

Remodeling involves two fundamental processes: bone resorption and bone formation. In normal young adults, these two processes are equivalent in magnitude. Formation is said to be coupled to resorption, and there is no net loss or gain of bone. It is bone formation that is coupled to bone resorption: resorption precedes formation during the remodeling cycle (Frost, 1966). This sequence will become a key issue when we explore the molecular mechanism of resorption cavity fill-in. The extent of formation after the resorption process determines the extent to which the two processes are coupled. Poor coupling is said to occur when there is an imbalance between bone resorption and bone formation, with a resulting net loss or gain of bone (Baylink and Jennings, 1993).

In order to understand the mechanism of coupling, we need to examine bone remodeling at the microscopic remodeling site. Remodeling at a site on a bone surface begins when osteoclasts appear. These cells subsequently produce an excavation cavity. Next, the cavity is filled in by osteoblastic activity. This constitutes the usual sequence of events during remodeling (Erikson et al., 1985). We should mention, however, that there are exceptions to this scheme. For example, bone formation may occur on a neutral surface when lining cells are stimulated to differentiate and begin to form bone. This can occur, for example, with parathyroid hormone (PTH) therapy in osteoporotic patients (Hodsman et al., 1993). In addition, when there is a change in the mechanical forces on bone, there may be situations where bone resorption is not followed by bone formation. Nonetheless, the formation of a resorption cavity is generally followed by a fill-in stage which completely fills in the excavation cavity. Disturbances in this fill-in phase can occur during aging and lead to the development of osteoporosis.

In terms of clinical significance, the phenomenon of coupling is thus relevant to the pathogenesis of osteoporosis, an age related disease characterized by a gradual loss of skeletal tissue. This disease affects over 35 million people in the United States alone (Chrischilles et al., 1994). Because loss of bone implies a malfunction in coupling between resorption and formation, studies in a number of laboratories have focused on identifying the potential signaling molecules which regulate coupling. The most promising candidates to date are growth factors and cytokines.

We present in this chapter a model to explain the coupling of formation to resorption and the factors that could influence the amount of resorption cavity fill-in and thereby determine whether there is a gain or a loss of bone. Our model will describe how growth factors, cytokines, and hormones act to influence the coupling of formation to resorption. We discuss the local sources of these factors (i.e., the cells involved in the production of growth factors and cytokines) as well as the systemic sources of hormones and growth factors that influence the coupling process.

II. A MODEL FOR COUPLING OF BONE FORMATION TO BONE RESORPTION

A model of coupling at a single remodeling site is depicted in Figure 1. This model is described in detail below in terms of the sources of growth factors on cytokines.

A. Local Source of Growth Factors and Cytokines That Affect Resorption Cavity Fill-in

Marrow cells and osteoblasts as a source of resorbing cytokines

Osteoclasts are formed and activated in response to cytokines produced by marrow cells (hematopoetic cells which include osteoclast precursors, osteoclasts, and stromal cells) or by osteoblasts. Cytokines which stimulate bone resorption include the interleukins (ILs), IL-1, IL-6, and IL-11, granulocyte-macrophage colony stimulating factor (GM-CSF), macrophage colony-stimulating factor (M-CSF), tumour necrosis factor-α (TNFα), and prostaglandins (Manologas, 1995). Osteoclasts themselves produce cytokines as evidenced by IL-6 production by osteoclasts from Paget's disease patients (Hoyland et al., 1994). Osteoblasts and stromal cells may also produce cytokines. For example, PTH increases bone resorption, in part, by stimulating the production of IL-6 by osteoblasts (Feyen et al., 1989).

Interestingly, at the menopause, when bone loss is known to be accelerated, the decrease in estrogen levels is thought to release the "brake" on the production of resorbing cytokines, thereby increasing bone resorption.

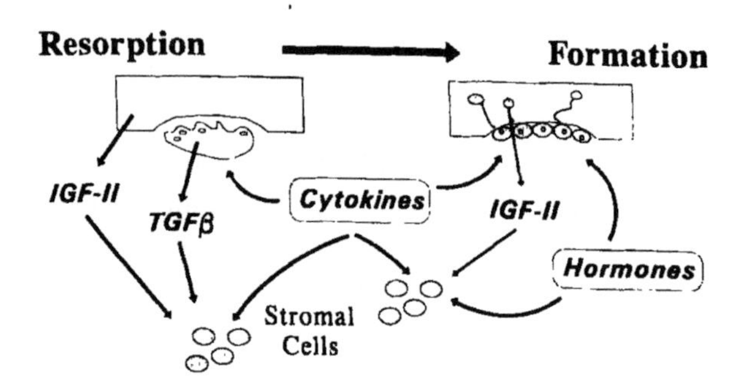

Figure 1. Model of the coupling of bone formation to bone resorption. This model depicts two phases of remodeling, bone resorption, on the left, which proceeds to the second stage of bone remodeling and bone formation, which is on the right. This model illustrates how the process of bone resorption leads to the production and release of messenger molecules, which act to promote resorption cavity fill-in by cells of the osteoblastic lineage. The process of bone resorption shown on the left is initiated by resorbing cytokines, which not only stimulate the osteoclasts but also may have negative or positive effects on the proliferation of cells of osteoblast lineage. Next, the osteoclasts, during the process of bone resorption, can secrete growth factors such as TGFβ and IGF-II. It is proposed that these growth factors are secreted in proportion to the resorptive activity of the osteoclast in order to ensure a proportionality between the extent of bone resorption and the final level of resorption cavity fill-in by osteoblasts. Another source of growth factor is the bone itself. During bone resorption by osteoclasts, several growth factors stored in bone, including IGF-II (which is abundant in bone), are released from bone in a biologically active state. These growth factors can then act to promote proliferation and differentiation of cells of the osteoblast lineage. Turning now to the regulation of osteoblastic activity during the formation phase, there is the release of IGF-II from osteocytes. It is proposed that, during mechanical loading, osteocytes detect changes in mechanical strain and also the presence of microdamage and, in response to these signals, produce growth factors themselves or send messages to nearby osteoblasts or lining cells to produce growth factors such as IGF-II. This mechanism, whereby mechanical forces influence growth factor production, is considered to be one of the most important determinants of the final level of resorption cavity fill-in. Another factor in the overall equation of resorption cavity fill-in by osteoblasts are circulating hormones and growth factors that can act on the process of osteoblastic bone formation. See text for further details. Reproduced with kind permission from Academic Press, Inc., San Diego.

Estrogen has been reported to suppress IL-1 and IL-6 production in osteoblasts, and its absence would lead to increased IL-1 and IL-6 production (Miyaura et al., 1995). Moreover, the administration of anti-IL-6 antibody to mice prevents the osteopenia produced by ovariectomy (Jilka et al., 1992). In a rat model, IL-1 receptor antagonist was shown to prevent the development of osteopenia following ovariectomy (Kimble et al., 1995) consistent with a crucial role of IL-1 in osteoporosis due to estrogen deficiency. In addition, Ammann et al. (1995), have shown that transgenic mice expressing high levels of soluble TNF receptor-1 are protected from bone loss caused by estrogen deficiency, suggesting a role for TNF in the pathogenesis of bone loss following estrogen deficiency.

It is now well documented that the aforementioned cytokines stimulate bone resorption, but do they also have an effect on the osteoblast and, therefore, bone formation? IL-1 stimulates osteoblast proliferation in normal human trabecular bone explants (Russell et al., 1990). Further, as IL-1 increases IL-6 and prostaglandin E_2 (PGE_2) production by osteoblasts (Feyen et al., 1989; Linkhart et al., 1991), it is possible that either IL-6 or PGE_2 may be responsible for increasing osteoblast proliferative activity during the resorption phase of bone remodeling. However, IL-6 has not been reported to affect human osteoblast proliferation (Littlewood et al., 1991). PGE_2, on the other hand, is a strong stimulator of bone resorption and, also, a strong mitogen for osteoblasts (Ke et al., 1992; Marks and Miller, 1993; Baylink et al., 1996).

We wish to emphasize that, although we think of cytokines as agents that promote bone resorption, it is possible these agents simultaneously increase the osteoclast population to excavate a bone cavity and initiate the proliferation of a preosteoblast population that are destined to fill-in this excavation cavity. It is also conceivable that certain resorbing cytokines may simultaneously increase resorption but decrease either proliferation or differentiation of preosteoblasts. This latter situation might take place in multiple myeloma, where one can see a large increase in bone resorption with a much smaller increase in bone formation (Valentin-Opran et al., 1982). This is particularly evident when skeletal radiographs are compared with bone scans. Accordingly, the radiograph may show a punched out lesion in the bone, reflecting an increase in net resorption, whereas in the same skeletal site there may be very little increase in the uptake of the bone seeking tracer, indicating an impairment in the coupling of bone formation to bone resorption.

Multiple myeloma is an example of poor coupling due to pathology. There are also physiologic situations that, of necessity, result in poor coupling. These include the poor coupling that occurs in calcium deficiency states. For bone to serve as a calcium reservoir, there must be molecular mechanisms that induce bone resorption without inducing an equivalent increase in bone formation. In this way, there is a net loss of calcium from bone, and, as a result, serum calcium can be maintained in the normal range, even in the face of a substantial deficit in calcium intake (Baylink and Jennings, 1993). Interestingly, once the subject is no longer exposed to a restricted calcium diet, there is a rapid replacement of the bone that was lost during the calcium depletion phase, a phenomenon which reflects the ability of skeletal tissue to maintain itself (Drivdahl et al., 1984).

Osteoclasts as a Source of Growth Factors

An interesting recent finding is that osteoclasts produce transforming growth factor-β (TGFβ), which is known to stimulate osteoblast progenitor proliferation (Bolander, 1992). Estrogen has been shown to regulate the production of TGFβ from chick osteoclasts, which were obtained from animals maintained on a low calcium diet (Robinson et al., 1996). Osteoclasts express mRNA for several TGFβs (Oursler, 1994), suggesting that TGFβ produced by oteoclasts may have important autocrine and paracrine effects within the bone microenvironment.

The action of estrogen in increasing the production of TGFβ could be important in bone resorption cavity fill-in. Previous studies have shown that TGFβ increases bone formation *in vivo* (Noda and Camilliere, 1989). Another possible regulator of bone formation is insulinlike growth factor-I (IGF-I), which has been demonstrated immunocytochemically in rat osteoclasts (Lazowski et al., 1994). The mRNA for IGF-I was not detected in these osteoclasts, possibly because of a lack of sensitivity of the detection method. It is reasonable to assume that osteoclasts produce sufficient amounts of growth factors so as to contribute to osteoblast proliferation and, thereby, resorption cavity fill-in; however, this issue needs further study. Such a function by osteoclasts would be analogous to activated tissue macrophages which produce growth factors which, in turn, act on nearby fibroblasts during wound healing (Greisler et al., 1993).

Osteoclasts also produce factors which inhibit osteoblast cell proliferation. Medium conditioned by either chicken osteoclasts, chicken giant cells, or human osteoclastoma cells (Galvin et al., 1994) was shown to inhibit collagen production and alkaline phosphatase activity in osteoblast cultures. The above information indicates that osteoclasts produce factors which are active on osteoblastlike cells. However, little is known whether these factors contribute significantly to the regulation of bone formation.

Bone As a Source of Growth Factors

It has been known for many decades that bone matrix contains growth factors (Urist, 1965). One possible function of these growth factors stored in bone is to regulate osteoblast proliferation in direct proportion to the amount of bone resorbed. Such a function would tend to ensure that any time bone is resorbed, a mechanism would be in place to guarantee that the resorbed bone was replaced by new bone. Thus, the storage of growth factors in bone would ensure bone replacement and bone repair. The major concern about whether such a mechanism could operate is that, if osteoclasts hydrolyze bone matrix during bone resorption, it would seem that this could degrade any growth factors present in bone. The following experiment addressed this important issue. Studies utilizing *in vitro* organ cultures of embryonic chick bones showed a positive correlation between the resorption rate and the amount of mitogenic activity released in the culture medium, even when cycloheximide was added in a concentration sufficient to block all new protein synthesis (Farley et al., 1987). Because the embryonic chick bones were cultured in a serum free medium and because cycloheximide would have been expected to inhibit new growth factor synthesis by either osteoblasts or osteoclasts, it is logical to assume that the growth factor appearing in the culture medium under these conditions represents its release during bone resorption. In further support of this conclusion, it was found that embryonic chick bone did, in fact, contain an abundance of extractable mitogenic activity.

Bone contains several growth factors, some in very high concentration (Figure 2). By far, the most abundant growth factor in bone is IGF-II, which is also the most abundant growth factor secreted by osteoblasts (Mohan and Baylink, 1991). *In vitro* studies have found that the IGF-II in bone could not be extracted until the bone was demineralized. The mechanism by which

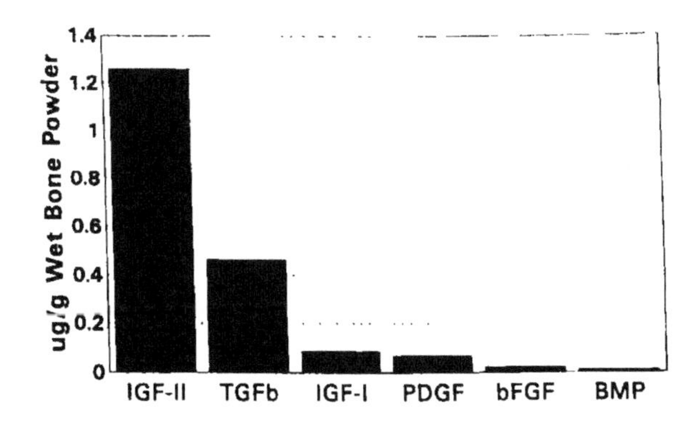

Figure 2. Human bone concentration of growth factors.

large amounts of IGF-II is stored in human bone can be explained as follows. Human bone cells in culture produce not only IGF-II, but also IGF-binding protein-5 (IGFBP-5), which has the unique ability to bind with high affinity not only to IGF-II but also to hydroxyapatite (Bautista et al., 1991). Thus, it can be envisioned that the secretion of IGFBP-5 by osteoblasts would lead to an accumulation in bone of IGF-II bound to IGFBP-5. This would help ensure the deposition of large amounts of IGF-II in bone for later use during the coupling of bone formation to bone resorption. Another interesting aspect of IGFBP-5 is that it not only sequesters IGF-I and II and fixes these IGFs in bone, but also enhances the mitogenic action of the two IGF mitogens (Mohan et al., 1995a).

Another growth factor that occurs at a high concentration in human bone is TGFβ (Centrella et al., 1994). Interestingly, TGFβ is also fixed in bone by a specific protein, in this case, a proteoglycan termed β-glycan which, it is believed, helps to ensure a relatively high deposition rate of TGFβ in human bone (Lopez-Casillas et al., 1991). In summary, growth factors can be released from bone due to osteoclastic resorption at sites where bone resorption cavity fill-in is required.

The concentrations of growth factors in bone can vary. Recently, it has been shown that the IGF-I and TGFβ content of human bone decreases with aging (Nicolas et al., 1994). Interestingly, estrogen deficiency and vitamin D deficiency have been shown to reduce the content of TGFβ in rat cortical bone (Finkelman et al,, 1991, 1992). Furthermore, TGFβ levels in the periosteum of rats is decreased during periods of reduced mechanical loading during space flight (Westerlind and Turner, 1995). These observations show that, at least in some circumstances, conditions which lead to diminished growth factor content in bone are associated with a poor coupling of bone formation to bone resorption.

With respect to the potential role of growth factors to promote excavation cavity fill-in, one might argue that the concentration of growth factors in bone cannot be acutely regulated in terms of growth factor deposition and subsequent release. Thus, this mechanism of growth factor release during bone resorption could not by itself efficiently regulate resorption cavity fill-in. For example, in a middle age male, the half-life of cortical bone is between 10 and 15 years and that of trabecular bone is between one and three years (Frost, 1961). Accordingly, it would appear impossible for contemporary osteoblasts, under the influence of contemporary stimuli, to deposit the appropriate amounts of growth factors in bone for release several years in the future. It therefore seems unlikely that a single source of growth factors (namely, that released from bone during bone resorption) is sufficient to ensure adequate coupling. It may be that coupling is such an important process that there is a redundancy in its regulatory systems. Such a view would suggest that there are several sources of growth factors that together ensure appropriate coupling of formation to resorption (Figure 1). Additional sources of growth factors are described below.

Osteocytes and Osteoblasts as a Source of Growth Factors

Culture medium conditioned by osteoblasts in culture contains a number of growth factors. Osteoblasts have been shown to produce IGF-I, IGF-II, platelet-derived growth factor (PDGF), epidermal growth factor (EGF), and TGFβ (Mohan and Baylink, 1991). Growth factors produced by osteoblasts may act in an autocrine manner or a paracrine manner to stimulate osteoblast proliferation. Thus, osteoblasts secrete growth factors that may do the following: (1) exert autocrine and paracrine actions and (2) be sequestered and for longterm storage in bone. All of these varous sources of growth factors could act to regulate resorption cavity fill-in (Figure 1).

Another cell that could be a source of growth factors for resorption cavity fill-in is the osteocyte, particularly those osteocytes that are located near such cavities. Osteocytes are located in strategic positions to sense and respond to mechanical strains and microdamage. It is also possible that bone lining cells are stimulated by mechanical strain directly or, indirectly, through signals from osteocytes. Several workers have provided indirect evidence that suggests that the osteocyte is the cell most likely to be responsible for transducing mechanical loads into biologic signals (Skerry et al., 1988; Lanyon, 1993; Weinbaum et al., 1995).

It has been known for many years that appropriate physical loading stimuli can result in large increases in bone formation. Furthermore, it has been shown that one of the cells responsible for growth factor production during mechanical loading is the osteocyte (Rawlinson et al., 1993). In addition, mechanical stimulation of canine bone cores has been shown to result in increased levels of IGF-I and IGF-II in both osteocytes and bone lining cells (Rawlinson et al., 1991) through a prostacyclin (PGI_2)-dependent mechanism. Increases in IGF-I have also been observed in rat caudal vertebrae after mechanical loading (Lean et al., 1994), as have increases in periosteal expression of IGF-I and TGFβ mRNA (Raab-Cullen et al., 1994).

It is now well established that mechanical loading can increase bone formation. For example, in studies using turkeys, mechanical loading was shown to result in a threefold increase in the mineral apposition rate (Rubin et al., 1995). Another mechanical loading study has shown approximately twofold increases in bone forming surface in rats (Forwood and Turner, 1994). In addition, a combination of aerobic and weight bearing exercise was found to result in increased bone mineral density in young women (Friedlander et al., 1995). Finally, in eumenorrheic young women (Taaffe et al., 1995), weight-bearing activity (gymnastics) was associated with increased whole body bone mineral density normalized for body mass when compared with nonweight bearing exercise (swimming).

There are two potential mechanisms through which mechanical forces could induce localized increases in growth factor concentration to regulate the extent of resorption cavity fill-in. First, mechanical loading, which causes a bone to bend could cause cell deformation directly. *In vitro* studies with isolated osteoblasts have shown that there is an increase in TGFβ release in response to physical deformation (Holbein et al., 1995). Secondly, mechanical loading which causes bones to bend could also lead to the generation of electrical fields (Bassett and Becker, 1962). Such fields are thought to be generated as a consequence of the production of streaming potentials in response to mechanical loading. It has been shown that applied electromagnet fields (EMFs) result in an increase in proliferation of these cells in culture that is associated with an increase in mitogenic activity in the culture medium. This conditioned culture medium also contains increased amounts of IGF-II (Fitzsimmons et al., 1992). Moreover, IGF-II blocking antibodies have been shown to inhibit the effects of EMFs in increasing osteoblast proliferation, emphasizing the pivotal role of IGF-II in the EMF effect (Fitzsimmons et al., 1995). Thus, IGF-II may be an important mediator of the effect of mechanical forces in regulating resorption cavity fill-in.

As the major function of the skeleton is to provide mechanical support, it is not surprising that the amount of bone present is determined critically by the amount of mechanical loading, and that the effect of mechanical forces to regulate bone formation is the only known feedback control system. These observations, together with the aforementioned mechanisms through which mechanical loading can regulate local growth factor production, emphasize the importance of mechanical forces in the regulation of resorption cavity fill-in. In this regard, it is now well known that resorption cavity fill-in decreases with age (i.e., wall thickness, which is a measure of resorption cavity fill-in, decreases progressively with age in both sexes; Baylink and Jennings, 1993). It seems likely that the decrease in physical activity that attends chronological aging is at partly responsible for the decrease in wall thickness that also accompanies aging.

B. Systemic Source of Growth Factors and Hormones that Affect Resorption Cavity Fill-in

Serum and extracellular fluids contain hormones, growth factors, and cytokines. There is evidence that sex hormones (e.g., estrogen, testosterone, and progesterone) may influence bone formation (Gray et al., 1989; Kasperk et al., 1990; Tremollieres et al., 1992). Growth hormone (Schiltz et al., 1992), glucocorticoids (Libanati and Baylink, 1992), calcium-regulating hormones such as PTH (Cheng et al., 1995), calcitonin (Farley et al., 1993), and 1,25 dihydroxyvitamin D_3 (Kyeyune-Nyombi et al., 1991), are also important regulators of bone remodeling. It seems likely that excavation cavity fill-in is influenced by the ambient hormone levels. For example, mean wall thickness, which is a measure of cavity fill-in, is depressed in patients treated with glucocorticoids (Dempster et al., 1983). This example raises the possibility that serum hormone levels could result in underfilling or overfilling of the resorption bay.

While hormones undoubtedly influence the final level of fill-in, they do not do so in a feedback manner where fill-in or bone density or some other bone parameter is an endpoint of feedback. In this regard, bone is merely a target organ of hormones and does not benefit from any feedback regulation which would assure an appropriate bone density. It is true that calcium regulating hormones act on bone in a feedback manner to sustain serum calcium. However, in this situation, it is the serum calcium rather than bone density that is being controlled and that, too, at the expense of bone density.

The only, and major, feedback system through which the amount of bone tissue is controlled is the prevailing mechanical strains. These act either to increase or to decrease the level of bone tissue and, thus, the bone's mechanical performance. As discussed above, this system involves locally produced growth factors that may be regulated according to the level of mechanical strain. In addition, growth hormone may operate in a feedback system to help regulate the amount of bone tissue. Thus exercise brings about both changes in local mechanical strains and rapid increases in circulating growth hormone levels (Cappon et al., 1994). The latter may act in concert with local growth factors to trigger changes in bone formation to meet the requirements dictated by mechanical strain (Figure 3). Finally, if mechanical strain is the only feedback mechanism in bone regulating the level of bone tissue, this means that the deleterious effects of hormones such as glucocorticoids on wall thickness (i.e., resorption cavity fill-in) occur despite the effects of the mechanical strain feedback system in bone that normally regulates resorption cavity fill-in.

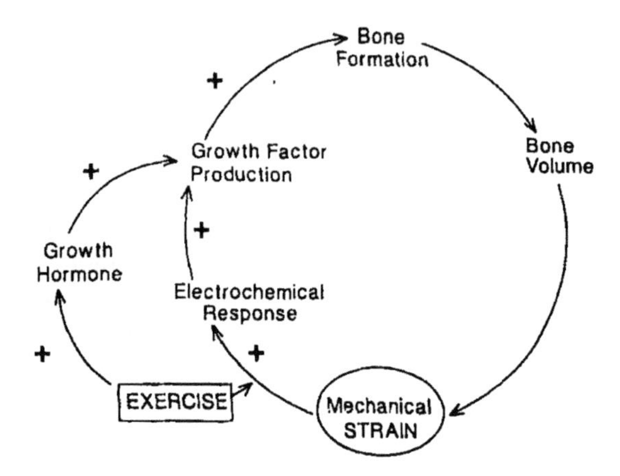

Figure 3. Model of the regulation of bone strength by mechanical strain. Exercise alters bone formation via both systemic and local factors. Among systemic factors, exercise increases growth hormone secretion which, in turn, could be expected to increase systemic levels of IGF-I, which then increases bone formation and, thus, bone volume. In addition, exercise increases local mechanical strain, which produces an electrical response which, in turn, increases the production of local growth factors such as the IGFs by bone cells and, thus, bone formation and bone volume. The increase in bone volume that occurs through this local mechanism decreases the mechanical strain and, thus, closes the feedback loop. Reproduced with permission.

Serum also contains growth factors and cytokines (Libanati et al., 1995). These include the IGFs and their corresponding binding proteins, as well as other growth factors (see above). Thus, bone cells are exposed to locally produced IGFs and their binding proteins, as well as to circulating IGFs and their binding proteins. Serum also contains cytokines, but the extent to which systemic levels of cytokines regulate local bone cell activities is not yet known. A number of studies suggest a possible physiological significance of serum cytokines. Thus, in studies of pubertal girls undergoing their acquisition of peak bone density, we found that the girls with the lowest levels of serum IL-6 had the highest levels of peak bone density and, also, that IL-6 levels correlated positively with bone resorption (Libanati et al., 1995). What was not determined was whether the bone resorption rate was reflected by the serum level or whether the serum level was a measure of locally produced IL-6. In any case, it seems likely that serum cytokines, as well as growth factors and hormones, will eventually be found to influence the process of resorption cavity fill-in.

Recently, one hormone has received special attention because of its remarkable anabolic effects on bone, namely, PTH. PTH has been thought to be a mediator of bone resorption or bone turnover (Breslau et al., 1983). However, PTH, given by daily injection, which results in an acute spike in serum concentration, produces more of an anabolic than a catabolic effect on bone (Hesp et al., 1981). It has been suggested that the normal spikes in PTH secretion are reduced in osteoporosis, and that this accounts for the relative decrease in bone formation that can occur in some osteoporotic patients (Prank et al., 1995). This could be a plausible explanation for their deficient bone formation. However, PTH secretion is not a plausible regulator of bone formation, at least in a feedback mechanism, since the parathyroid glands cannot sense directly how much bone formation or how much bone density is needed. On the other hand, it seems quite possible that parathyroid hormone-related protein (PTHrP), which, like PTH, has anabolic actions on bone (Fermor and Skerry, 1995; Goltzman, 1995) and which is produced locally by bone cells, could act in an anabolic manner. Thus, PTHrP should be considered along with the other aforementioned growth factors and cytokines as a possible regulator of osteoblast proliferation during resorption cavity fill-in.

III. CLINICAL CORRELATES OF IMPAIRED FORMATION-RESORPTION COUPLING

An impairment in the coupling phenomenon is pivotal to the pathogenesis of osteoporosis, since osteoporosis could never occur if resorption cavity fill-in was equal to the extent of the resorption depth. In this regard, we discuss two examples of secondary osteoporosis, for which there is evidence that decreased resorption cavity fill-in is the cause of the osteoporotic bone loss.

A. Secondary Hyperparathyroidism in the Elderly

Serum PTH tends to increase with age, even in those patients who have normal serum levels of 1,25-hydroxyvitamin D_3 and, thus, are not deficient in vitamin D (Wiske et al., 1979). The cause of the increase in PTH is not entirely settled, but it is thought to result from impaired enteral calcium absorption (Eastell et al., 1991). This increase in serum PTH is clinically important since it is probably associated with decreased resorption cavity fill-in. Accordingly, calcium deficient rats with secondary hyperparathyroidism due to dietary restriction of calcium show an impairment in endosteal resorption cavity fill-in (Stauffer et al., 1973).

We recently explored a potential mechanism to explain impaired bone formation in secondary hyperparathyroidism in a group of elderly subjects with hip fracture. Our study demonstrated that serum PTH levels in hip fracture patients, when compared to age-matched control subjects without hip fracture, was significantly increased (Rosen et al., 1992). Because PTH is known to increase the *in vitro* production of IGFBP-4 (an inhibitory IGF-binding protein), we measured the serum level of IGFBP-4 in the control group and the hip fracture group. Serum IGFBP-4 was significantly higher in the hip fracture group, and data pooled from both groups showed a positive correlation between serum PTH and serum IGFBP-4 (Rosen et al., 1992). Because the IGFs are strong anabolic agents for bone, and because IGFBP-4 is an inhibitory binding protein which can obliterate the actions of both IGF-I and IGF-II, we have speculated that the increased production of IGFBP-4 was part of the mechanism for the relative reduction in bone formation and resorption cavity fill-in in secondary hyperparathyroidism. The effect of PTH to increase IGFBP-4 under conditions where serum calcium is low or below normal could explain the depression of bone formation that occurs under physiological conditions where the organism needs to utilize the bone mineral reservoir to maintain a normal serum calcium, such as in lactating mothers. Because poor coupling is now recognized as being the main determinant of the pathogenesis of osteoporosis, it seems probable that, in the near future, additional molecular mechanisms will be discovered to explain how calcium deficiency leads to bone loss in both physiological and pathological situations.

In senile osteoporosis, there are probably other factors besides secondary hyperparathyroidism that contribute to decreased resorption cavity fill-in. For example, there might well be a defect in some aspect or aspects of the feedback mechanism that increases or decreases bone tissue in response to changes in mechanical loading. Consequently, in osteoporosis, the bone density drops so low that patients sustain spontaneous fractures. The implication of this observation is that the strains occurring on the small amount of bone tissue remaining in the osteoporotic patient could be considerably greater than normal, yet bone formation is not increased in patients with severe osteoporosis. Thus, despite an increase in mechanical strain in the severely osteoporotic patient, there is no obvious increase in bone formation. In contrast, in young animals subjected to large increases in strain, there is an increase in bone formation. Interestingly, this change is less marked in older animals (Turner et al., 1995). Thus, the effect of mechanical loading on bone formation and also, perhaps, on bone resorption in the osteoporotic individual is somehow impaired. This is particularly important given the fact that mechanical loading is the only known feedback mechanism that regulates the volume of bone tissue.

B. Pathogenesis of Glucocorticoid-Induced Osteoporosis

As mentioned earlier, it has now been established that excess glucocorticoid treatment results in a decrease in the mean wall thickness, which is a measure of resorption cavity fill-in (Dempster et al., 1983). Glucocorticoid induced osteoporosis illustrates the importance of the coupling phenomenon in the pathogenesis of osteoporosis. Recent work has offered insights into the molecular mechanisms involved in glucocorticoid induced osteoporosis. Thus, highly significant changes have been observed in the IGF system in response to glucocorticoid therapy. For example, in a group of patients with chronic obstructive lung diseases who were treated with large doses of glucocorticoids, there were statistically significant reductions in the serum levels of IGF-I, IGF-II, IGFBP-3, and IGFBP-5, all of which are anabolic for human bone cells *in vitro*. Note that IGFBP-3 and IGFBP-5 tend to enhance the anabolic effects of IGF-I and IGF-II on bone cells (Mohan et al., 1995). Thus, glucocorticoid therapy appears to downregulate the stimulatory components of the IGF system. This response could impair resorption cavity fill-in and, thus, contribute to the bone loss typical of glucocorticoid therapy. With respect to the molecular mechanism of action of glucocorticoids on IGF binding proteins, we found that, in human bone cells *in vitro,* dexamethasone caused a dose-dependent decrease in the mRNA level of IGFBP-3 and IGFBP-5 and, also, corresponding decreases in IGFBP-3 and IGFBP-5 protein levels in the conditioned medium from human bone cells *in vitro* (Chevalley et al., 1996). Thus, it appears that the effects of glucocorticoids on the IGF system are at least, in part, mediated through systemic actions of glucocorticoids on the local bone cell production of at least two components of the IGF system, IGFBP-3 and IGFBP-5.

The coupling phenomenon is not only important in the pathogenesis of osteoporosis, but is also potentially important in the regeneration of the skeleton which is required for the treatment of established osteoporosis. In patients with large deficits in skeletal tissue, an elimination of the risk for fracture requires that such patients have large increases in bone density. We now know that such increases in bone density can be brought about by the over-filling of resorption cavities. This results in a net positive balance at each remodeling site, which, if large, can lead to large increases in bone density in a short time frame. For example, fluoride therapy can result in a doubling of trabecular bone density of the spine within a year of commencing therapy (Farley et al., 1992). The anabolic effect of fluoride results in increases in wall thickness (Erikson et al., 1985). Another anabolic agent that can cause large increases in bone density is PTH(1-34). PTH also has been shown to increase wall thickness (Meunier et al., 1984). In addition, PTH acts by stimulating the differentiation and synthetic activity of cells that line neutral surfaces (Hodsman et al., 1993). Thus, PTH exhibits a repertoire of mechanisms that can rapidly increase bone formation, bone formation on neutral surfaces, and an increased filling-in of resorption cavities. Because of these positive skeletal responses, it would seem that the future holds considerable promise with respect to the development of drugs which could promote the regeneration of the skeleton in patients with a severe skeletal deficit.

IV. SUMMARY

One of the major functions of bone is to serve the mechanical needs of the body. The control mechanism which determines the extent to which bone accomplishes this function is the coupling of bone formation to bone resorption; this, in turn, regulates bone mass. Under normal steady-state conditions during the bone remodeling process, the level of resorption cavity fill-in by osteoblasts is identical to the size of the resorption cavity and, as such, bone formation is said to be coupled to bone formation. Because of the importance of this bone mass regulatory mechanism, we have developed a model of the coupling of bone formation to resorption. In this model, the emphasis is on the mechanism that determines resorption cavity fill-in, a mechanism which involves cytokines and growth factors. There are several local sources of cytokines and growth factors that are relevant to the proliferation and differentiation of osteoblasts, cells whose synthetic activity is required for resorption cavity fill-in. Firstly, resorbing cytokines not only regulate bone resorption, but also either increase or decrease osteoblast proliferation. These cytokines can be produced by either marrow cells, osteoblasts, or osteoclasts. Secondly, osteoclasts can produce the IGFs and TGFβ and probably other growth factors. It is postulated that these growth factors are secreted in proportion to the resorptive activity of the osteoclast, thereby assuring an adequate bone formation response to osteoclastic bone resorption. Thirdly, the bone tissue itself is a storage depot for many growth factors. These bone growth factors are released during osteoclastic bone resorption and then act on cells of the osteoblast lineage to promote resorption cavity fill-in. Fourth, osteocytes and osteoblasts have been shown to produce growth factors in response to several conditions, including mechanical forces. Indeed, mechanical forces regulate the coupling of bone formation to bone resorption by regulating the secretion of growth factors from cells of the osteoblast lineage in a feedback mechanism. The function of the latter mechanism is to ensure a bone mass which is appropriate to mechanical needs. An example of poor coupling due to local factors is the bone loss that occurs with immobilization. Resorption cavity fill-in can also be influenced by systemic sources of hormones and growth factors. An example of poor coupling in osteoporosis that occurs from systemic factors is the osteoporosis seen in response to glucocorticoid therapy, where there is an inadequate fill-in of resorption cavities and, thus, net bone loss. Another cause of poor coupling is related to the fact that the body also needs a mechanism to subvert the coupling mechanism during calcium deficiency where bone calcium is needed for serum calcium regulation. Thus, the PTH-calcium regulatory mechanism can induce poor coupling and, in this way, use the bone as a calcium reservoir to maintain serum calcium. Poor coupling of this type over prolonged periods can result in osteoporosis. Indeed, it is apparent that osteoporosis cannot occur unless there is poor coupling. Moreover, therapeutic agents which are capable of increasing bone density and restoring skeletal mass in osteoporosis do so by influencing the coupling mechanism. Accordingly, both fluoride and PTH, which are known to increase bone density, do so at least in part by increasing resorption cavity fill-in. Thus, the coupling of formation to resorption is an important mechanism in health and disease and is also the target of therapeutic agents to replete an osteoporotic skeleton.

REFERENCES

Ammann, P, Garcia, I., Rizzoli, R., Meyer, J-M., Vassali, P., and Bonjour, J-P. (1995). Transgenic mice expressing high levels of soluble tumor necrosis factor receptor-1 fusion protein are protected from bone loss caused by estrogen deficiency. J. Bone Miner. Res. 10 (Suppl. 1), S139.

Bassett, C.A.L. and Becker, R.O. (1962). Generation of electric potentials by bone in reponse to mechanical stress. Science 137, 1063-1064.

Bautista, C.M., Baylink, D.J., and Mohan, S. (1991). Isolation of a novel insulinlike growth factor (IGF) binding protein from human bone: A potential candidate for fixing IGF-II in human bone. Biochem. Biophys. Res. Commun. 176, 756-763.

Baylink, D.J. and Jennings, J.C. (1993). Calcium and bone homeostasis and changes with aging. In: Principles of Geriatic Medicine and Gerontology, 3rd. ed. (Hazzard, W.R., Bierman, E.L., Blass, J.P., Ettinger, W.H., Halter, J.B. Eds.), pp. 879-896. McGraw-Hill.

Baylink, T., Mohan, S., Fitzsimmons, R.J., and Baylink, D.J. (1996). Evaluation of signal transduction mechanisms for the mitogenic effects of prostaglandin E2 in normal human bone cells, in vitro. J. Bonė Miner. Res. 11, 1413-1418.

Bolander, M.E. (1992). Regulation of fracture repair by growth factors. Proc. Soc. Exp. Biol. Med. 200, 165-170.

Breslau, N.A., Moses, A.M., and Pak, C.Y. (1983). Evidence for bone remodeling but lack of calcium mobilization response to parathyroid hormone in pseudohypoparathyroidism. J. Clin. Endocrinol. Metab. 57, 638-644.

Cappon, J., Brasel, J.A., Mohan, S., and Cooper, D.M. (1994). Effect of brief excercise on circulating IGF-I. J. Applied Physiol. 76, 1418-1422.

Centrella, M., Horowitz, M.C., Wozney, J.M., and McCarthy, T.L. (1994). Transforming growth factor-β gene family members in bone. Endocr. Rev. 15, 27-39.

Cheng, P.T., Chan, C., and Muller, K. (1995). Cyclical treatment of osteopenic ovariectomized adult rats with PTH (1-34) and pamidronate. J. Bone Miner. Res. 10, 119-126.

Chevalley, T., Stong, D.D., Mohan, S., Baylink, D.J., and Linkhart, T.A. (1996). Evidence for a role of IGF binding proteins in glucocorticoid inhibition of normal human osteoblastlike cell proliferation. Eur. J. Endocrinol. 134, 591-601.

Chrischilles, E., Shireman, T., and Wallace, R. (1994). Costs and health effects of osteoporotic fractures. Bone 15, 377-386.

Dempster, D., Arlot, M., and Meunier, P. (1983). Mean wall thickness and formation periods of trabecular bone packets in corticosteriod-induced osteoporosis. Calcif. Tiss. Int. 35, 410-417.

Drivdahl, R.H., Liu, C.C., and Baylink, D.J. (1984). Regulation of bone repletion in rats subjected to varying low-calcium stress. Am. J. Physiol. 246 (2), R190-R196.

Eastell, R., Yergey, A.L., Vieira, N., Cedel, S.L., Kumar, R., and Riggs, B.L. (1991). Interrelationship among vitamin D metabolism, true calcium absorption, parathyroid function, and age in women: Evidence of an age-related intestinal resistance to $1,25(OH)_2D$ action. J. Bone Miner. Res. 6, 125-132.

Erikson, E.F. et al. (1985). Effect of sodium fluoride, calcium, phosphate, and vitamin D2 on trabecular bone balance and remodeling in osteoporotics. Bone 6, 381-389.

Farley, J.R., Tarbaux, N., Murphy, L.A., Masuda, T., and Baylink, D.J. (1987). In vitro evidence that bone formation may be coupled to resorption by release of mitogen(s) from resorbed bone. Metabolism 36, 314-321.

Farley, J.R., Wergedal, J.E., and Baylink, D.J. (1993). A review of evidence for anabolic actions of calcitonin on bone and new evidence for anabolic calcitonin-fluoride interactions. In: The Proc. of Fourth International. Symposium on Osteoporosis and Concensus Development Conference. (Christiansen, C. and Riis, B. Eds.), p. 348. Handelstry Kkeriet Aalborg, ApS, Aalborg, Denmark.

Farley, S.M., Wergedal, J.E., Farley, J.R., Javier, G.N., Schulz, E.E., Talbot, J.R., Libanati, C.R., Lindegren, L., Bock, M., Goette, M.M., Mohan, S.S., Kimball-Johnson, P., Perkel, V.S., Cruise, R.J., and Baylink, D.J. (1992). Spinal fractures during fluoride therapy for osteoporosis: Relationship to spinal bone density. Osteoporosis International 2, 213-218.

Fermor, B. and Skerry, T.M. (1995). PTH/PTHrP receptor expression on osteoblasts and osteocytes but not resorbing bone surfaces in growing rats. J. Bone Miner. Res. 10 (12), 1935-1943.

Feyen, J.H.M., Elford, P., Di Padova, F.E., and Trechsel, U. (1989). Interleukin-6 is produced by bone and modulated by parathyroid hormone. J. Bone Miner. Res. 4, 633-638.

Finkelman, R.D., Linkhart, T.A., Mohan, S., Lau, KHW., Baylink, D.J., and Bell, N.H. (1991). Vitamin D deficiency causes a selective reduction in the deposition of transforming growth factor β in rat bone: Possible mechanism for impaired osteoinduction. Proc. Nat. Acad. Sci. USA 88, 3657-5660.

Finkelman, R.D., Bell, N.H., Strong, D.D., Demers, L.M., and Baylink, D.J. (1992). Ovariectomy selectively reduces the concentrations of transforming growth factor in rat bone: Implications for estrogen deficiency-associated bone loss. Proc. Nat. Acad. Sci. USA 89, 12190-12193.

Fitzsimmons, R.J., Strong, D.D., Mohan, S., and Baylink, D.J. (1992). Low-amplitude, low-frequency electric field-stimulated bone cell proliferation may in part be mediated by increased IGF-II release. J. Cellular Physiol. 150, 84-89.

Fitzsimmons, R.J., Ryaby, J.T., Mohan, S., Magee, F.P., and Baylink, D.J. (1995). Combined magnetic fields increase IGF-II in one cell cultures. Endocrinology. 136, 3100-3106.

Forwood, M.R. and Turner, C.H. (1994). The response of rat tibiae to incremental bouts of mechanical loading: A quantum concept for bone formation. Bone 15, 603-609.

Frost, H.M. (1961). Human osteoblastic activity: 2. Measurement of the biological half-life of bones with the aid of tetracyclines. Henry Ford Hosp. Med. Bull. 9, 80-96.

Frost, H.M. (1966). Relation between bone tissue and cell-population dynamics, histology, and tetracycline labeling. Clin. Orthop. 49, 65-75.

Galvin, R.J.S., Cullison, J.W., Avioli, L.V., and Osdoby, P.A. (1994). Influence of osteoclasts and osteoclastlike cells on osteoblast alkaline phosphatase activity and collagen synthesis. J. Bone Miner. Res., 9, 1167-1178.

Goltzman, D. (1995). Interaction of parathyroid hormone and of parathyroid hormone-related peptide with target cells in the skeleton. J. Bone Miner. Res. 13, 57-60.

Gray, T.K., Mohan, S., Linkhart, T.A., and Baylink, D.J. (1989). Estradiol stimulates in vitro the secretion of insulinlike growth factors by the clonal osteoblastic cell line, UMR106. Biochem. Biophys. Res. Comm. 158, 407-412.

Greisler, H.P., Henderson, S.C., and Lam, T.M. (1993). Basic fibroblast growth factor production in vitro by macrophages exposed to Dacron and polyglactin 910. J. Biomaterials Sci., Polymer Edition. 4 (5), 415-430.

Hesp, R., Hulme, P., Williams, D., and Reeve, J. (1981). The relationship between changes in femoral bone density and calcium balance in patients with involutional osteoporosis treated with human PTH fragment 1-34. Metab. Bone Dis.Rel. Res. 2, 331-334.

Hodsman, A.B., Fraher, L.J., Ostbye, T., Adachi, J.D., and Steer, B.M. (1993). An evaluation of several biochemical markers for bone formation and resorption in a protocol utilizing cyclical parathyroid hormone and calcitonin therapy for osteoporosis. J. Clin. Invest. 91, 1138-1148.

Holbein, O., Niedlinger-Wilke, C., Suger, G., Kinzl, L., and Claes, L. (1995). Ilizarov callus distraction produces systemic bone cell mitogens. J. Orthop. Res. 13, 629-638.

Hoyland, J.A., Freemont, A.J., and Sharpe, P.T. (1994). Interleukin-6, IL-6 receptor, and IL-6 nuclear factor gene expression in paget's disease. J. Bone Miner. Res. 9, 75-80.

Jilka, R.L., Hangoc, G., Girasole, G., Passeri, G., Williams, D.C., Abrams, J.S., Boyce, B., Broxmeyer, H., and Manologas, S.C. (1992). Increased osteoclast development after estrogen loss: Mediation by interleukin-6. Science 257, 88.

Kasperk, C., Fitzsimmons, R.J., Stong, D., Mohan, S., Jennings, J., Wergedal, J., and Baylink, D.J. (1990). Studies of mechanism by which androgens enhance mitogenesis and differentiation in bone cells. J. Clin. Endocrinol. Metab. 71, 1322-1329.

Ke, H.Z., Li, M., and Jee, W.S.S. (1992). Prostaglandin E_2 prevents ovariectomy-induced cancellous bone loss in rats. Bone and Mineral 19, 45-62.

Kimble, R.B., Matayoshi, A.B., Vannice, J.L., Kung, V.T., Williams, C., and Pacifici, R. (1995). Simultaneous block of interleukin-1 and tumor necrosis factor is required to completely prevent bone loss in the early postovariectomy period. Endocrinology 136, 3054-3061.

Kyeyune-Nyombi, E., Lau, K-HW, Baylink, D.J., and Strong, D.D. (1991). 1,25-dihydroxyvitamin D_3 stimulates both alkaline phosphatase gene transcription an dmRNA stability in human bone cells. Arch. Biochem. Biophys. 291 (2), 316-325.

Lanyon, L.E. (1993). Osteocyte, strain detection, bone modeling and remodeling. Calc. Tiss. Int. 53, S102.

Lazowski, D.A., Fraher, L.J., Hodsman, A., Steer, B., Modrowski, D., and Han, V.K.M. (1994). Regional variation of insulinlike growth factor-I gene expression in mature rat bone and cartilage. Bone 15, 563-576.

Lean, J.M., Jagger, C.J., Chambers, T.J., and Chow, J.W.M. (1994). Increased insulinlike growth factor-I mRNA expression in osteocytes precedes the increase in bone formation in response to mechanical stimulation. J. Bone Miner. Res. 9, S142.

Libanati, C.R. and Baylink, D.J. (1992). Prevention and treatment of glucocorticoid-induced osteoporosis: A pathogenic perspective. Chest 102 (5), 1426-1435.

Libanati, C.R., Lee, J.E.S., Lois, E., and Baylink, D.J. (1995). The development of peak bone density in pubertal girls is associated with decreased bone resorption and decreased serum interleukin-6. J. Bone Miner. Res. 10 (Suppl. 1), S344.

Littlewood, A.J., Aarden, L.A., Evans, D.B., Russel, R.G.G., and Gowen, M. (1991). Human osteoblastlike cells do not respond to interleukin-6. J. Bone Miner. Res. 6, 141-148.

Lopez-Casillas, F., Cheifetz, S., Doody, J., Andres, J.L., Lane, W.S., and Massague, J. (1991). Structure and expression of the membrane proteoglycan betaglycan; a component of the TGF β receptor system. Cell 67, 785-795.

Manolagas, S.C. (1995). Role of cytokines in bone resorption. Bone 17 (2), 63S-67S.

Marks, Jr., S.C. and Miller, S.C. (1993). Prostaglandins and the skeleton: The legacy and challenges of two decades of research. Endocrine Journal 1, 337-334.

Miyaura, C., Kusano, K., Masuzawa, T., Chaki, O., Onoe, Y., Aoyagi, M., Sasaki, T., Tamura, T., Koishihara, Y., Ohsugi, Y., and Suda, T. (1995). Endogenous bone-resorbing factors in estrogen deficiency: Cooperative effects of IL-1 and IL-6. J. Bone Miner. Res. 10, 1365-1373.

Mohan, S. and Baylink, D.J. (1991). Bone growth factors. Clin. Orthopaedics Rel. Res. 263, 30-48.

Mohan, S., Nakao, Y., Honda, Y., Landale, E., Leser, U., Dony, C., Lang, K., and Baylink, D.J. (1995a). Studies on the mechanisms by which insulinlike growth factor binding protein (IGFBP-4) and IGFBP-5 modulate IGF actions in bone cells. J. Biol. Chem. 270, 20424-20431.

Mohan, S., Libanati, C., Chevalley, T., Linkhart, T., Dony, C., Lang, K., and Baylink, D.J. (1995b). Reduced serum levels of IGF-I, IGF-II, IGFBP-3, and IGFBP-5 in COPD patients after acute glucocorticoid treatment. Bone 16 (Suppl.), 1885, 411 (Abstract).

Meunier, P.J. et al. (1984). Treatment of primary osteoporosis with drugs that increase bone formation: Sodium fluoride, hPTH 1-34, AFDR concept. In: Osteoporosis, International Symposium on Osteoporosis. (Christiansen, C., Arnaud, C.D., Nordin, B.E.C., Parfitt, A.M., Peck, W.A., and Riggs, B.A., Eds.), pp 595-602. Copenhagen, Denmark.

Nicolas, V., Prewett, A., Bettica, P., Mohan, S., Finkelman, R.D., Baylink, D.J., and Farley, J.F. (1994). Age-related decreases in insulinlike growth factor-I and transforming growth factor β in femoral cortical bone from both men and women: Implications for bone loss with aging. J. Clin. Endocrinol. Metab. 78, 1011-1016.

Noda, M. and Camilliere, J.J. (1989). In vivo stimulation of bone formation by transforming growth factor β. Endocrinology 124, 2991-2994.

Oursler, M.J. (1994). Osteoclast synthesis and secretion and activation of latent transforming growth factor β. J. Bone Miner. Res. 9, 443-452.

Prank, K., Nowland, S.J., Harms, H.M., Kloppstech, M., Brabant, G., Hesch, R.D., and Sejnowski, TJ. (1995). Time series prediction of plasma hormone concentration. Evidence for differences in predictability of parathyroid hormone secretion between osteoporotic patients and normal controls. J. Clin. Invest. 95 (6), 2910-2919.

Raab-Cullen, D.M., Thiede, M.A., Kimmel, D.B., and Recker, R.R. (1994). Mechanical loading stimulates rapid changes inperiosteal gene expression. Calcif. Tissue Int. 55, 473-478.

Rawlinson, S.C.F., El-Haj, A.J., Minter, S.L., Tavares, I.A., Bennett, A., and Lanyon, L.E. (1991). Loading-related increases in prostaglandin production in cores of adult canine cancellous bone in vitro: A Role for prostacyclin in adaptive bone remodeling. J. Bone Miner. Res. 6, 1345-1351.

Rawlinson, S.C.F., Mohan, S., Baylink, D.J., and Lanyon, L.E. (1993). Exogenous prostacyclin, but not prostaglandin E2, produces similar responses in both G6PD activity and RNA production as mechanical loading, and increases IGF-II, release in adult cancellous bone in culture. Calcif. Tiss. Int. 53, 324-329.

Robinson, J.A., Riggs, T.C., Spelsberg, T.C., and Oursler, M.J. (1996). Osteoclasts and transforming growth factor-β: Estrogen-mediated isoform-specific regulation of production. Endocrinology 137, 615-621.

Rosen, C., Donahue, L.R., Hunter, S., Holick, M., Kavookjian, H., Kirshenbaum, A., Mohan, S., and Baylink, D.J. (1992). The 24/25–kDa serum insulinlike growth factor binding protein is increased in elderly osteoporotic women. J. Clin. Endocrinol. Metab. 74, 24-27.

Rubin, C.T., Gross, T.S., McLeod, K.J., and Bain, S.D. (1995). Morphologic stages in lamellar bone formation stimulated by a potent mechanical stimulus. J. Bone Miner. Res. 10, 488-495.

Schilitz, P.M., Ohta, T., Glass, D., Mohan, S., and Baylink, D.J. (1992). Growth hormone stimulates cortical bone formation in immature hypophysectomized rats. Endocrine Res. 18, 19-30.

Skerry, T., Bitenshky, L., Chayen, J., and Lanyon, L. (1988). Early stain-related changes in enzyme activity in osteocytes following bone loading in vivo. J. Bone Miner. Res. 4, 783-788.

Stauffer, M., Baylink, D.J., Wergedal, J., and Rich, C. (1973). Decreased bone formation and mineralization, and enhanced resorption in calcium-deficient rats. Am. J. Physiol. 225, 269-276.

Taaffe, D.R., Snow-Harter, C., Connolly D.A., Robinson, T.L., Brown, M.D., and Marcus, R. (1995). Differential effects of swimming versus weight-bearing activity on bone mineral status of eumenorrheic athletes. J. Bone Miner. Res . 10 (4), 586-593.

Tremollieres, R., Stong, D.D., Baylink, D.J., and Mohan, S. (1992). Progesterone and progmestone stimulate human bone cell proliferation: Evidence that the mechanism involves increased insulinlike growth factor-II production. Acta Endocrinol. (Copenhagen) 126, 329-337.

Turner, C.H., Yakano, Y., and Owan, I. (1995). Aging changes mechanical loading thresholds for bone formation in rats. J. Bone Miner. Res. 10, 1544-1549.

Urist, M.R. (1965). Bone: Formation by autoinduction. Science 150, 893-899.

Valentin-Opran, A., Charhon, S.A., Meunier, P.J., Edouard, C.M., and Arlot, M.E. (1982). Quantitative histology of myeloma-induced bone changes. Br. J. Haematol. 52, 601-610.

Weinbaum, S., Cowin, S.C., and Zeng, Y. (1995). A model for the excitation of osteocytes by mechanical loading induced by mechanical fluid shear stresses. J. Biomechanics 27, 3399360.

Westerlind, K.C., and Turner, R.T. (1995). The skeletal effects of spaceflight in growing rats: Tissue specific alterations in mRNA levels for TGF β. J. Bone and Miner. Res. 10, 843-848.

Wiske, P.S., Epstein, S., Bell, N.H., Queener, S.F., Edmonson, J., and Johnston, C.C. (1979). Increases in immunoreactive parathyroid hormone with age. N. Eng. J. Med. 300, 1419-1421.

MECHANOTRANSDUCTION IN BONE

Elisabeth H. Burger, Jenneke Klein-Nulend,
and Stephen C. Cowin

I. Introduction . . . 123
II. The Syncytium of Osteocytes, Bone-Lining Cells, and Osteoblasts . . . 125
III. The Osteocyte as the Sensor of Mechanical Loading . . . 126
IV. Mechanical Stimulation of the Osteocyte . . . 128
V. Response of the Osteocyte to Fluid Flow and Pressure . . . 129
VI. Osteocyte to Bone Surface Cell Communication . . . 130
VII. Conclusion . . . 131
VIII. Summary . . . 132
Acknowledgments . . . 132

I. INTRODUCTION

It is well-known that bone tissue is able to adapt its mass and structure to the prevailing mechanical loads resulting from gravity and muscle function. This phenomenon, first described by Wolff (1986) and Roux (1881) is

Advances in Organ Biology
Volume 5A, pages 123-136.

ISBN: 0-7623-0390-5

called functional or mechanical adaptation. The mechanical adaptation of bone requires a biological system that senses the applied mechanical loading and communicates the loading information to effector cells. These cells then set about accomplishing the structural changes in the bone tissue that lead to altered mechanical characteristics of the bone organ. Several hypotheses have been proposed concerning the substance and nature of the biological system that accomplishes theses changes. The mechanosensory system that is currently basic to the approach of the authors consists of the following mechanisms:

1. The mechanical loads applied to the bone from normal activity (walking, running, lifting) cause flow of interstitial bone fluid in the lacunar canalicular porosity. This occurs because the load-induced straining causes the volume of some pores to decrease slightly and the volume of other pores to increase slightly, creating differences in bone fluid pressure which are then equalized by the movement of the bone fluid from the high pressure pores to the low pressure pores. While the flow of bone fluid is important in the lacunar canalicular porosity, it is negligible in the Haversian and Volkmann channels because the pores of those channels are much larger (30,000 times) and the pressure is more uniform as it must be almost the same as the blood pressure.
2. The fluid flowing past an osteocytic cell process in a canaliculus is sensed by the osteocyte.
3. The osteocyte, having sensed the fluid movement due to the applied loading, communicates this information through the syncytium of osteocytes to the connected bone cells on the bone surface, the bone-lining cells, and the osteoblasts.
4. The bone surface cells that receive the signal from the syncytium may then organize a group of cells to effect resorption and/or deposition of bone tissue.

This bone mechanosensory hypothesis is partially sustained by experimental evidence and model calculations that will be described. The questions addressed by these studies include the following: (a) Which cells (osteoblasts, osteocytes, bone-lining cells) are the bone mechanosensors? (b) What mechanical stimulus activates the mechanosensor? And (c) How is a local mechanical signal translated into an anabolic or catabolic event? Over the last decade, important progress has been made related to these questions, that will be reviewed here.

II. THE SYNCYTIUM OF OSTEOCYTES, BONE-LINING CELLS, AND OSTEOBLASTS

The cells that lie directly on the bony surfaces are bone-lining cells, osteoblasts, and osteoclasts. Osteoblasts and osteoclasts are the effector cells of bone formation and bone resorption, respectively, but in the adult human skeleton only some 5% of the bone surface is covered by osteoblasts and roughly 1% by osteoclasts. By far the greatest part of the bone surface, some 94%, is covered by bone-lining cells. Bone-lining cells are in many ways similar to osteocytes. Unlike osteoblasts, they are flat cells with few organelles and a condensed nucleus, that are not actively engaged in protein synthesis or other energy-consuming processes. They are connected to the superficial osteocytes via cell processes extending into canaliculi, and gap junctions are present between adjacent bone-lining cells and between bone-lining cells and osteocytes (Jee, 1988). The bone cells that are buried in the extracellular bone matrix are the osteocytes. Each osteocyte, enclosed within its mineralized lacuna, has many (perhaps as many as 80) cytoplasmic processes. These processes are approximately 15 μm long and are arrayed three-dimensionally in a manner that permits them to interconnect with similar processes of up to 12 neighboring cells (Palumbo et al., 1990a,b). These processes lie within mineralized bone matrix channels called canaliculi. The small space between the cell process plasma membrane and the canalicular wall is filled with bone fluid and macromolecular complexes of a slightly different composition as the mineralized interlacunar matrix. In particular, large proteoglycans are more prevalent in the pericellular and canalicular space than in the calcified, interlacunar matrix (Jande, 1971; Sauren et al., 1992). However, the sheath of unmineralized matrix is easily penetrated by macromolecules such as albumin (Owen and Triffit, 1976) and peroxidase (Doty and Schofield, 1972; Tanaka and Sakano, 1985). The bone-lining cells, osteoblasts, and osteocytes (i.e., all bone cells except osteoclasts) are extensively interconnected by cell processes, thereby forming a syncytium (Cowin et al., 1991; Moss, 1991).

The touching cell processes of two neighboring bone cells contain gap junctions (Bennett and Goodenough, 1978; Doty, 1981, 1989; Jones and Bingmann, 1991; Schirrmacher et al., 1992; Gourdie and Green, 1993; Jones et al., 1993; Civitelli, 1995). A gap junction is a group of channels connecting two cells. The walls of a channel consist of matching rings of proteins piercing the membrane of each cell, and when the rings associated with two cells connect with each other, the cell-to-cell junction is formed. This junction allows ions and compounds of low molecular weight to pass between the two cells without passing into the extracellular space. The proteins making up a gap junction are called connexins; in bone as in the heart, the protein is connexin 43 (the number refers to the size of the proteins calculated in kilodaltons) (Minkoff et al., 1994). Gap junctions connect superficial osteocytes to periosteal and endosteal bone-lining cells and osteoblasts. All bone surface cells are similarly interconnected laterally on a bony surface. Gap junctions are found where the plasma membranes of a pair of markedly overlapping canalicular processes meet (Rodan, 1992). In compact bone, canaliculi cross the cement lines that form the outer boundary of osteons. Thus extensive communication exists between osteons and interstitial regions (Curtis et al., 1985).

Live bone cells allow the active intercellular transmission of ions and small molecules; gap junctions exhibit both electrical and fluorescent dye transmission (Schirrmacher et al., 1993; Moreno et al., 1994; Spray, 1994). In a physical sense, the bone cell syncytium represents the hard-wiring (Moss, 1991; Nowak, 1992) of bone tissue.

III. THE OSTEOCYTE AS THE SENSOR OF MECHANICAL LOADING

There is only circumstantial evidence that the osteocyte is the primary mechanosensory cell in bone tissue. A list of that evidence is as follows:

1. The placement and distribution of osteocytes in the three-dimensional labyrinthine syncytium is architecturally well-suited to sense deformation of the mineralized tissue encasing them (Lanyon, 1993). Further, the syncytium provides an intracellular as well as an extracellular route for rapid passage of ions and signal molecules. A contrary argument is that, if the bone cell syncytium does not sense mechanical loading, what does it do?
2. The only other candidates for the role of the primary mechanosensory cell in bone tissue are the osteoblasts, the bone lining cells, and the osteoclasts. The osteoclasts may be eliminated directly because they are only present in the bone tissue when they are accomplishing their resorption function. Bone-lining cells should probably be considered as surface osteocytes, because they likely represent the last group of osteoblasts on a (re)modeling bone surface, that have ceased activity and flattened out because the bone surface was complete. Thus the only other serious candidate is the

osteoblast. It should be anticipated that, if the osteocyte has some mechanosensory capacity, then so should the osteoblast because the osteoblast is the progenitor of the osteocyte. However the location of the osteoblasts on bone surfaces means that they must generally sense strain through their supporting substrate and, since the strain in the bone is small (0.2%) this requires very great cell sensivity (Cowin et al., 1991). Furthermore, there is generally a layer of osteoid between the surface cell and the mineralized matrix, compromizing the contact. The osteocyte, on the other hand, directly senses the bone strain fluid movement as we describe in the following section.

3. The osteocyte has been shown to be extremely sensitive to fluid shear stress, but not to compressive stress (Klein-Nulend et al., 1995a). Chicken osteocytes were shown to be the most stress-sensitive cells of bone, capable of rapidly transducing mechanical stress into a release of chemical messengers such as prostaglandins (Klein-Nulend et al., 1995a) and nitric oxide (Klein-Nulend et al., 1995b; Pitsillides et al., 1995).
4. Lastly, a computer-simulation study based on the assumption that osteocytes are the bone mechanosensors, instructing osteoblasts and osteoclasts to adapt bone structure, found that such a model indeed produces structures resembling actual trabecular architecture, which aligns with the actual principal stress orientation according to Wolff's trajectorial hypothesis (Mullender and Huiskes, 1995).

If osteocytes are involved in the transduction of mechanical signals into chemical signals regulating bone remodeling, hormones might modulate the osteocytic response to mechanical strain. Experimental evidence for such a role for hormones or other humoral or local factors is still very limited. Parathyroid hormone (PTH) receptors have been demonstrated on isolated osteocytes (Van der Plas et al., 1994). Receptors for 1,25-dihydroxyvitamin D_3 (1,25-$(OH)_2D_3$) were also shown on osteocytes (Boivin et al., 1987). Indirect evidence for prostaglandin receptors derives from the study by Lean et al. (1995). As osteocytes do release prostaglandins in response to stress (Klein-Nulend et al. 1995a,b), this suggests a role for prostaglandins as amplifiers of a mechanical signal. Finally, Braidman et al. (1995) identified osteocytes as target cells for estrogen. This might implicate a role for estrogen in the mechanosensory function of the osteocyte (Frost, 1992), but nothing definitive has been shown experimentally.

IV. MECHANICAL STIMULATION OF THE OSTEOCYTE

The stimulus for bone remodeling is defined as that particular aspect of the bone's stress or strain history that is employed by the bone to sense its mechanical load environment and to signal for the deposition, maintenance, or resorption of bone tissue. The case for strain rate as a remodeling stimulus has been building over the last quarter century. The animal studies of Hert and his co-workers (Hert et al., 1969, 1971, 1972) suggested the importance of strain rate. Others (O'Connor et al., 1982; Lanyon, 1984; Rubin and Lanyon, 1984, 1987; Goldstein et al., 1991; Turner et al., 1994; Rubin and McLeod, 1996) have quantified the importance of strain rate over strain as a remodeling stimulus. The studies of Weinbaum et al. (1991, 1994), Zeng et al. (1994), Cowin et al. (1995), and Zhang et al. (1996) directed at the understanding of the cellular mechanism for bone remodeling, have suggested that the prime mover is the bone strain rate driven motion of the bone fluid, whose signal is transduced by osteocytes. It was proposed that the osteocytes are stimulated by relatively small fluid shear stresses acting on the membranes of their osteocytic processes.

A hierarchical model of bone tissue structure which related the cyclic mechanical loading applied to the whole bone to the fluid shear stress at the surface of the osteocytic cell process has been presented (Weinbaum et al., 1994). In this model the sensitivity of strain detection is a function of frequency. In the physiological frequency range (1–20 Hz) that is associated with either locomotion (1–2 Hz) or the maintenance of posture (15–30 Hz), the fluid shear stress is nearly proportional to the product of frequency and strain. Thus if bone cells respond to strains of the order of 0.1% at frequencies of one or two Hz, they will also respond to strains of the order of 0.01% at frequencies of 20 Hz. The fluid shear stresses will also strain the macromolecular mechanical connections between the cell and the extracellular bone matrix mentioned in the section above; thus fluid shear stress is also potentially capable of transmitting information from the strained matrix to the bone cell membrane. Extracellular matrix macromolecules connect via integrins in the cell membrane to the cytoskeleton. As the cytoskeleton has an important role in the transduction of information from outside the cell to the cell nucleus (Lazarides, 1980; Wang et al., 1993; Banes et al., 1995), this role allows for an efficient regulation of genomic functions.

Skeletal muscle contraction is a typical bone loading event and has been suggested (Moss, 1969, 1978) and implicated (Rubin and McLeod, 1996) as a stimulus of bone cell activity. Frequency is one of the critical parameters of the muscle stimulus and it serves to differentiate this stimulus from the direct mechanical loads of ambulation which occur at a frequency of one to two Hz. The frequency of contracting muscle in tetanus is from 15 Hz to a maximum of 50–60 Hz in mammalian muscle (McMahon, 1984). It has been observed (McLeod and Rubin, 1992; Rodriquez et al., 1993) that these higher order frequencies, significantly related to bone adaptational responses, are present within the muscle contraction strain energy spectra regardless of animal or activity (Rubin et al., 1993). The close similarity of muscle stimulus frequencies to bone tissue response frequencies is discussed below.

V. RESPONSE OF THE OSTEOCYTE TO FLUID FLOW AND PRESSURE

It has recently been shown that osteocytes, but not periosteal fibroblasts, are extremely sensitive to fluid flow, and that this results in increased prostaglandin as well as nitric oxide (NO) production (Klein-Nulend et al., 1995a,b). Three different cell populations, namely osteocytes, osteoblasts, and periosteal fibroblasts, were subjected to two stress regimes, pulsatile fluid flow (PFF) and intermittent hydrostatic compression (IHC) (Klein-Nulend et al., 1995a). IHC was applied at 0.3 Hz with a 13 kPa peak pressure. PFF was a fluid flow with a mean shear stress of 0.5 Pa with cyclic variations of 0.02 Pa at 5 Hz. The maximal hydrostatic pressure rate was 130 kPa/sec and the maximal fluid shear stress rate was 12 Pa/sec. Under both stress regimes, osteocytes appeared more sensitive than osteoblasts, and osteoblasts more sensitive than periosteal fibroblasts. However, despite the large difference in peak stress and peak stress rate, PFF was more effective than IHC. Osteocytes, but not the other cell types, responded to one hour PFF treatment with a sustained prostaglandin E_2 upregulation lasting at least one hour after the PFF was terminated. By comparison, IHC needed six hours treatment to elicit a response. These results suggested that osteocytes are more sensitive to mechanical stress than osteoblasts, which are, in turn, more sensitive than periosteal fibroblasts. Furthermore, osteocytes appeared particularly sensitive to fluid shear stress, rather than to hydrostatic stress.

These conclusions are in remarkable agreement with the theory developed by Cowin's group (Cowin et al., 1991; Weinbaum et al., 1994) that osteocytes are the "professional" mechanosensory cells of bone, and that they detect mechanical loading events by the canalicular flow of interstitial fluid which results from that loading event. Weinbaum et al. (1994) used Biot's porous media theory to relate loads applied to a whole bone to the flow of canalicular interstitial fluid past the osteocytic processes. Their calculations predict fluid induced shear stresses of 0.8–3 Pa, as a result of peak physiological loading regimes. The findings that bone cells *in vitro* actually respond to fluid shear stress of 0.2–6 Pa (Reich et al., 1990; Williams et al., 1994; Hung et al., 1995; Klein-Nulend et al., 1995a,b) lend experimental support to their theory.

Osteocytes also rapidly release NO in response to stress (Pitsillides et al., 1995; Klein-Nulend et al., 1995b) and this NO response seems to be required for the stress-related prostaglandin release (Klein-Nulend et al., 1995b). Therefore, the behavior of osteocytes is comparable to that of endothelial cells which regulate the flow of blood through the vascular system. They also respond to fluid flow of 0.5 Pa with increased prostaglandin and NO production (Hecker et al., 1993). The response of endothelial cells to shear stress is likely related to their role in mediating an adaptive remodeling of the vasculature, so as to maintain constant endothelial fluid shear stress throughout the arterial site of the circulation (Kamiya et al., 1984). Mutatis mutandis, osteocytes would mediate the adaptive remodeling of bones, to maintain constant strain and, thus, constant canalicular fluid shear stress throughout the skeletal system.

VI. OSTEOCYTE TO BONE SURFACE CELL COMMUNICATION

From a communications viewpoint, the syncytium is a multiply noded (each osteocyte is a node) and a multiply connected network. Each osteocytic process is a connection between (at least) two osteocytes, and each osteocyte is multiply connected to a number of osteocytes that are near neighbors. In order to transmit a signal over the syncytium one osteocyte must be able to signal to a neighboring osteocyte which will then pass the signal on until it reaches bone cells on the bone surface. There are a variety of means of chemical and electrical cell-to-cell communication (De Mello, 1987). The passage of chemical signals, such as Ca^{2+}, from cell to cell appears to occur at a rate that would be too slow to respond to the approximately 30 Hz signal associated with muscle firing. Accordingly, we focus here on electrical cell-to-cell communication. Zhang et al. (1996) have formulated a cable model for cell-to-cell communication in an osteon. The spatial distribution of intracellular electric potential and current from the cement line to the lumen of an osteon was estimated as the frequency of the loading and conductance of the gap junction were altered. In this model the intracellular potential and current are driven by the mechanically induced strain generated streaming potentials (SGPs) produced by the cyclic mechanical loading of bone. The model differs from earlier studies (Harrigan and Hamilton, 1993) in that it pursues a more physiological approach in which the microanatomical dimensions of the connexon pores, osteocytic processes, and the distribution of cellular membrane area and capacitance are used to quantitatively estimate the leakage of current through the osteoblast membrane, the time delay in signal transmission along the cable, and the relative resistance of the osteocytic processes and the connexons in their open and closed states.

The cable model predicts that the connected osteocytic processes function as a high-pass, low-pass filter. The generation of the streaming potentials is a high-pass filter because the SPG generation rises from zero at zero frequency to a plateau with respect to frequency. The decay of the signal along the connected osteocytic processes functions as a low-pass filter because higher frequencies are not propagated. The theory also predicts that the pore pressure relaxation time for the draining of the bone fluid into the osteonal canal has the same order as the characteristic diffusion time for the spread of current along the membrane of the osteocytic processes. This coincidence of characteristic times produced a spectral resonance in the cable at 30 Hz. Thus there is a large amplification of the intracellular potential and current in the surface bone cells which could serve as the initiating signal for a remodeling response. This voltage amplification might also explain why live bone appears to be selectively responsive to the mechanical loading in a specific frequency range (15–60 Hz), as has been experimentally demonstrated for several species (Rubin and McLeod, 1996).

The primacy of electrical signals is suggested here, since while bone cell transduction may also use small biochemical molecules that can pass through gap junctions, the time-course of mechanosensory processes is believed to be too rapid for the involvement of secondary messengers (French, 1992; Carvalho et al., 1994). As we noted above, the passage of chemical signals, such as Ca^{2+}, from cell to cell appears to occur at a rate that would be too slow to respond to the approximately 30 Hz signal associated with muscle firing.

VII. CONCLUSIONS

Although many details of the mechanosensory system in bone are still unclear, important progress has been made over the last decade. Theoretical and experimental studies in animals and cultured bone cells agree that the network of osteocytes in conjunction with the bone-lining cells provides the three-dimensional cellular structure that allows the detection and integration of mechanical signals. The flow of fluid resulting from stress, through the vast lacunar-canalicular porosity of bone, likely provides the mechanical signal that activates the osteocytes. Electrical cell-to-cell communication may provide the means for transmitting fast mechanical signals over the osteocyte lining-cell syncytium.

VIII. SUMMARY

It is becoming increasingly evident that mechanical strain is an important regulator of bone homeostasis. However, the mechanism whereby bone tissue detects the strain in a bone organ during mechanical loading, and how mechanical signals are transduced into local anabolic or catabolic responses, is only partially understood. We briefly review current theoretical and experimental evidence which suggests that osteocytes are the principal mechanosensor cells of bone, that they are activated by shear stress from fluid flowing through the osteocyte canaliculi, and that the electrically coupled three-dimensional network of osteocytes and lining cells provides the physiological basis for a geometrically meaningful coordinated response.

ACKNOWLEDGMENTS

The work of S.C.C. was performed while on sabbatical leave, supported by a Fogarty Senior International Fellowship, and the Netherlands Organization for Scientific Research NWO.

REFERENCES

Banes, A.J., Tsuzaki, M., Yamamoto, J., Fischer, T., Brigman, B., Brown, T., and Miller, L. (1995). Mechanoreception at the cellular level: the detection, interpretation, and diversity of responses to mechanical signals. Biochem. Cell Biol. 73, 349-365.

Bennett, M.V.L. and Goodenough, D.A. (1978). Gap junctions. Electronic coupling and intercellular communication. Neurosci. Res. Prog. Bull. 16, 373-485.

Boivin, G., Mesguich, P., Pike, J.W., Bouillon, R., Meunier, P.J., Haussler, M.R., Dubois, P.M., and Morel, G. (1987). Ultrastructural immunocytochemical localization of endogenous 1,25-dihydroxyvitamin D_3 and its receptors in osteoblasts and osteocytes from neonatal mouse and rat calvariae. Bone Miner. 3, 125-136.

Braidman, J.P., Davenport, L.K., Carter, D.H., Selby, P.L., Mawer, E.B., and Freemont, A.J. (1995). Preliminary in situ identification of estrogen target cells in bone. J. Bone Miner. Res. 10, 74-80.

Carvalho, R.S., Scott, J.E., Suga, D.M., and Yen, E.H.K. (1994). Stimulation of signal; transduction pathways in osteoblasts by mechanical strain potentiated by parathyroid hormone. J. Bone Miner. Res. 9, 999-1011.

Civitelli, R. (1995). Cell-cell communication in bone. Calcif. Tissue Int. 56, S29-S31.

Cowin, S.C., Moss-Salentijn, L., and Moss, M.L. (1991). Candidates for the mechanosensory system in bone. J. Biomech. Engin. 113, 191-197.

Cowin, S.C., Weinbaum, S., and Zeng, Y. (1995). A case for bone canaliculi as the anatomical site of strain generated potentials, J. Biomech. 28, 1281-1296.

Curtis, T.A., Ashrafi, S.H., and Weber, D.F. (1985). Canalicular communication in the cortices of human long bones. Anat. Rec. 212, 336-344.

De Mello, W.C. (1987). The ways cells communicate. In: *Cell-to-Cell Communication.* (de Mello, W.C., Ed.) pp. 1-20. Plenum Press, New York.

Doty, S.B. and Schofield, B.M. (1972). Metabolic and structural change within osteocytes of rat bone. In: Calcium, parathyroid hormone and the calcitonins. (Talmage, B.V. and Munson, P.L., Eds.), pp. 353-365, Exerpta Medica, Amsterdam.

Doty, S.B. (1981). Morphological evidence of gap junctions between bone cells. Calcif. Tissue Int. 33, 509-512.

Doty, S.B. (1989). Cell-to-cell communication in bone tissue. In: The Biological Mechanism of Tooth Eruption and Root Resorption. (Davidovitch, Z., Ed.), pp. 61-69, EBSCO Media, Birmingham, AL.

French, A.S. (1992). Mechanotransduction. Ann. Rev. Physiol. 54, 135-152.

Frost, H.J. (1992). The role of changes in mechanical usage set points in the pathogenesis of osteoporosis. J. Bone Miner. Res. 7, 253-261.

Goldstein, S.A., Matthews, L.S., Kuhn, J.L., and Hollister, S.J. (1991). Trabecular bone remodeling: An experimental model. J. Biomech. 24, 135-150.

Gourdie, R. and Green, C. (1993). The incidence and size of gap junctions between bone cells in rat calvaria. Anat. Embryol. 187, 343-352.

Harrigan, T.P. and Hamilton, J.J. (1993). Bone strain sensation via transmembrane potential changes in surface osteoblasts: Loading rate and microstructural implications. J. Biomech. 26, 183-200.

Hecker, M., Mülsch, A., Bassenge, E., and Busse, R. (1993). Vasoconstriction and increased flow: Two principal mechanisms of shear stress-dependent endothelial autacoid release. Am. J. Physiol. 265 (Heart Cir. Physiol. 34), H828-H833.

Hert, J., Liskova, M., and Landgrot, B. (1969). Influence of the long-term continuous bending on the bone. An experimental study on the tibia of the rabbit. Folia Morphologia 17, 389-399.

Hert, J., Liskova, M., and Landa, J. (1971). Reaction of bone to mechanical stimuli. Part I. Continuous and intermittent loading of tibia in rabbit. Folia Morphologia 19, 290-300.

Hert, J., Pribylova, E., and Liskova, M. (1972). Reaction of bone to mechanical stimuli. Part 3. Microstructure of compact bone of rabbit tibia after intermittent loading. Acta Anatomica 82, 218-230.

Hung, C.T., Pollack, S.R., Reilly, T.M., and Brighton, C.T. (1995). Real-time calcium response of cultured bone cells to fluid flow. Clin. Orthop. Rel. Res. 313, 256-269.

Jande, S.S. (1971). Fine structural study of osteocytes and their surrounding bone matrix with respect to their age in young chicks. J. Ultrastr. Res. 37, 279-300.

Jee, W.S.S. (1988). The skeletal tissues. In: Cell and Tissue Biology. (Weiss, L., Ed.), pp. 213-253, Urban and Schwarzenberg, Munich.

Jones, D.B. and Bingmann, D. (1991). How do osteoblasts respond to mechanical stimulation? Cells Materials 1, 329-340.

Jones, S.J., Gray, C., Sakamaki, H., Arora, M., Boyde, A., Gourdie, R., and Green, C. (1993). The incidence and size of gap junctions between bone cells in rat calvaria. Anat. Embryol. 187, 343-352.

Kamiya, A., Bukhari, R., and Togawa, T. (1984). Adaptive regulation of wall shear stress optimizing vascular tree function. Bull. Math. Biol. 46, 127-137.

Klein-Nulend, J., Van der Plas, A., Semeins, C.M., Ajubi, N.E., Frangos, J.A., Nijweide, P.J., and Burger, E.H. (1995a). Sensitivity of osteocytes to biomechanical stress in vitro. FASEB J. 9, 441-445.

Klein-Nulend, J., Semeins, C.M., Ajubi, N.E., Nijweide, P.J., and Burger, E.H. (1995b). Pulsating fluid flow increases nitric oxide (NO) synthesis by osteocytes but not periosteal fibroblasts—correlation with prostaglandin upregulation. Biochem. Biophys. Res. Commun. 217, 640-648.

Lanyon, L.E. (1984). Functional strain as a determinant for bone remodeling. Calcif. Tissue Int. 36, S56-S61.

Lanyon, L.E. (1993). Osteocytes, strain detection, bone modeling and remodeling. Calcif. Tissue Int. 53, S102-S106.

Lazarides, E. (1980). Intermediate filaments as mechanical integrators of cellular space. Nature (London) 283, 249-256.

Lean, J.M., Jagger, C.J., Chambers, T.J., and Chow, J.W. (1995). Increased insulinlike growth factor I mRNA expression in rat osteocytes in response to mechanical stimulation. Am. J. Physiol. 268, E318-E327.

McLeod, K.J. and Rubin, C.T. (1992). The effect of low-frequency electrical fields on osteogenesis. J. Bone Joint Surg. 74A, 920-929.

McMahon, T.A. (1984). Muscles, Reflexes, and Locomotion. Princeton University Press, Princeton, NJ.

Minkoff, R., Rundus, V.R., Parker, S.B., Hertzberg, E.L., Laing, J.G., and Beyer, E. (1994). Gap junction proteins exhibit early and specific expression during intramembranous bone formation in the developing chick mandible. Anat. Embryol. 190, 231-241.

Moreno, A.P., Rook, M.B., Fishman, G.I., and Spray, D.C. (1994). Gap junction channels: Distinct voltage-sensitive and -insensitive conductance states. Biophys. J. 67, 113-119.

Moss, M.L. (1969). A theoretical analysis of the functional matrix. Acta Biotheoret. 18, 195-202.

Moss, M.L. (1978). The Muscle-Bone Interface: An Analysis of a Morphological Boundary. (Monograph). pp. 39-72. Center for Human Growth and Development, Ann Arbor, MI.

Moss, M.L. (1991). Bone as a connected cellular network: Modeling and testing. In: Topics in Biomedical Engineering (Ross, G., Ed.), pp. 117-119, Pergamon Press, New York.

Mullender, M.G., and Huiskes, R. (1995). Proposal for the regulatory mechanism of Wolff's Law. J. Orthopaed. Res. 13, 503-512.

Nowak, R. (1992). Cells that fire together, wire together. J. NIH Res. 4, 60-64.

O'Connor, J.A., Lanyon, L.E., and MacFie, H. (1982). The influence of strain rate on adaptive bone remodeling, J. Biomech. 15, 767-781.
Owen, M. and Triffit, J.T. (1976). Extravascular albumin in bone tissue. J. Physiol. 257, 293-307.
Palumbo, C., Palazzini, S, and Marotti, G. (1990a). Morphological study of intercellular functions during osteocyte differentiation. Bone 11, 401-406.
Palumbo, C., Palazzini, S., Zaffe, D., and Marotti, G. (1990b). Osteocyte differentiation in the tibia of newborn rabbit: An ultrastructural study of the formation of cytoplasmic processes. Acta Anat. 137, 350-358.
Pitsillides, A.A., Rawlinson, S.C.F., Suswillo, R.F.L., Bourrin, S, Zaman, G, and Lanyon, L.E. (1995). Mechanical strain-induced NO production by bone cells—A possible role in adaptive bone (re)modeling. FASEB J. 9, 1614-1622.
Reich, K.M, Gay, C.V., and Frangos, J.A. (1990). Fluid shear stress as a mediator of osteoblast cyclic adenosine monophosphate production. J. Cell. Physiol. 143, 100-104.
Rodan, G. (1992). Introduction to bone biology. Bone 13, S3-S6.
Rodriquez, A.A., Agre, J.C., Knudtson, E.R., Franke, T.M., and Ng, A.V. (1993). Acoustic myography compared to electromyography during isometric fatigue and recovery. Muscle Nerve 16, 188-192.
Roux, W. (1881). Der Kamkpf der Teile im Organismus. Leipzig, Engelmann.
Rubin, C.T., Donahue, H.J., Rubin, J.E., and McLeod, K.J. (1993). Optimization of electric field parameters for the control of bone remodeling: Exploitation of an indigenous mechanism for the prevention of osteopenia. J. Bone Miner. Res. 8, S573-S581.
Rubin, C.T. and Lanyon, L.E. (1984). Regulation of bone formation by applied dynamic loads. J. Bone Joint Surg. 66A, 397-415.
Rubin, C.T. and Lanyon, L.E. (1987). Osteoregulatory nature of mechanical stimuli: Function as a determinant for adaptive bone remodeling. J. Orthop. Res. 5, 300-310.
Rubin, C.T. and McLeod, K.J. (1996). Inhibition of osteopenia by biophysical intervention. In: Osteoporosis (Marcus, R., Feldman, D., and Kelsey, J., Eds.), pp. 351-371, Academic Press, New York.
Sauren, Y.M.H.F., Mieremet, R.H.P., Groot, C.G., and Scherft, J.P. (1992). An electron microscopic study on the presence of proteoglycans in the mineralized matrix of rat and human compact lamellar bone. Anat. Rec. 232, 36-44.
Schirrmacher, K., Schmitz, I., Winterhager, E., Traub, O., Brummer, F., Jones, D., and Bingmann, D. (1992). Characterization of gap junctions between osteoblastlike cells in culture. Calcif. Tissue Int. 51, 285-290.
Schirrmacher, K., Brummer, F., Dusing, R., and Bingmann, D. (1993). Dye and electric coupling between osteoblastslike cells in culture. Calcif. Tissue Int. 53, 53-60.
Spray, D.C. (1994). Physiological and pharmacological regulation of gap junction channels. In: Molecular Mechanisms of Epithelial Cell Junctions: From Development to Disease (Chi, S., Ed.), pp. 195-215, RG Landes, Austin, TX.
Tanaka, T. and Sakano, A. (1985). Differences in permeability of microperoxidase and horseradish peroxidase into alveolar bone of developing rats. J. Dent. Res. 64, 870-876.
Turner, C.H., Forwood, M.R., and Otter, M.W. (1994). Mechanotranduction in bone: Do bone cells act as sensors of fluid flow? FASEB J. 8, 875-878.
Van der Plas, A., Aarden, E.M., Feyen, J.H.M., de Boer, A.H., Wiltink, A., Alblas, M.J., De Ley, L., and Nijweide, P.J. (1994). Characteristics and properties of osteocytes in culture. J. Bone Miner. Res. 9, 1697-1704.

Wang, N., Butler, J.P., and Ingber, D.E. (1993). Mechanotransduction across the cell surface and through the cytoskeleton. Science 260, 1124-1127.

Weinbaum, S., Cowin, S.C., and Zeng, Y. (1991). A model for the fluid shear stress excitation of membrane ion channels in osteocytic processes due to bone strain. In: Advances in Bioengineering. (Vanderby, Jr, R., Ed.), pp. 317-320. American Society of Mechanical Engineers, New York.

Weinbaum, S., Cowin, S.C. and Zeng, Y. (1994). Excitation of osteocytes by mechanical loading-induced bone fluid shear stresses. J. Biomech. 27, 339-360.

Williams, J.L., Iannotti, J.P., Ham, A., Bleuit, J., and Chen, J.H. (1994). Effects of fluid shear stress on bone cells. Biorheol. 31, 163-170.

Wolff, J. (1986). The Law of Bone Remodeling. (Translated by Maquet, P. and Furlong, R.), Springer-Verlag, Berlin.

Zeng, Y., Cowin, S.C., and Weinbaum, S. (1994). A fiber matrix model for fluid flow and streaming potentials in the canaliculi of an osteon. Ann. Biomed. Engin. 22, 280-292.

Zhang, D., Cowin, S.C., and Weinbaum, S. (1996). Electrical signal transmission and gap junction regulation in bone cell network: A cable model for an osteon. Ann. Biomed. Engin., (in press).

VASCULAR CONTROL OF BONE REMODELING

Ted S. Gross and Thomas L. Clemens

I. Introduction . . . 138
II. Importance of the Vasculature to Bone Development and Remodeling . . . 138
III. Vasoactive Agents and Bone Cell Activity . . . 142
A. Endothelins . . . 142
B. Thrombin . . . 145
C. Parathyroid Hormone-Related Protein . . . 145
D. Calcitonin Gene-Related Peptide . . . 147
E. Vascular Endothelial Growth Factor . . . 149
F. The Natriuretic Peptides . . . 149
G. Nitric Oxide and Small Oxygen Radicals . . . 150
H. Catecholamines . . . 152
IV. Role of Blood Vessels and Their Products in Bone Pathophysiology . . . 153
V. Summary . . . 154

Advances in Organ Biology
Volume 5A, pages 137-160.

ISBN: 0-7623-0390-5

I. INTRODUCTION

The coordinated regulation of cell activity within bone tissue is achieved through an extensive array of local and systemic factors. Factors made by endothelial and vascular smooth muscle cells have been shown to modulate the activity of osteoblasts and osteoclasts under a variety of conditions. Reciprocally, specific bone cell-derived factors elicit distinct actions on endothelial and vascular smooth muscle cells. A wide variety of soluble molecules and peptides are produced in the vasculature-bone microenvironment. Among these are numerous growth factors and cytokines. Prostaglandins and other ecosanoids have been extensively studied and reviewed elsewhere (Raisz, 1996) and will therefore not be addressed here. The primary purpose of this chapter is to briefly review the recent literature on the effects on bone tissue of the less well-known vasoactive substances. To provide context at the *in vivo* level, we also discuss the physiologic roles that endothelial cells and the vascular system as a whole are thought to play in bone physiology and pathophysiology.

II. IMPORTANCE OF VASCULATURE TO BONE DEVELOPMENT AND REMODELING

Adult bone is highly vascularized. It has been estimated that approximately 5% of resting cardiac output passes through the adult skeleton (Guyton and Hall, 1996). Mature long bones are vascularized by a nutrient artery and an intricate vascular network of vessels, capillaries, and blood sinusoids (Figure 1) (Brooks, 1971). The nutrient artery arises from the systemic circulation, enters the diaphysis, and then branches into ascending and descending medullary arteries within the marrow cavity. These vessels are then further subdivided into arterioles which penetrate the endocortical surface to form the primary supply of the diaphyseal cortex (Rhinelander, 1974). The high degree of vascularity of bone is visually evident on any cross-section of mammalian cortical bone (Figure 2). The three-dimensional lattice of Volkmann and Haversian canals each represents a vessel populated by endothelial cells. The osteocytes surrounding each canal are ideally located to communicate with the resident endothelial cells. It is the close spatial relationship between endothelial cells and bone cells during bone development, modeling, and remodeling that suggested the potential for interaction between these cell populations (Trueta, 1963).

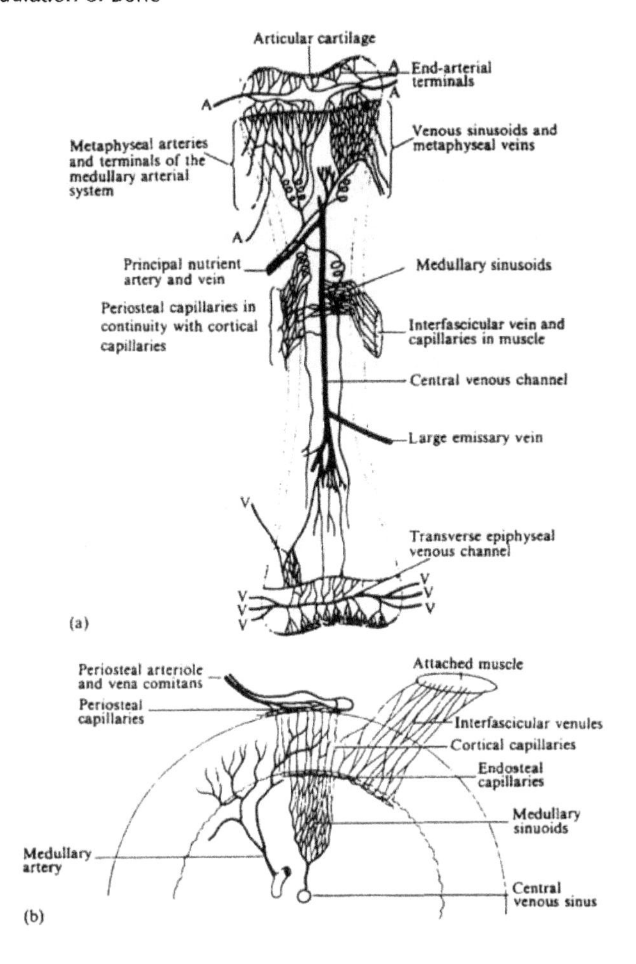

Figure 1. Panel **A**, Diagrammatic illustration of the vasculature of a long bone based on the work of Brooks and Harrison. Panel **B**: Diagram of a transverse section of a long bone showing the anatomy of the main vasculature. Reproduced with permission from Brooks, M. (1971). The Blood Supply of Bone. Butterworths, London.

In a simple analogy, the circulatory system serves as a highway upon which O_2, nutrients, and hormones are transported to cells, and CO_2 and metabolic end products are cleared from the tissue. Rather than passively observing this exchange, there is now a substantial body of evidence indicating that endothelial cells actively mediate physiologic processes required by nutrient and cytokine exchange (Shireman and Pearce, 1996). Within the skeletal system, endothelial cells are associated with two processes that potentially affect bone cell populations: angiogenesis and vasoregulation.

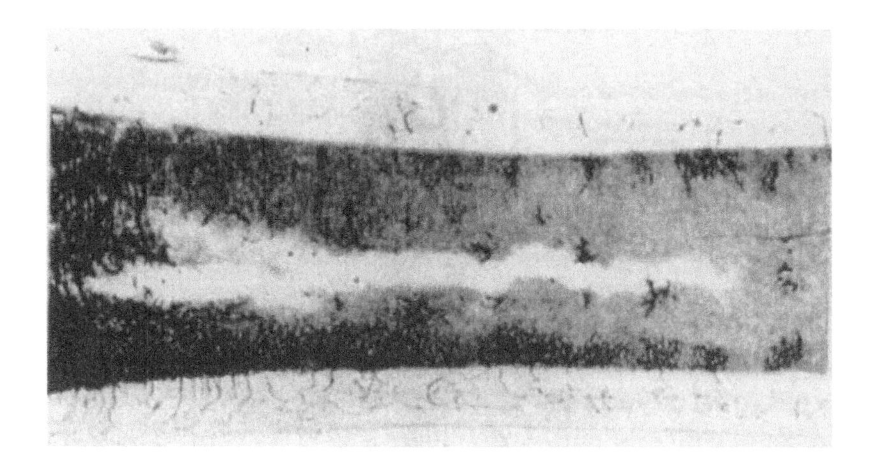

Figure 2. India-ink gelatin profusion of the femoral diaphysial shaft illustrating the cortical bone canals and the "bushlike" sinusoid formations in the marrow. Reproduced with permission from De Saint-Georges, L. and Miller, S.C. (1992). Anat. Rec. 233:169–177.

Intramembranous and endochondral ossification and intracortical remodeling occur in close association and proximity to capillary in-growth or angiogenesis. Intramembranous ossification is characterized by invasion of capillaries into the mesenchymal zone, emergence of preosteoblasts from the developing mesenchyme and, finally, differentiation of osteoblasts which deposit osteoid. In endochondral bone formation, mesenchymal chondrocyte precursors differentiate into chondroblasts which then secrete a cartilaginous matrix. During this stage, the matrix is avascular, possibly due to the actions of antiangiogenic substances. With progressive chondrocyte proliferation and enlargement of the extracellular matrix, chondrocytes become hypertrophic and ultimately undergo terminal differentiation and apoptosis. Chondroclasts then invade this region and form lacunae into which blood vessels and associated perivascular osteoprogenitor cells take residence and differentiate into mature, mineralizing osteoblasts (Trueta, 1963). The process of vasculogenesis in embryonic tissues appears to coordinate limb development by providing structural support for bone-forming osteoblasts (Caplan et al., 1983). Calcification of bone matrix is accompanied by in-growth of capillary endothelial cells and the rate of new bone formation and blood flow appear to be tightly coupled (Lewinson and Silberman, 1992).

Once skeletal growth ceases, bone mass is controlled by the activity of basic multicellular units (Frost, 1989). Remodeling is characterized by activation and resorption by osteoclasts coupled with osteoblastic refilling of excavated bone cavities (see Chapter 2 for details on the remodelling process). Each of these intracortical remodeling events is accompanied by vascular budding (Geiser and Trueta, 1958). As the number of remodeling events accumulates with age, the number of vascular channels within bone are also elevated in the elderly (Laval-Jeantet et al., 1983).

Skeletal vasoregulation, as with other tissues, is accomplished by the interaction of systemic and local feedback control (Mellander, 1970; Duling and Klitzman, 1980). Systemic hormonal alterations, such as those caused by ovariectomy (Egrise et al., 1992) and castration (Kapitola et al., 1995) alter bone blood flow. The metabolic state of the tissue has been postulated as being the primary stimulus for local vasoregulation (Adair et al., 1990). Within this framework, any condition that alters the metabolic state of a tissue should precipitate local vasoregulation. The altered bone blood flow observed in response to exercise (Tøndevold and Bülow, 1983) and disuse (Semb, 1969; Hardt, 1972, Gross et al., submitted) suggests that bone conforms to this supposition. It is therefore reasonable to propose that the mechanism by which local vasoregulation is achieved in bone is similar to that described for other tissues such as muscle.

Within muscle, endothelial cells are connected chemically, electrically, and mechanically with upstream and downstream endothelial cells and vascular smooth muscle cells (Guyton and Hall, 1996). As such, endothelial cells within the tissue are ideally located to monitor the physiologic demands of the tissue and initiate the signal for local vasoregulation (Segal, 1994). Interestingly, endothelial cells appear to interact with bone cell populations in a variety of manners. Recent *in vitro* studies have begun to identify specific factors responsible for coupling of endothelial cells and bone cells. Endothelial cells modulate local blood flow by releasing vasoactive substances that act upon the smooth muscle cell population (Davies, 1995). Endothelial cell released vasoconstrictors (e.g., endothelins; Kuchan and Frangos, 1993) and vasodilators (e.g., nitric oxide; Miller and Burnett, 1992; and prostaglandins; Wilson and Kapoor, 1993) both modulate bone cell activity (Collins and Chambers, 1991; Alam et al., 1992; Ralston et al., 1994). Further, it appears that endothelial cells are capable of stimulating osteoclastic adhesion via the insulinlike growth factor-I (IGF-1) pathway (Formigli et al., 1995). While these studies support the hypothesis that endothelial cells are capable of dynamically regulating bone cell activity (Zaidi et al., 1993), these pathways have not yet been confirmed at the *in vivo* level (Figure 3).

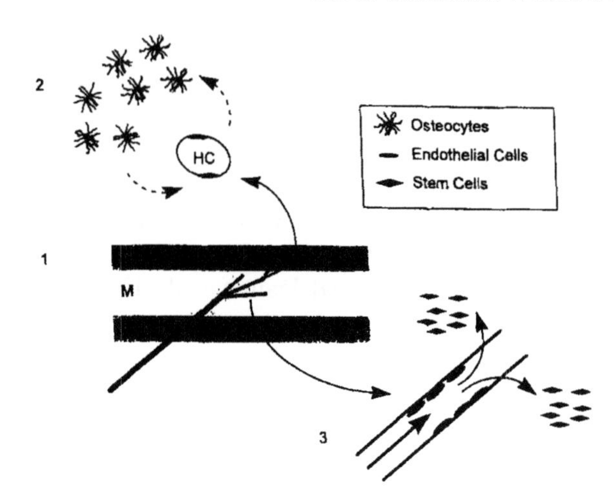

Figure 3. Schematic illustration of two postulated sites of interaction between endothelial cell and bone cell populations. The descending vascular supply of the long bone penetrates the cortex (C), extensively branches within the marrow cavity (M), and then penetrates the endocortical surface to supply the diaphyseal cortex (1). Within the cortex (2), osteocytes surround the endothelial populated Haversian canals (HC). At this level, the close juxtaposition of these cell populations would create an ideal means of monitoring and responding to tissue metabolic demands. Within marrow arterioles (3), endothelial cell mediated tissue vasoregulation is accomplished by substances that have potent effects on bone cell populations. Given that this process occurs adjacent to stem cell populations, the potential for interaction is high.

III. VASOACTIVE AGENTS AND BONE CELL ACTIVITY

A. Endothelins

Endothelins are 21 amino acid long peptides which are similar structurally to reptilian neurotoxins and are among the most potent vasoconstrictors known (Rubanyi and Polokoff, 1994; Levin, 1995). Three different isopeptides differ structurally by only a few amino acids and are cleaved from their larger proendothelins and each mature form exhibits an intrachain disulfide bridge. Endothelin (ET)-1 is made by endothelial cells and a large number of other mesenchymal and epithelial cells, whereas the sites of production of ET-2 and ET-3 are more restricted. The endothelins bind and activate two distinct classes of G-protein coupled receptor subtypes called ET_a and ET_b. ET-1 and ET-2 bind the ET_a receptor with equal affinity whereas ET-3 has reduced affinity. Each of the three endothelins bind the ET_b receptor with approximately equal affinity. Activation of ET receptors is associated with stimulation of multiple signal transduction second messenger pathways in a cell-specific fashion. ET_a receptors are expressed in vascular smooth muscle cells and cardiac myocytes and signal through a linked G-protein to activate phospholipase C (PLC), K^+ channels, cAMP-dependent protein kinases and nitric oxide (NO) formation. ET_b receptors are expressed predominantly in endothelial cells and like the ET_a receptor signal through the phosphoinositide-Ca^{2+} cascade. In some cells the ET_b receptor activates G_i which leads to inhibition of cAMP and activation of the Na^+/H^+ antiporter.

An increasing body of *in vitro* evidence supports a role for endothelin in the regulation of bone cell activity. As mentioned above, the importance of functional vasculature in bone development and repair, together with the close proximity of bone osteoblasts, osteoclasts and stromal cells to the ET producing endothelium and vascular smooth muscle cells provides circumstantial evidence that ET functions locally in bone. In addition, ET has been detected in rat osteoblasts and osteoclasts *in situ* (Sasaki and Hong, 1993) and osteoblasts express both ET_a and ET_b receptors (Sakurai et al., 1990).

The properties of the ET receptors in bone have been studied by determining the second messengers produced following ligand activation (Stern et al., 1995). Thus, ET-1 evokes a calcium transient and stimulates formation of inositol phosphates in several different osteoblast cell types (Takuwa et al., 1989; Lee and Stern, 1995). ET_b receptors in osteoblast-like cells (see below) are coupled to G_i and their activation inhibits cAMP production. In addition, ET-1 treatment of osteoblastic cells causes rapid desensitization to thrombin and epidermal growth factor (EGF) (Tatrai and Stern, 1993) but enhances calcium transients elicited by parathyroid hormone (PTH) (Figure 4) (Lee and Stern, 1995). Therefore, functional diversity of ET action is amplified in part by activation of multiple signal transduction pathways.

ET-1 has weak mitogenic activity in MC3T3-E1 mouse osteoblasts (Takuwa et al., 1989; Schvartz et al., 1992). The mitogenic effects of ET-1 in these cells appears to require tyrosine phosphorylation. ET-1 also regulates the expression and elaboration of several extracellular matrix proteins. In MC3T3-E1 cells ET-1 inhibits alkaline phosphatase activity. In neonatal calvaria ET-1 stimulates both non-collagenous and collagenous protein synthesis (Tatrai et al., 1992). ET-1 and ET-2, but not ET-3, stimulate expression of osteopontin and osteocalcin mRNA in ROS 17/2.8 cells, suggesting that ET actions in osteoblasts are mediated through the ET_a receptor. In addition, ET stimulates the production of interleukin-6 (IL-6) in rat bone marrow-derived stromal cells (Agui et al., 1994).

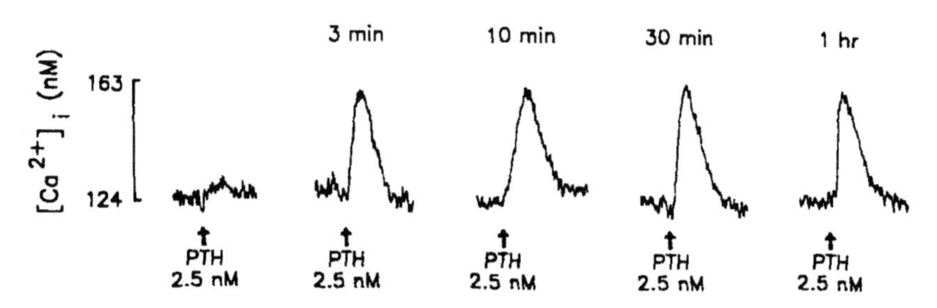

Figure 4. Potentiation of PTH-induced calcium transients by endothelin 1. Endothelin 1 (1 nM) was added to UMR-106 cells three minutes to one hour before parathyroid hormone (2.4 nM). Cytosolic calcium was determined using fluo-3. Reprinted with permission from Stern, P.H., et al. (1995). J. Nutrition, 125: 2028S–2023S.

The effects of ET-1 on bone resorption are less well defined and appear to depend on the type of bone cell culture preparation. For instance, ET-1 causes resorption in fetal mouse calvaria which is abolished by treatment with indomethacin suggesting that these effects are secondary to stimulation of prostaglandin production. However, in fetal long bone cultures, ET-1 has no effect on resorption (Tatrai et al., 1992). By contrast, in disaggregated osteoclasts, ET-1 inhibits both the motility and resorptive activity suggesting a direct inhibitory activity of ET on this cell type (Alam et al., 1992). The type of ET receptor expressed on the osteoclast is not known, but the ability of ET-1 to inhibit osteoclast activity is not associated with a change in intracellular calcium suggesting that ET-1 affects osteoclast activity through different second messengers than those operating in osteoblasts. The established antiresorptive action of NO (see below) raises the possibility that this signaling molecule might mediate the osteoclast inhibitory action of ET.

To date there have been no *in vivo* studies on the effect of exogenous ET administration on bone and mineral metabolism. However, transgenic mice with targeted ablation of ET-1 die at birth from respiratory failure and display severe maldevelopment of craniofacial tissues including aberrant zygomatic and temporal bones and absent auditory ossicles (Kurihara et al., 1994). It is unclear whether or to what extent the developmental abnormalities seen in these animals are due to lack of normal ET actions on developing bone or cartilage cells. Targeted disruption of the mouse ET_b receptor gene produces megacolon and spotted coat color similar to that seen in the natural mouse piebald-lethal mutation of the ET_b receptor (Hosoda et al., 1994), but no skeletal abnormalities were reported in these mouse models.

B. Thrombin

Alpha thrombin is a circulating serine protease and is the principal end product of the coagulation cascade (Fenton, 1986). Thrombin elicits the release reaction and platelet aggregation and also has potent mitogenic effects on a variety of different cell types. Thrombin is chemotactic for monocytes and is mitogenic for lymphocytes, fibroblasts, and vascular smooth muscle cells. Thrombin induces biological responses following its association with a specific receptor which is a member of the seven transmembrane domain receptor family. However, unlike other more classical peptide ligands, thrombin signals by a unique proteolytic process referred to as tethered liganding (Coughlin, 1995). Thrombin binds to its receptor's N-terminal extension and then cleaves off an inactive fragment to unmask a new amino terminus. This "unmasked" portion (referred to as a tethered peptide) then directly activates the receptor by binding to a putative binding pocket. Activation of the thrombin receptor stimulates phosphoinositide hydrolysis and inhibits cAMP production by virtue of its interaction with both pertussis toxin sensitive and insensitive G proteins.

A potential role for thrombin in bone stems from initial observations that fibrin depositions in certain inflammatory lesions such as rheumatoid arthritis are often associated with bone degradation. In this process, macrophages initiate the coagulation activity and deposit the vitamin K dependent coagulation factors and thromboplastin (factor II). Alpha-thrombin increases bone resorption as measured in fetal and neonatal long bone cultures (Lerner and Gustafson, 1988). Since calcitonin inhibits the thrombin-induced bone resorptive activity in fetal mouse calvarial bones, these effects appear to be mediated by actions on osteoclasts. Moreover, the resorptive effects of thrombin are most likely indirect and involve both prostaglandin-dependent and independent pathways (Lerner and Gustafson, 1988). The ability of thrombin to increase intracellular calcium in UMR osteoblast-like cells were first reported by van Leeuwen et al. (1988). More recently, Babich and co-workers (Babich et al., 1990) demonstrated that thrombin increased inositol phosphate production and intracellular calcium concentrations in UMR-106-H5 cells. Subsequent studies (Babich et al., 1991) demonstrated that the mechanism by which thrombin mobilized intracellular calcium in UMR-106 cells were distinct from those mediating PTH-induced calcium transients.

C. Parathyroid Hormone-Related Protein

Parathyroid hormone-related protein (PTHrP) was identified from tumor cells originally derived from patients with the syndrome of humoral hypercalcemia of malignancy (for a review, see Mosely and Martin, 1996). PTHrP and PTH share limited N-terminal sequence homology which enables both PTH and PTHrP proteins to activate the same G-protein linked receptor in bone and kidney (also see Abou-Samra's chapter in volume 5A). There is also direct evidence that several different cell types indeed produce a mid-region PTHrP fragment with the N-terminus beginning at amino acid 37 and probably extending to a putative cleavage site at amino acids 102–106 (Wu et al., 1996). Novel C-terminal fragments are also produced but have yet to be conclusively identified. These different cleavage products, which lack the PTH-like N-terminal region, are postulated to activate receptors distinct for the PTH/PTHrP receptor and have a biological profile different from N-terminal PTHrP peptides. In support of this are studies demonstrating that synthetic N-terminal fragments or recombinant PTHrP-(1–141) usually exhibit biological effects similar to PTH, whereas mid-region PTHrP peptides uniquely stimulate transplacental calcium transport (Care et al., 1990). In addition, C-terminal fragments have been shown to affect osteoclastic activity. Since its discovery in tumors, PTHrP has emerged as an important paracrine regulator in many fetal and adult tissues including vascular smooth muscle, endothelial cells, and bone.

The cardiovascular actions of systemically delivered PTH have been known for decades (Mok et al., 1989). The hormone exerts acute vasodilatory actions on both conductance and resistance vessels. In addition, it produces both positive inotropic and chronotropic effects on the heart. Gastrointestinal, urogenital, and reproductive smooth muscle are also relaxed by PTH (Mok et al., 1989). The discovery of PTHrP, which is produced in abundance in vascular smooth muscle (Hongo et al., 1991), suggests that the well-documented effects of PTH on the cardiovascular system could be subserved by the local production and action of PTHrP.

In addition to the dramatic bone-resorbing effects of PTHrP in bone of patients with humoral hypercalcemia of malignancy, more recent evidence indicates that it also functions locally in developing and adult bone cells. PTHrP is produced within the skeleton in chondrocytes, in osteoblasts, and perhaps in marrow hematopoietic, lymphoid, and stromal cells (Moseley and Martin, 1996). The local production of PTHrP within the skeleton appears to subserve a number of increasingly well-defined roles. Perhaps the most revealing illustration of PTHrPs role in skeletal development is seen in mice in which either the PTHrP gene (Karaplis et al., 1994) or the PTH/PTHrP receptor gene (Lanske et al., 1996) have been inactivated by homologous recombination. Both knockouts are lethal, with the peptide knockout leading to death at the time of delivery, and the receptor knockout leading to death even earlier in gestation. In addition, both knockout mouse models display a striking form of accelerated skeletal mineralization which leads to dwarfism, and an abnormally small rib cage. This defect apparently leads to death shortly after birth from respiratory failure. Thus PTHrP plays a fundamental role in skeletal development, particularly in the regulation of chondrocyte maturation and skeletal mineralization.

Studies in bone cells culture systems have provided evidence for additional actions of PTHrP. A C-terminal peptide of PTHrP, PTHrP (107–139), has been shown to inhibit bone resorption by isolated osteoclasts (Fenton et al., 1991a,b). This activity has been shown subsequently to lie within the sequence PTHrP (107–111) and requires cleavage from the parent molecule. Other investigators (Sone et al., 1992) have failed to confirm this action of PTHrP. Studies using N-terminal PTHrP fragments indicate that PTHrP has actions identical to PTH in bone and bone cells *in vitro,* and both peptides activate common signal transduction pathways with similar potency and *in vivo* (Martin et al., 1991; Moseley and Gillespie, 1995). Thus, PTHrP (1–34), (1–84), and (1–141) all stimulate bone resorption in organ culture (Raisz et al., 1990; Pilbeam et al., 1993) and in osteoclasts resorption systems (Evely et al., 1991). The ability of PTHrP and PTH to stimulate osteoclast resorption *in vitro* appears to require the presence of osteoblasts. In addition, autoradiographic studies have demonstrated binding of labeled PTHrP only to osteoblasts in such co-cultures (Evely et al., 1991). Similarly, PTHrP is equipotent with PTH in its ability to stimulate osteoclast generation in co-cultures or osteoblasts and bone marrow cells (Rakopoulos, unpublished data). Finally, studies *in vivo* suggest that the anabolic actions of PTH can be mimicked by PTHrP containing the PTH-like N-terminal domain (Hock et al., 1989). However, most studies to date have assessed the effects of exogenous addition of synthetic PTHrP fragments and it remains to be determined to what extent local production of PTHrP contributes to osteoblast and osteoclast function.

D. Calcitonin Gene-Related Peptide

Calcitonin gene-related peptide (CGRP) is a 37 amino acid peptide generated by alternative processing of the calcitonin gene (reviewed in Cooper, 1994). Calcitonin is the major product made in parafollicular cells of the thyroid, whereas CGRP is made predominantly in the nervous system. An additional CGRP, (CGRP-2), which differs by only three amino acids in humans and by a single amino acid in the rat, is encoded by a separate gene. Both have a common six amino acid ring at the N-terminus which is formed by a disulfide bridge between amino acids two and seven. CGRPs share structural homology with amylin which is produced by the β cells of the pancreas.

CGRP is one of the most potent vasodilators yet identified and acts through a novel G-protein linked receptor. Although the exact molecular structure(s) for the CGRP receptor(s) are currently unknown, binding and signaling by CGRP has been demonstrated throughout the cardiovascular system. In smooth muscle, CGRP appears to exert direct vasodilatory activity whereas in endothelial cells it causes a release in a diffusible mediator, believed to be NO. CGRP is also made in sensory nerves which are widely distributed throughout the cardiovascular and skeletal systems suggesting its importance in regulation of blood flow in peripheral vascular beds. It seem likely that the epiphyseal and periosteal regions, which are known to be extensively innervated, would be exposed to higher concentrations of CGRP than those which are known to circulate. The intimate association of these nerves with blood vessels suggests they may also have a role in regulating blood flow to the sites of fracture repair or growth (see below).

In vitro studies provide evidence that CGRP modulates the activity of both osteoclasts and osteoblasts (reviewed in Reid and Cornish, 1996). In neonatal mouse calvarial cultures CGRP inhibited the basal and PTH-stimulated release of prelabeled ^{45}Ca (Yamamoto et al., 1986), however, these effects were seen at CGRP concentrations 500-fold greater than those observed for calcitonin in the same culture system. Similarly, in disaggregated neonatal rat osteoclasts, CGRP inhibits motility and resorptive activity but only at concentrations 100-fold greater than those achieved with calcitonin (Zaidi et al., 1987).

CGRP also effects osteoblast proliferation (Bernard and Shih, 1990), an activity not replicated by calcitonin. Specific CGRP binding sites (Datta et al., 1990) and cAMP responses (Tamura et al., 1992) have also been demonstrated in several normal or transformed osteoblast-like cell lines. In fetal rat osteoblasts CGRP stimulates the production of IGF-1 through a cAMP-dependent mechanism (Vignery and McCarthy, 1996).

Despite these results suggesting specific actions of CGRP on bone cells *in vitro,* studies *in vivo* have generally failed to provide a clear picture of its activity (Reid and Cornish, 1996). Early studies showed that CGRP caused hypocalcemia in the rat but at concentrations 100- to 1,000-fold higher than those produced by calcitonin. However, in the rabbit, injections of large concentrations of CGRP resulted in hypercalcemia. In the chicken, the peptide causes only hypercalcemia, with a fall in serum calcium not occurring at any dose. Clearly, further clarification of any physiological role for CGRP in bone metabolism will depend on a better definition of its receptor and separation of potential effects of this peptide from those seen by interactions with the calcitonin receptor.

E. Vascular Endothelial Growth Factor

Vascular endothelial growth factor (VEGF) is a novel heparin-binding glycoprotein which induces endothelial cell proliferation, angiogenesis, and capillary permeability (Leung et al., 1989; Ferrara et al., 1992). Four different isoforms are derived from alternative splicing of VEGF mRNA. VEGF is distinguished from other endothelial cell mitogens such as fibroblast growth factor in that VEGF is a secreted protein and uniquely alters vascular permeability. Vascularized tissues including kidney, heart, lung, and brain abundantly express VEGF. Studies of its temporal and spatial expression into these tissues during development strongly suggest an imminent role in angiogenesis.

Recent studies have shown that VEGF is a normal product of osteoblasts. Harada et al. (1994) showed that expression of VEGF mRNA was induced in RCT-3 osteoblast-like cells by prostaglandin E_2 (PGE_2) and PGE_1. The induction in these cells was shown to be dependent on the production of cAMP and could be inhibited by dexamethasone. VEGF mRNA was also expressed in normal rat tibia. Based on these findings these authors speculated that the ability of PGE to stimulate bone formation may depend in part on expression of VEGF.

F. The Natriuretic Peptides

The natriuretic peptide family are a group of peptides which activate several membrane bound guanylyl cyclases leading to increased intracellular cGMP. Atrial natriuretic peptide (or atriopeptin) is a 28 amino acid peptide which is produced by atrial cadiocytes and exerts potent actions on renal salt and water balance and vascular smooth muscle contractility (Drewett and Garbers, 1994) Two other family members are brain natriuretic peptide (BNP) and C-type natriuretic peptide. Each of these peptides exhibit natriuretic-diuretic and vasorelaxant properties which lower blood pressure. These activities are mediated by activation a series of related membrane guanylyl cyclase receptors. Atrial natriuretic peptide is the principal ligand for the type A guanylyl cyclase whereas C-type natriuretic peptide is the natural ligand for the type C receptor.

The possible involvement of intracellular cGMP in control of bone remodeling was first suggested by Rodan et al. (1976) who reported that mechanical compression of embryonic chick epiphyseal bone cells stimulated accumulation of both cAMP and cGMP. More recently, guanylyl cyclase activity has been localized histochemically to the surface of osteoblasts but was not detected in osteoclasts (Fukushima and Gay, 1991). The first studies to directly investigate the effects of ANP on bone cells were those of Fletcher et al. (1986) who reported the presence of ANP binding sites on both newborn rat osteoblasts and in UMR-106 osteosarcoma cells. In addition, these investigators demonstrated that a synthetic Ile-ANF-26 stimulated cGMP accumulation in these cells consistent with activation of the receptor. In a separate study (Vargus et al., 1989), ANP blunted the PGE stimulated bone resorption as assessed in fetal rat long bones but had no direct effect on basal or PTH-stimulated bone resorption. A recent study by Holliday and co-workers (Holliday et al., 1995) has provided evidence for a role of C-type peptide (CNP) on bone resorption. Based on knowledge that CNP and its receptor were expressed in bone marrow and chondrocytes, these investigators used an *in vitro* osteoclast formation model involving bone marrow osteoclast precursors to demonstrate that two different indices of resorption were specifically enhanced by CNP. These "osteoclasts" expressed both CNP and the C-type receptor suggesting the possibility that this natriuretic factor normally functions in the local control of bone resorption. To date there have been no studies that have directly assessed the effect of ANP on bone cell activity *in vivo*. However, mice null for both ANP and the guanylyl type A receptor exhibit hypertension, but no bone abnormalities.

G. Nitric Oxide and Small Oxygen Radicals

Nitric oxide (NO) is a short-lived molecule produced in endothelial cells from L-arginine (Moncada and Higgs, 1993), and is a potent endogenous vasodilator which is now understood to mediate the relaxant effects of acetylcholine and bradykinin. In addition to its established role in regulation of blood pressure and hemodynamics, NO is increasingly recognized as an important ubiquitous signal transduction effector (Stefanovic-Racic et al., 1993). In vascular endothelium and nerve cells, NO is produced by a constitutively active calcium-calmodulin dependent nitric oxide synthase (NOS). NO liberated by this enzyme activates a soluble guanylyl cyclase which increases cGMP. In macrophages, neutrophils, and bone marrow cells, NO is produced through an inducible nitric oxide synthase (iNOS) which is stimulated by endotoxin and several cytokines including IL-1, tumor necrosis factor alpha (TNFα) and gamma interferon (IFNγ). These cytokines are known to be activated in inflammatory states such as rheumatoid arthritis and osteoarthritis, which are associated with increased bone resorption (Gowen et al., 1986).

Several lines of evidence suggest that NO inhibits osteoclastic bone resorption. Studies by MacIntyre et al. (1991) first showed that NO or NO-releasing agents such as sodium nitroprusside (SNP) decreased spreading and resorptive ability of disaggregated rat osteoclasts and *in vitro*. This effect was apparently not dependent on cGMP production as cGMP analogues did not affect osteoclast activity. These observations led to the speculation that the action of NO in osteoclasts might involve activity on substrate adhesion molecules or cytoskeletal components. However, subsequent work using a bone organ culture system found that SNP inhibited bone resorption while increasing cGMP production (Stern and Diamond, 1992).

The sites of NO production have been studied by Schmidt et al. (Schmidt et al., 1992), who showed that NOS activity was localized in bone at sites of osteoclastic activity. Other recent experiments clearly demonstrated that NADPH-dependent diaphorase staining (Figure 5), an indicator of NOS activity, was evident in osteoclasts of chicken long bones and was increased when osteoclast activity was experimentally increased by lowering the dietary calcium (Kasten et al., 1994). It still unclear, however, whether the observed NADPH-dependent activity derived from NOS rather than from other potential sources such as NADPH oxidases or the oxidoreductases. These authors also demonstrated that chicken osteoclast activity was increased when NO levels were experimentally raised by treatment with nitroprusside. Alternatively, osteoclastic bone resorption was stimulated when NOS was inhibited by treatment with N-nitro-L-

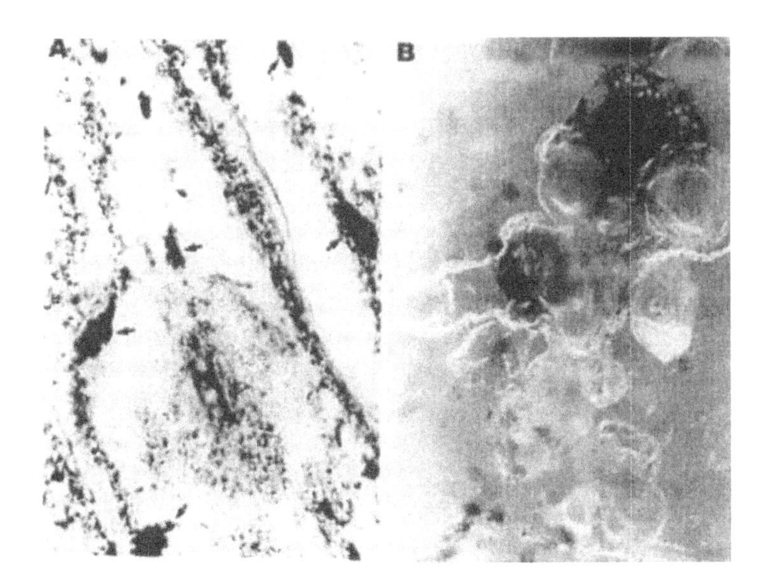

Figure 5. Localization of nitric oxide synthase (NOS) in bone. Panel **A**: Photomicrograph of a frozen section chicken tibia from an animal maintained on a low calcium diet for four weeks. Osteoclasts (arrowheads) were observed in close proximity to bone trabeculae. Panel **B**: Localization of NOS using the diaphorase stain in isolated osteoclasts cultured on cortical bone slices. Osteoclasts (indicated by the arrows) are closely associated with resorption pits. Reproduced with permission from Kasten, T.P., et al., (1994). Proc. Natl. Acad. Sci:. USA, 91:3569–3573.

arginine methyl ester or aminoguanidine. In addition, the enhancement of osteoclastic activity in rats following ovariectomy was augmented by treatment with aminoguanidine. These data support the concept that NO production dampens osteoclastic activity, but it is not impossible to exclude other NO-independent effects of aminoguanidine. In fact, other studies reported either inhibition or unaltered osteoclastic activity when these cells were treated with inhibitors of NOS. These apparently conflicting studies might be explained by considering the possibility that the relative amount of NO produced by osteoclasts determines activity, such that constitutive levels are necessary for normal activity and amplified NO levels are inhibitory.

The involvement of NO in osteoblast function is suggested by several studies. Both primary human osteoblasts and osteosarcoma osteoblast-like cells produce NO in response to stimulation with cytokines IL-1, TNF and IFNγ (Damoulis and Hauschka, 1994; Lowik et al., 1994; Ralston et al., 1994; Riancho et al., 1995). Since these cytokines are established inhibitors of bone formation, it was suggested that NO might mediate their activity. Thus despite the uncertainty of the precise physiological role of NO in bone, there is growing enthusiasm for an interaction of this critical signaling molecule with local cytokines, which are established regulators of both osteoblastic and osteoclastic activity.

In addition to NO, other small reactive oxygen molecules, which are known to be produced by endothelial cells, are believed to influence the activity of bone cells. Treatment of osteoclasts with H_2O_2 stimulated bone resorption. Zaidi and co-workers (Zaidi et al., 1993) have postulated that in the local hemivacuolar osteoclast environment, H_20_2 (and, theoretically, other short-lived free radical species) could be generated from osteoclast-derived H_20_2 through the action of a superoxide dismutase. In this environment, H_20_2 could provide an excitatory signal that would enable the osteoclast to increase its motility following a resorptive episode and move to a new resorption site.

H. Catecholamines

The catecholamines, norepinephrine and isoproterenol, have also been shown to modulate activity of bone cells. Beta adrenergic agonists are potent regulators of heart rate, airway tone and blood pressure and selectively activate different subtypes of G-protein coupled receptors, which are expressed in variable abundance in different target tissues (Caron et al., 1993). The different adrenergic receptor subtypes have been cloned and their signaling properties have been extensively characterized. Activation by receptor-selective ligands leads to formation of intracellular cAMP and subsequent activation of protein kinase A. Studies by Rodan and Rodan (1981) demonstrated that isoproterenol stimulated adenylate cyclase and specifically bound to cell surface receptors in clonal rat osteosarcoma (ROS) cells. In addition, dexamethasone was shown to augment isoproterenol stimulated adenylate cyclase activity and increase receptor binding sites in ROS 17/2.8 cells (Rodan and Rodan, 1986). In a more recent study, Moore et al. (1993) clearly demonstrated the existence of β-2-receptors on both ROS 17/2.8 rat osteosarcoma cells and on human osteosarcoma cells. In addition, these investigators showed that both norepinephrine and isoproterenol stimulated bone resorption in neonatal mouse calvariae via a cAMP dependent mechanism.

IV. ROLE OF BLOOD VESSELS AND THEIR PRODUCTS IN BONE PATHOPHYSIOLOGY

Fracture healing illustrates the frequently symbiotic relation between the vascular system and bone. Fracture disrupts the normal afferent blood supply of bone (Rhinelander, 1974; Smith et al., 1990). After fracture, compensatory flow through small periosteal arterioles is elevated via an endothelial cell mediated process (Swiontkowski and Senft, 1992; Triffitt et al., 1993). The initial hematoma is filled with growth factors that enhance cell recruitment and differentiation, (Assoian and Sporn, 1986). When a cartilaginous callus is formed to link the exposed bone ends, an oxygen deficient environment must be maintained (Brighton and Krebs, 1972). The subsequent mineralization of the cartilage substrate, however, requires capillary invasion. The initial stimulus for this angiogenesis is unclear but may be derived from endothelial cells (Brown and McFarland, 1992). Impotent or abnormal vascularization during fracture healing is associated with delayed union and non-union (Mohanti and Mahakul, 1983; Smith et al., 1992; Fernandez and Eggli, 1995).

Numerous bone pathologies are associated with disruption or alteration of the tissue's blood supply or vasculature. Rheumatoid arthritis is characterized by increased blood flow to the joint capsule (Tamai et al., 1994) and overexpression of vascular endothelial growth factor by the synovium (Nagashima et al., 1995). Osteoarthritis is associated with vascular invasion of cartilage and the growth plate (Harrison et al., 1953; Farkas et al., 1987). The collapse of the femoral head concomitant with avascular necrosis clearly illustrates that a healthy vascular system is required for a successful skeleton. The specific role of the vasculature in the pathoetiologies of these diseases, however, is only beginning to be elucidated.

V. SUMMARY

Current literature suggests that an expanding array of vasoactive molecules derived from endothelial cells, vascular smooth muscle cells, and bone marrow can influence bone cell populations. It is worth noting, however, that the majority of the cited studies provide only circumstantial evidence for the biological role of specific factors in bone. Importantly, the demonstration that a vasoactive agent is capable of modulating bone cell gene expression *in vitro* is not sufficient evidence for invoking a substantive role *in vivo*. This consideration not withstanding, there is every reason to believe that vasoactive agents do exert important functions in bone. The intimate temporal and spatial relation between the vasculature and bone during development and the common mesenchymal and cell surface antigens of vascular and osteogenic cells suggest that vascular-derived factors do modulate bone cell functions under both normal and pathologic conditions. An improved understanding of this syncytium will therefore enhance our ability to successfully intervene in bone pathologies.

REFERENCES

Adair, T.H., Gay, W.J., and Montani, J.P. (1990). Growth regulation of the vascular system: Evidence for a metabolic hypothesis. Am. J. Physiol. 259, R393-404.

Agui, T., Xin, X., Cai, Y., Sakai, T., and Matsumoto K. (1994). Stimulation of interleukin-6 production by endothelin in rat bone marrow-derived stromal cells. Blood 84, 2531-2538.

Alam, A.S.M.T., Gallagher, A., Shankar, V., Ghatei, M.A., Datta, H.K., Huang, CL-H., Moonga, B.S., Chambers, T.J., Bloom, S.R., Zaidi, M. (1992). Endothelin inhibits osteoclastic bone resorption by a direct effect on cell motility: Implications for the vascular control of bone resorption. Endocrinology 130, 3617-3624.

Assoian, R.K., Sporn, M.B. (1986). Type-β transforming growth factor in human platelets: Release during platelet degranulation and action on vascular smooth muscle cells. J. Cell Biol. 102, 1217-1223.

Babich, M., King, K., and Nissenson, R.A.. (1990). Thrombin stimulates inositol phosphate production and intracellular free calcium by a pertussis toxin-insensitive mechanism in osteosarcoma cells. Endocrinology 126, 948-954.

Babich, M., Choi, H., Johnson, R.M., King, K.L., Alford, G.E., and Nissenson, R.A. (1991). Thrombin and parathyroid hormone mobilize intracellular calcium in rat osteosarcoma cells by distinct pathways. Endocrinology 129, 1463-1470.

Bernard, G.W. and Shih, C. (1990). The osteogenic stimulating effect of neuroactive calcitonin gene-related peptide. Peptides 11, 625-632.

Brooks, M. (1971). The Blood Supply of Bone. Butterworths, London.

Brighton, C.T. and Krebs, A.G. (1972). Oxygen tension of healing fractures in the rabbit. J. Bone Jt. Surg. 54A, 323-332.

Brown, R.A. and McFarland, C.D. (1992). Regulation of growth plate cartilage degradation in vitro: Effects of calcification and a low molecular weight angiogenic factor (ESAF). Bone Miner. 17, 49-57.

Burkhardt, B., Kettner, G., Bohm, W., Schmidmeier, M., Shlag, R., Frisch, B., Mallman, B., Eisenmeyer, N. and Gilg, T. (1987). Changes in trabecular bone, hematopoiesis and bone marrow vessels in aplastic anemia, primary osteoporosis and old age: A comparative histomorphometric study. Bone 8, 157-164.

Caplan, A., Syftestad, G., and Osdoby P. (1983). The development of embryonic bone and cartilage in tissue culture. Clin. Ortho. Rel. Res. 174, 243-261.

Care, A.D., Abbas, S.K., Pickard, D.W., Barri, M., Drinkhill, M., Findlay, J.B, White, I.R., and Caple, I.W. (1990). Stimulation of ovine placental transport of calcium and magnesium by mid-molecule fragments of human parathyroid hormone-related protein. Exp. Physiol. 75, 605-608.

Caron, M.G., Le, S., and Kowitz, R.J. (1993). Catecholamine receptors: Structure, function, and regulation. Rec. Prog. Horm. Reg. 48, 1.

Collin-Osdoby, P. (1994). Role of vascular endothelial cells in bone biology. J. Cell. Biochem. 55, 304-309.

Collins, D.A. and Chambers, T.J. (1991). Effect of prostaglandins E_1, E_2, and F_{2a} on osteoclast formation in mouse bone marrow cultures. J. Bone Min. Res. 6, 157-164.

Cooper, G.J. (1994). Amylin compared with calcitonin gene-related peptide-structure, biology, and relevance to metabolic disease. Endocr. Rev. 15, 163-201.

Coughlin, S.R. (1995). Molecular mechanisms of thrombin signaling. Semin. Hematol. 31, 270-277.

Damoulis, P.D. and Hauschka, P.V. (1994). Cytokines induce nitric oxide production in mouse osteoblasts. Biochem. Biophys. Res. Commun. 201, 924-931.

Datta, H.K., Rafter, P.W., Ohri, S.K., MacIntyre, I., and Wimalawansa, S.J. (1990). Amylin-amide competes with CGRP binding sites on osteoblastlike osteosarcoma cells. J. Bone Min. Res. 5, S229.

Davies, P.F. (1995). Flow-mediated endothelial mechanotransduction. Physiol. Rev. 75, 519-560.

Drewett, J.G. and Garbers, D.L. (1994). The family of gyanylyl cyclase receptors and their ligands. Endocr. Rev. 15, 135-162.

Duling, B.R., Klitzman, B. (1980). Local control of microvascular function: Role in tissue oxygen supply. Ann. Rev. Physiol. 42, 373-382.

Egrise, D., Martin, D., Neve, P., Vienne, A., Verhas, M., and Schoutens, A. (1992). Bone blood flow and in vitro proliferation of bone marrow and trabecular bone osteoblastlike cells in ovariectomized rats. Calcif. Tiss. Int. 50, 336-341.

Evely, R.S., Bonomo, A., Schneider, H.G., Moseley, J.M., Gallagher, J., and Martin, T.J. (1991). Structural requirements for the action of parathyroid hormone–related protein (PTHrP) on bone resorption by isolated osteoclasts. J. Bone Min. Res. 6, 85-93.

Farkas, T., Boyd, R.D., Schaffler, M.B., Radin, E.L., Burr, D.B. (1987). Early vascular changes in rabbit subchondral bone after repetitive impulsive loading. Clin. Orthop. Rel. Res. 219, 259-267.

Fenton, I.I. (1986). Thrombin. Ann. NY Acad. Sci. 485, 5.

Fenton, A.J., Kemp, B.E., Hammonds, R.G., Mitchelhill, K., Martin, T.J., and Nicholson, G.C. (1991a). A potent inhibitor of osteoclastic bone resorption within a highly conserved pentapeptide region of parathyroid hormone-related protein: PTHrP 107-111. Endocrinology 129, 1762-1768.

Fenton, A.J., Kemp, B.E., Kent, G.N., Moseley, J.M., Zheng, L-H., Rowe, D.J., Britto, J.M., Martin, T.J., and Nicholson, G.C. (1991b). A carboxy-terminal fragment of parathyroid hormone-related protein inhibits bone resorption by osteoclasts. Endocrinology 129, 1762-1768.

Fernandez, D.L. and Eggli, S. (1995). Non-union of the scaphoid. Revascularization of the proximal pole with implantation of a vascular bundle and bone-grafting. J. Bone Jt. Surg. 77, 883-893.

Ferrara, N., Houck, K., and Leung, D.W. (1992). Molecular and biological properties of the vascular endothelial growth factor family of proteins. Endocr. Rev. 13, 18-32.

Fletcher, A.E., Allan, E.H., Casley, D.J., and Martin, T.J. (1986). Atrial natriuretic factor receptors and stimulation of cyclic GMP formation in normal and malignant osteoblasts. FEBS Letters 208, 263-268.

Formigli, L., Orlandini, S.Z., Benvenuti, S., Masi, L., Pinto, A., Gattei, V., Bernabei, P.A., Robey, P.G., Collin-Osdoby, P., Brandi, M.L. (1992). In vitro structural and functional relationships between preosteoclastic and bone endothelial cells: A juxtracrine model for migration and adhesion of osteoclast precursors. J. Cell Phys. 162, 199-212.

Frost, H.M. (1989). Some ABC's of skeletal pathophysiology III: Bone balance and the ΔB.BMU. Calcif. Tiss. Int. 45, 131-133.

Fukushima, O. and Gay, C.V. (1991). Ultrastructural localization of guanylate cyclase in bone cells. J. Histochem. Cytochem. 39, 529-535.

Geiser, M. and Trueta, J. (1958). Muscle action, bone rarefaction and bone formation. J. Bone Jt. Surg. 40B, 282-311.

Gowen, M., MacDonald, B.R., Hughes, D.E. et al. (1986). Immune cells and bone resorption. Exp.Biol. Med. 208, 261-273.

Gross, T.S., Damjii, A.A., Judex, S., Bray, R.C., and Zernicke, R.F. Bone hyperemia precedes disuse induced bone resorption. J. Appl. Phys., submitted.

Guyton, A.C. and Hall, J.E. (1996). The textbook of medical physiology, Ninth ed. p. 200. W.B. Saunders Company, Philadelphia.

Harada, S., Nagy, J.A., Sullivan, K.A., Thomas, K.A., Endo, N., Rodan, G.A., and Rodan, S.B. (1994). Induction of vascular endothelial growth factor expression by prostaglandin E_2 and E_1 in osteoblasts. J. Clin. Invest. 93, 2490-2496.

Hardt, A.B. (1972). Early metabolic responses of bone to immobilization. J. Bone Jt. Surg. 54A, 119-124.

Harrison, M.H.M., Schajowicz, F., and Trueta, J. (1953). Osteoarthritis of the hip: A study of the nature and evolution of the disease. J. Bone Jt. Surg. 35B, 598-626.

Hock, J.M., Fonseca, J., Guness-Hey, Kemp, M., and Martin, T.J. (1989). Comparison of the anabolic effects of synthetic parathyroid hormone-related protein (PTHrP) 1-34 and PTH 1-34 on bone in rats. Endocrinology 125, 2022-2027.

Holliday, S.L., Dean, A.D., Greenwald, J.E., and Gluck, S.L. (1995). C-type natriuretic peptide increases bone resorption in 1,25-dihydroxyvitamin D_3-stimulated mouse bone marrow cultures. J. Biol. Chem. 270, 18983-18989.

Hongo, T., Kupfer, J., Enomoto, H., Sharifi, B., Giannella-Neto, D., Forrester, J.S., Singer, F.R., Goltzman, D., Hendy, G.N., Pirola, C., et al. (1991). Abundant expression of parathyroid hormone-related protein in primary rat aortic smooth muscle cells accompanies serum-induced proliferation. J. Clin. Invest. 88, 1841-1847.

Honig, C.R., Odoroff, C.L., and Frierson, J.L. (1980). Capillary recruitment in exercise: Rate, extent, uniformity, and relation to blood flow. Am. J. Physiol. 238, H862-H868.

Hosoda, K., Hammer, R.E., Richardson, J.A., Baynash, A.G., Cheung, J.C., Gaid, A., and Yanagisawa, M. (1994). Targeted and natural (piebald-lethal) mutations of the endothelin-B receptor gene produce megacolon associated with spotted coat color in mice. Cell 79, 1267-1276.

Jespersen, S.M., Hoy, K., Christensen, K.O., Lindbad, B.E., He, S.Z., Bunger, C., Hansen, E.S. (1994). Axial and peripheral skeletal blood flow at rest and during exercise. Trans. Orthop. Res. Soc. 19, 293.

Jones, A.R., Clark, C.C., and Brighton, C.T. (1995). Microvessel endothelial cells and pericytes increase proliferation and repress osteoblast phenotypic markers in rat calverial bone cell cultures. S. Orthop. Res. 13, 553-561.

Kapitola, J., Kubícková, J., and Andrle, J. (1995). Blood flow and mineral content of the tibia of female and male rats: Changes following castration and/or administration of estradiol or testosterone. Bone 16, 69-72.

Karaplis, A.C., Luz, A., Glowacki, J., Bronson, R.T., Rybulewicz, V.L., Kronenberg, H.M., and Mulligan, R.C. (1994a). Lethal skeletal dysplasia from targeted disruption of the parathyroid hormone-related peptide gene. Genes Dev. 8, 277-289.

Karaplis, A., Luz, A., Mulligan, R., and Kronenberg, H. (1994b). Characterization of mice homozygous for the PTH/PTHrP receptor gene null mutation. J. Bone Min. Res. 9, S121.

Kasten, T.P., Collin-Osdoby, P., Patel, N. Krukowski, M., Misko, T.P., Settle, S.L., Currie, M.G., and Nickols, G.A. (1994). Potentiation of osteoclastic bone resorption activity by inhibition of nitric oxide synthase. Proc. Nat. Acad. Sci. (USA) 93, 3569-3573.

Kuchan, M.J. and Frangos, J.A. (1993). Shear stress regulates endothelin-1 release via protein kinase C and cGMP in cultured endothelial cells. Am. J. Phys. 264, H150-156.

Kurihara, Y., Kurihara, H., Suzuki, H., Kodama, T., Maemura, K., Nagai, R., Oda, H., Kuwaki, T., Cao, W.H., Kamada, N., Jishage, K., Ouchi, Y., Azuma, S., Toyada, Y., Ishikawa, T., Kamada, M., and Yazaki, Y. (1994). Elevated blood pressure and craniofacial abnormalities in mice deficient in endothelin-1. Nature 368, 703-710.

Lanske, B., Karaplis, A.C., Lee, K., Luz, A., Vortkamp, A., Pirro, A., Karperien, M., Defize, L.H.K., Ho, C., Mulligan, R.C. et al. (1996). PTH/PTHrP receptor in early development and Indian hedgehog-regulated bone growth. Science 273, 663-666.

Laval-Jeantet, A.M., Bergot, C., Carroll, R., and Garcia-Schaefer, F. (1983). Cortical bone senescence and mineral bone density of the humerus. Calcif. Tiss. Int. 35, 268-272.

Lee, S.K. and Stern, P.H. (1995). Endothelin$_B$ receptor activation enhances parathyroid hormone-induced calcium signals in UMR-106 cells. J. Bone Min. Res. 10, 43-1351.

Lerner, U.H. and Gustafson, G.T. (1988). Coagulation and bone metabolism, some characteristics of the bone resorptive effect of thrombin in mouse calvaria. Biochimica et Biophysica Acta 964, 309-318.

Leung, D.W., Cachianes, G., Kuang, W-J, Goeddle, D.W., and Ferrara, N. (1989). Vascular endothelial growth factor is a secreted angiogenic mitogen. Science 246, 1306-1309.

Levin, E.R. (1995). Endothelins. N. Engl. J. Med. 333, 356-363.

Lewinson, D. and Silberman, M. (1992). Chondroclasts and endothelial cells collaborate in the process of cartilage resorption. Anat. Rec. 233, 504-514.

Lowik, C.W.G.M., Nibbering, P.H., Van der Ruit, M., and Papapoulos, S.E. (1994). Inducible production of nitric oxide is osteoblastlike cells and in fetal mouse bone explants is associated with suppression of osteoclastic bone resorption. J. Clin. Invest. 93, 1465-1472.

MacIntyre, I., Zaidi, M., Towhidul Alum, A.S.M., Datta, H.K., Moonga, B.S., Lidbury, P.S, Hecker, M., and Vane, J.M. (1991). Osteoclast inhibition: An action of nitric oxide not mediated by cGMP. Proc. Natl. Acad. Sci. USA 88, 2936-2940.

Martin, T.J., Moseley, J.M., Gillespie, M.T. (1991). Parathyroid hormone-related protein: biochemistry and molecular biology. (Review). Crit. Rev. Biochem. Mol. Biol. 26, 377-395.

Mellander S. (1970). Systemic circulation: Local control. Ann. Rev. Physiol. 32, 313-344.

Miller, V.M. and Burnett, J.C. (1992). Modulation of NO and endothelin by chronic increases in blood flow in canine femoral arteries. Am. J. Phys. 263, H103-108.

Mohanti, R.C. and Mahakul, N.C. (1983). Vascular response in fractured limbs with and without immobilization: An experimental study on rabbits. Intl. Orthop. 7, 173-177.

Mok, L.L., Nickols, G.A., Thompson, J.C., and Cooper, C.W. (1989). Parathyroid hormone as a smooth muscle relaxant. (Review). Endocr. Rev. 10, 420-436.

Moseley, J.M. and Martin, T.J. (1996). Parathyroid Hormone-Reland Protein; Physiological Actions. In: *Principals of Bone Biology*. (Bitezikian, J.P., Raisz, L.G., and Rodon, GA., Eds.), pp. 363-376. Academic Press, New York.

Mosellev, J.M. and Gillespie, M.T. (1995). Parathyroid hormone-related protein. Crit. Rev. Clin. Lab. Sci. 32, 299-343.

Moncada, S. and Higgs, A. (1993). The L-arginine-nitric oxide pathway. New Eng. J. Med. 239, 2002-2012.

Moore, R.E., Smith, C.K., Baily, C.S., Voelkel, E.F., and Tashjian, Jr., A.H. (1993). Characterization of β-adrenergic receptors on rat and human osteoblastlike cells and demonstration that β-receptor agonists stimulate bone resorption in organ culture. Bone Miner. 23, 301-315.

Nagashima, M., Yoshino, S., Ishiwata, T., and Asano, G. (1995). Role of vascular endothelial growth factor in angiogenesis of rheumatoid arthritis. J. Rheum. 22, 1624-1630.

Owen, M. (1988). Marrow stromal stem cells. J. Cell Sci. 10, 63-76S.

Pilbeam, C.C., Alander, C.B., Simmons, H.A., and Raisz, L.G. (1993). Comparison of the effects of various lengths of synthetic human parathyroid hormone-related peptide (hPTHrP) of malignancy on bone resorption and formation in organ culture. Bone 14, 717-720.

Ralston, S.H., Ho, L-P., Helfrich, M.H., Grabowski, P.S., Johnston, P.W., and Benjamin, N. (1995). Nitric Oxide: A cytokine-induced regulator of bone resorption. J. Bone Min. Res. 10, 1040-1049.

Raisz, L. Principals of Bone Biology. (Bilezikian, J.P., Raisz, L.G., and Rodon, G.A., Eds.), Academic Press, New York.

Raisz, L.G., Simmons, H.A., Vargas, S.J., Kemp, B.E., and Martin, T.J. (1990). Comparison of the effects of amino-terminal synthetic parathyroid hormone-related peptide (PTHrP) of malignancy and parathyroid hormone on resorption of cultured fetal rat long bones. Calcif. Tissue Int. 46, 233-238.

Ralston, S.H., Todd, D., Helfrich, M.H., Benjamin, N., and Grabowski, P. (1994). Human osteoblastlike cells produce nitric oxide and express inducible nitric oxide synthase. Endocrinology 135, 330-336.

Reid, I.R. and Cornish, J. (1996). Amylin and CGRP. In: Principals in Bone Biology. (Bilezikian, J.P., Raisz, L.G., and Rodan, G.A., Eds.), pp. 495-505. Academic Press, New York.

Rhinelander, F.W. (1974). Tibial blood supply in relation to fracture healing. Clin. Orthop. Rel. Res. 105, 34-81.
Riancho, J.A., Salas, E., Zarrabeitia, M.T., et al. (1995). Expression and functional role of nitric oxide synthase in osteoblastlike cells. J. Bone Min. Res. 10, 439-446.
Rodan, G.A., Bourret, L.A., Harvey, A., and Mensi, T. (1976). Cyclic AMP and cyclic GMP: Mediators of the mechanical effects on bone remodeling. Science 189, 467-469.
Rodan, S.B. and Rodan, G.A. (1981). Parathyroid hormone and isoproterenol stimulation of adenylate cyclase in rat osteosarcoma clonal cells. Hormone competition and site heterogeneity. Biochimica Biophysica Acta 673, 46-54.
Rodan, S.B. and Rodan, G.A. (1986). Dexamethazone effects on β-adrenergic receptors and adenylate cyclase regulatory proteins G_i and G_s in ROS 17/2.8 cells. Endocrinology 118, 2510-2518.
Rubanyi, G.M. and Polokoff, M.A. (1994). Endothelins: Molecular biology, biochemistry, pharmacology, physiology, and pathophysiology. Pharmacological Rev. 46, 324-415.
Sakurai, T., Yanagisawa, M., Takuwa, Y., Miyazaki, H., Kimura, S., Goto, K., and Masaki, T. (1990). Cloning of a cDNA encoding a non-isopeptide selective subtype of the endothelin receptor. Nature 348, 732-735.
Sasaki, T. and Hong, M.H. (1993). Localization of endothelin-1 in osteoclasts. J. Electron Microsc. 42, 193-196.
Schmidt, H.H., Gagne, G.D., Nakane, M., Pollock, J.S., and Miller, M.F. (1992). Mapping of neural nitric oxide synthase in rat suggests frequent co-localization with NADPH diaphorase but not with soluble guanylyl cyclase, and novel paraneural functions for nitrinergic signal transduction. J. Histochem. Cytochem. 40, 1439-1456.
Schvartz, I., Ittoop, O., Davidai, G., and Hazum, E. (1992). Endothelin rapidly stimulates tyrosine phosphorylation in osteoblastlike cells. Peptides 13, 159-163.
Segal, S.S. (1994). Cell-to-cell communication coordinates blood flow control. Hypertension 23, 1113-1120.
Semb, H. (1969). Experimental limb disuse and bone blood flow. Acta Orthop. Scand. 40, 552-562.
Shireman, P.K., Pearce, W.H. (1996). Endothelial cell function: Biologic and physiologic functions in health and disease. Am. J. Radiology 166, 7-13.
Smith, J.W., Arnoczky, S.P., and Hersh, A. (1992). The intraosseous blood supply of the fifth metatarsal: Implications for proximal fracture healing. Foot Ankle. 13, 143-52.
Smith, S.R., Bronk, J.T., Kelly, P.J. (1990). Effect of fracture fixation on cortical bone blood flow. J. Orthop. Res. 8, 471-478.
Sone, T., Kohno, H., Kikuchi, H., Ikeda, T., Kasai, R., Kikuchi, Y., Takeuchi, R., Konishi, J., and Shigeno, C. (1992). Human parathyroid hormone-related peptide (107-111) does not inhibit bone resorption in neonatal mouse calvariae. Endocrinology 131, 2742-2746.
Stefanovic-Racic, M., Stadler, J., and Evans, C.H. (1993). Nitric oxide and arthritis. Arthritis and Rheum. 36, 1036-1044.
Stern, P.H., Tatrai, A., Semler, D.E., Lee, S.K., Lakatos, P., Strieleman, P.J., Tarjan, G., and Sanders, J.L. (1995). Endothelin receptors, second messengers, and actions in bone. J. Nutr. 125, 2028S-2032S.
Stern, P.H. and Diamond, J. (1992). Sodium nitroprusside increases cyclic GMP in fetal rat long bone cells and inhibits resorption of fetal rat limb bones. Res. Comm. Chem. Path. Clin. Pharm. 75, 19-28.

Swiontkowski, M.R. and Senft, D. (1992). Cortical bone microperfusion: Response to ischemia and changes in major arterial blood flow. J. Orthop. Res. 10, 337-343.

Takuwa, Y., Ohue, Y,. Takuwa, N., and Yamashita, K. (1989). Endothelin-1 activates phospholipase C, and mobilizes Ca^{2+} from extra- and intracellular pools in osteoblastic cells. Am. J. Physiol. 257, E797-E803.

Tamai, K., Yamato, M., Yamaguchi, T., and Ohno, W. (1994). Dynamic magnetic resonance imaging for the evaluation of synovitis in patients with rheumatoid arthritis. Arthritis and Rhuem. 8, 1151-1157.

Tamura, T., Miyaura, C., Owan, I., and Suda, T. (1992). Mechanism of action of amylin in bone J. Cell. Physiol. 153, 6-14.

Tatrai, A., Foster, S., Lakatos, P., Shanker, G., and Stern, P.H. (1992). Endothelin-1 actions on resorption, collagen and noncollagen protein synthesis, and phosphatidylinositol turnover in bone organ cultures. Endocrinology 131, 603-607.

Tatrai, A. and Stern, P.H. (1993). Endothelin-1 modulates calcium signaling by epidermal growth factor, a thrombin, and prostaglandin E_1 in UMR-106 osteosarcoma cells. J. Bone Min. Res. 8, 943-952.

Tøndevold, E. and Bülow, J. (1983). Bone blood flow in conscious dogs at rest and during exercise. Acta Orthop. Scand. 54, 53-57.

Triffitt, P.D., Cieslak, C.A., and Gregg, P.J. (1993). A quantitative study of the routes of blood flow to the tibial diaphysis after an osteotomy. J. Orthop. Res. 11, 49-57.

Trueta, J. (1963). The role of vessels in osteogenesis. J. Bone Joint Surg. 45, 402-418.

Van Leeuwen, J.P.T.M., Bos, M.P., Lowik, C.W.G.M., and Herrmann-Erlee, M.P.M. (1988). Effect of parathyroid hormone and parathyroid hormone fragments on the intracellular ionized calcium concentration in an osteoblast cell line. Bone Miner. 4, 177.

Vargus, S.J., Holden, S.N., Fall, P.M., and Raisz, L.G. (1989). Effects of atrial natriuretic factor on cyclic nucleotides, bone resorption, collagen and deoxyribonucleic acid synthesis, and prostaglandin E2 production in fetal rat bone cultures. Endocrinology 125, 2527-2531.

Vaughan, J.M. (1975). Oxford University Press, Oxford.

Vignery, A. and McCarthy, T.L. (1996). The neuropeptide calcitonin gene-related peptide stimulates insulinlike growth factor I production by primary fetal rat osteoblasts. Bone 18, 331-335.

Wilson, J.R. and Kapoor, S.C. (1993). Contribution of prostaglandins to exercise-induced vasodilation in humans. Am. J. Phys. 264, H171-H175.

Wimalawansa, S.J. (Ed.) Calcitonin Gene-Related Peptide and Its Receptors: Molecular Genetics, Physiology, Pathophysiology, and Therapeutic Potentials. Dept. of Internal Medicine, University of Texas Medical Branch at Galveston, Galveston, TX.

Wu, T.L., Vasavada, R.C., Yang, K., Massfelder, T., Ganz, M., Abbas, S.K., Care, A.D., Stewart, A.F. (1996). Structural and physiological characterization of the mid-region secretory species of parathyroid hormone-related protein. J. Biol. Chem. 271, 24371-24381.

Yamamoto, I., Kitamura, N., Aoki, J., Shigeno, C., et al. (1986). Human calcitonin gene-related peptide possesses weak inhibitory potency of bone resorption in vitro. Calcif. Tissue Int. 38, 339-341.

Zaidi, M., Breimer, L.H., and MacIntyre, I. (1987). Biology of peptides from the calcitonin gene. Quart. J. Exp. Physiol. 72, 371-408.

Zaidi, M., Alam, A.S.M.T., Bax, B.E., Shanker, V.S., Bax, C.M.R., Gill, J.S., Pazianas, M., Huang, C.L-H., Sahinoglu, T., Moonga, B.S., Stevens, C.R., and Blake, D.R. (1993). Role of the endothelial cell in osteoclast control: New perspectives. Bone 14, 97-102.

PARATHYROID HORMONE AND ITS RECEPTORS

Abdul B. Abou-Samra

I. Introduction 162
II. Parathyroid Hormone 162
III. Parathyroid Hormone-Related Peptide 164
IV. PTH/PTHrP Receptor 166
A. Molecular Cloning of the PTH/PTHrP Receptor 166
B. Ligand Binding Properties of the PTH/PTHrP Receptor 169
C. Signaling Properties of the PTH/PTHrP Receptor 170
D. Regulation of the PTH/PTHrP Receptor 171
E. Homologous PTH Receptor Downregulation and Desensitization of Cellular Responsiveness to PTH 171
F. Heterologous Regulation of the PTH/PTHrP Receptor 172
V. The PTH/PTHrP Receptor Gene 173
VI. PTH_2 Receptor 175
VII. Summary 175

Advances in Organ Biology
Volume 5A, pages 161-185.

ISBN: 0-7623-0390-5

I. INTRODUCTION

Maintenance of calcium ion homeostasis is a vital physiological process that requires coordination between the effects of parathyroid hormone (PTH) and vitamin D on bone and kidneys. Deficiency in the functioning of the parathyroid glands leads to decreased calcium concentrations. In contrast, overactivity of the PTH system leads to increased levels of extracellular calcium. Thus, the functioning of the parathyroid system is essential for the maintenance of blood calcium concentrations. PTH acts on its target tissues through specific receptors located on the cell membrane of the target cells. Although several tissues have been described to contain receptors for PTH, most of the physiological actions of PTH on calcium homeostasis are mediated by specific receptors in bone and kidneys. PTH stimulates calcium release from bone and inhibits the urinary excretion of calcium from the kidneys. Additionally, PTH increases the synthesis of the active metabolite of vitamin D ($1,25(OH)_2D_3$) by stimulating the renal 1-α hydroxylase that converts 25-OHD_3 into $1,25(OH)_2D_3$. In turn, $1,25(OH)_2D_3$ acts on the intestine to increase calcium absorption.

Parathyroid hormone-related peptide (PTHrP) was characterized from malignant tumors that cause hypercalcemia in cancer patients. PTHrP has significant sequence homology to PTH that is limited to its 13 amino-terminal residues. However, this limited sequence homology allows high binding affinity and full activation of the PTH receptor by PTHrP. The molecular cloning of a single receptor from bone and kidney that equivalently binds PTH and PTHrP ultimately proved that one single receptor molecule binds both ligands. The cloned receptor was, therefore, named the PTH/PTHrP receptor.

I. PARATHYROID HORMONE

PTH is synthesized in the parathyroid cells as a pre-pro-hormone comprised of 115 amino acids (Aurbach et al., 1972; Keutmann et al., 1975; Goltzman et al., 1976; Kronenberg et al., 1977, 1979; Habener et al., 1981). The signal peptide of PTH allows translocation of the polypeptide into the secretory vesicles, a property that is shared with most secreted proteins (Habener et al., 1981). Cleavage of the signal peptide results in the generation of Pro-PTH, a 90 amino acid polypeptide, that eventually matures to the 84 amino acid intact PTH molecule (Goltzman et al., 1976). PTH is then secreted in the blood stream.

The synthetic amino terminal fragment, PTH(1–34), was recognized early on to have full biological activity *in vitro* and *in vivo* (Potts et al., 1971). PTH(1–34), however, does not correspond to a metabolic product of PTH. The bioactive PTH molecule in the circulation is the 84 amino acid intact hormone (Bringhurst et al., 1982, 1988). Carboxy-terminal fragments of PTH were also detected in the blood stream, however, these fragments lack bioactivity (Bringhurst et al., 1988). The presence of inactive PTH fragments in the circulation had made it difficult to interpret the blood level of PTH using the old radioimmunoassay techniques (Segre et al., 1974, 1975; Habener et al., 1976; Sharp and Marx, 1985). However, the immunoradiometric assay (IRMA) in which the PTH molecule is sandwiched between a carboxyl-terminal antiserum immobilized on beads and an ^{125}I-labeled amino-terminal antiserum, allowed a more accurate determination of plasma levels of the active PTH (Nussbaum et al., 1987).

The molecular cloning of the PTH gene from human, rat, bovine, porcine, and chicken species has revealed an extraordinary conservation of the PTH sequences across these species (Kronenberg et al., 1977, 1979, 1986; Vasicek et al., 1983; Heinrich et al., 1984; Khosla et al., 1988). Surprisingly, both the amino-terminus and the carboxy-terminus are highly conserved (Figure 1). The high degree of conservation of the carboxyl regions of the

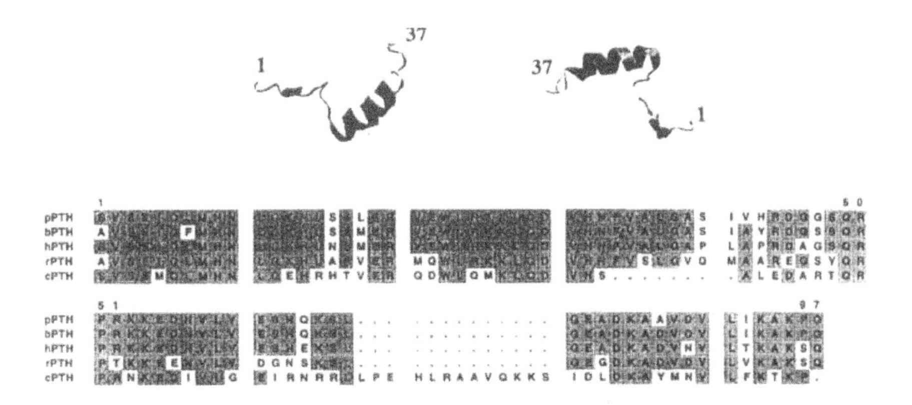

Figure 1. Comparison between the PTH sequence from porcine, bovine, human, rat, and chicken species. Identical sequences in three or more species are shaded. Notice that the chicken PTH has a nine amino acid deletion from the sequence within the region 34–41 and a 13 amino acid insert in the region 68–80. The NMR structure of human PTH(1–37) is shown above. The atomic coordinates and structure factors are available in the Protein Databank, Brookhaven National Laboratory, Upton, NY (Marx et al., 1995). Reproduced with kind permission from Journal of Biological Chemistry.

PTH molecule suggest that these regions may subserve distinct biologic functions. *In vitro* data have in fact revealed specific binding sites for the carboxyl-terminal region of PTH that are distinct from those that bind the amino-terminal fragments (Inomata et al., 1995).

Analysis of the structural requirement for PTH action has revealed that the amino-terminus of PTH(1–34) is essential for bioactivity. Progressive amino terminal truncation of PTH(1–34) led to PTH molecules with low bioactivity (Rosenblatt, 1981). For example, PTH(3–34), PTH(5–34), and PTH(7–34) are potent competitive antagonists *in vitro*. However, only PTH(7–34) maintained this property *in vivo*. Therefore, it was hypothesized that the N-terminus of PTH is the receptor activating region whereas the C-terminus of PTH(1–34) is the receptor binding region. This hypothesis was further supported by extensive *in vitro* binding and activation data (Nussbaum et al., 1980).

Nuclear magnetic resonance (NMR) analysis of the amino-terminal PTH fragments has revealed a rich secondary structure (Bundi et al., 1976; Klaus et al., 1991; Barden and Cuthbertson, 1993; Barden and Kemp, 1993; Wray et al., 1994). The secondary structure of hPTH(1–37) (Marx et al., 1995) shows two α helices formed of residues 5 to 10 and 17 to 28 (Figure 1). The two helices are connected with a flexible link (residues 12 and 13) and a turn region (residues 14 to 17) (Figure 1). These two helices may play important roles in receptor recognition and activation.

III. PARATHYROID HORMONE-RELATED PEPTIDE

PTHrP was characterized from tumor tissues that cause hypercalcemia in cancer patients (Moseley et al., 1987; Stewart et al., 1987; Suva et al., 1987). This protein is now appreciated to be a paracrine or autocrine factor that plays a role both in fetal development and in adult physiology. Many different tissues and cell types produce PTHrP, including brain, pancreas, heart, lung, mammary tissues, placenta, endothelial cells, and smooth muscles (Martin et al., 1989; Martin and Ebeling, 1990).

PTH and PTHrP, although distinct proteins and products of different genes, share considerable structural and functional similarity. Nine out of residues 1–13 are identical in PTH and PTHrP (Figure 2). The remaining residues (14–84 of PTH and 14–141, 14–139, or 14–173 of PTHrP) do not share any significant sequence homology. However, amino-terminal fragments of PTH and PTHrP bind equivalently to one single receptor (Juppner et al., 1991; Abou-Samra et al., 1992), and the carboxy-terminal regions of

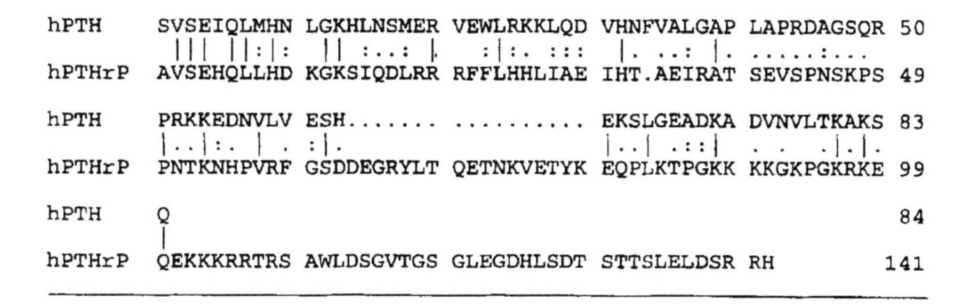

Figure 2. Comparison between the sequences of human PTH(1–84) and human PTHrP(-141). Notice the high degree of sequence identity in the first 13 residues.

PTH(1–34) and PTHrP(1–36) fully compete with the PTH and PTHrP radioligands for binding to rat osteosarcoma (ROS) 17/2.8 cells (Abou-Samra et al., 1989c). Therefore, the secondary structure of these peptides must be sufficiently similar to allow their interaction with one single receptor. In that regard NMR analysis of [Ala^{26}]PTHrP(1–34) and hPTH(1–34) in solution revealed similar structural features (Barden and Kemp, 1989, 1994; Ray et al. 1993). These analyses indicate two segments of α-helix extending from Glu^{4} to Lys^{13} and from Leu^{27} to Thr^{33}, with two turns from Gln^{16} to Arg^{19} and Phe^{22} to His^{25} (Barden and Kemp, 1989, 1994; Ray et al., 1993). A salt-bridge appears likely between Arg^{20} and Glu^{30} which may be critical for holding the receptor-binding domain together (Barden and Kemp, 1989, 1994; Ray et al., 1993). Hybrid PTH(1–34)/PTHrP(1–34) ligands, constructed at break points around residues 14, 15, and 16, revealed interactions between the amino-terminus and the carboxy-terminus of the ligands that is important for receptor recognition (Gardella et al., 1995).

Three alternatively-spliced forms of PTHrP were described, PTHrP(1–141), PTHrP(1–139), and PTHrP(1–173). The varying molecular forms have potential internal cleavage sites to generate multiple hormonal products (Stewart et al., 1987; Philbrick et al., 1996). It is quite possible that PTHrP is a prohormone that is processed to several biologically active peptides (Burtis et al., 1990; Martin and Ebeling, 1990; Soifer et al., 1992). The biologic role of the various molecular forms and of the potential cleavage products is not yet clear (Yang et al., 1994). Gene knock-out techniques revealed that PTHrP is essential for life (Karaplis et al., 1994). Mice with homozygous deletion of the PTHrP gene died either at term or shortly after delivery. These animals have a striking deformity in their growth plates and shortening and thickening of the long bones. Transgenic animals expressing

PTHrP under control of collagen type II promoter, which directs PTHrP expression to chondrocytes, resulted in chondrocytic proliferation from the periphery of bone and decreased mineralization (Weir et al., 1996). These data clearly indicate an important role for PTHrP in bone development.

IV. PTH/PTHrP RECEPTOR

PTH action on its target cells involves interaction with a specific G-protein-linked receptor that lies on the cell membrane. The development of an oxidation resistant PTH analogue, [$Nle^{8,18}$, Tyr^{34}]bPTH(1–34)amide (NlePTH), that can be radioiodinated to an optimal specific activity, i.e., one iodine atom per one molecule of PTH, had facilitated the biochemical characterization of the PTH receptor (Segre et al., 1979). High affinity binding sites were described in canine renal membranes (Segre et al., 1979) and on intact osteosarcoma (ROS 17/2.8, UMR 106, SaOS-02, and MC_3T_3) and opossum kidney (OK) cell lines. Photo affinity cross-linking of the iodinated PTH analogue revealed that the PTH receptor has a molecular weight of 80–90 kDa on SDS-PAGE and that 30% of its mass is attributable to N-linked glycosylation (Shigeno et al. 1988a,b). Synthetic PTHrP fragments were also used to probe the PTHrP binding sites on osteoblastic and renal tubular cells. Surprisingly, PTHrP fully displaces PTH from its binding sites on ROS 17/2.8 cells and renal membranes, and vice-versa (Juppner et al., 1988; Abou-Samra et al., 1989c).

Attempts to purify the PTH/PTHrP receptor from renal membranes and osteosarcoma cell lines were not successful. The main obstacle was the inability to solubulize large quantities of functional PTH receptor molecules. The successful isolation of several receptor cDNAs by functional screening of cDNA libraries expressed into COS cells or xenopus oocytes prompted investigators in the PTH field to use expression cloning technology. These techniques have led to isolation of several rat and opposum cDNA clones encoding a G protein-linked receptor that equivalently binds the amino-terminal fragments of PTH and PTHrP (Juppner et al. 1991; Abou-Samra et al., 1992).

A. Molecular Cloning of the PTH/PTHrP Receptor

The expression cloning system does not require previous knowledge of the sequence of the receptor and does not require purification or solubulization of the receptor molecule. However, this technique requires a highly sensitive and specific screening method. Since radio receptor assays for the PTH/PTHrP receptor were both sensitive and specific, these assays were used to screen cDNA libraries expressed in COS-7 cells. ROS 17/2.8 and OK cell cDNA libraries were screened in pools, each representing 10,000 independent clones. Each pool was transfected into COS-7 cells that were plated on a small slide. Three days after transfection, radio receptor assays were performed on these slides. After extensive rinsing, the slides were fixed and subjected to photoemulsion autoradiography. Each slide was inspected under dark-field microscopy to identify the cDNA pool that contained the positive clone. Three such pools were characterized; they were subdivided into smaller pools until one single clone from each pool was isolated: R15B from ROS 17/2.8 cells and OK-H and OK-O from OK cells.

Sequence analysis of the rat and opossum PTH/PTHrP receptor cDNA showed no homology to any other G protein-coupled receptor known at the time. However, the porcine calcitonin receptor (Lin et al., 1991), showed a great degree of sequence homology with the PTH/PTHrP receptor. Thus a novel G protein-linked receptor family, represented by the PTH/PTHrP and calcitonin receptors, was predicted (Juppner et al., 1991; Lin et al., 1991). Expression cloning technology and polymerase chain reaction (PCR) amplification of cDNA libraries using primers based on the conserved sequences in this receptor family have resulted in the molecular cloning of many other members of this receptor family including the receptors for secretin (Titus et al., 1991), glucagon (Jelinek et al., 1993), glucagon-like peptide (Thorens, 1992), vasoactive intestinal peptide (Ishihara et al., 1992; Lutz et al., 1993), pituitary adenylate cyclase stimulating peptide (Hashimoto et al., 1993; Spengler et al., 1993), growth hormone releasing factor (Lin et al., 1992; Gaylinn et al., 1993), corticotropin-releasing factor (Chen et al., 1993; Lovenberg et al., 1995), and an insect diuretic hormone (Reagan, 1994). The active fragments of these hormones are simple polypeptides of an intermediate length (20–40 amino acids). These receptors are characterized by an amino-terminal extracelluar domain that is intermediate in length (100–200 amino acids) and that contains six highly conserved extracellular cystein residues.

The cDNA of the rat PTH/PTHrP receptor was used as a probe to screen cDNA libraries constructed from human kidney and SaOS cells (Schipani et al., 1993), murine cell lines (Karperien et al., 1994), porcine kidney (Smith et al., 1995), UMR 106-01 cells (Pausova et al., 1994), and *Xenopus* kidney (Bergwitz et al., 1994). Analysis of the PTH/PTHrP receptor sequence from different species revealed an extraordinary conservation across the species

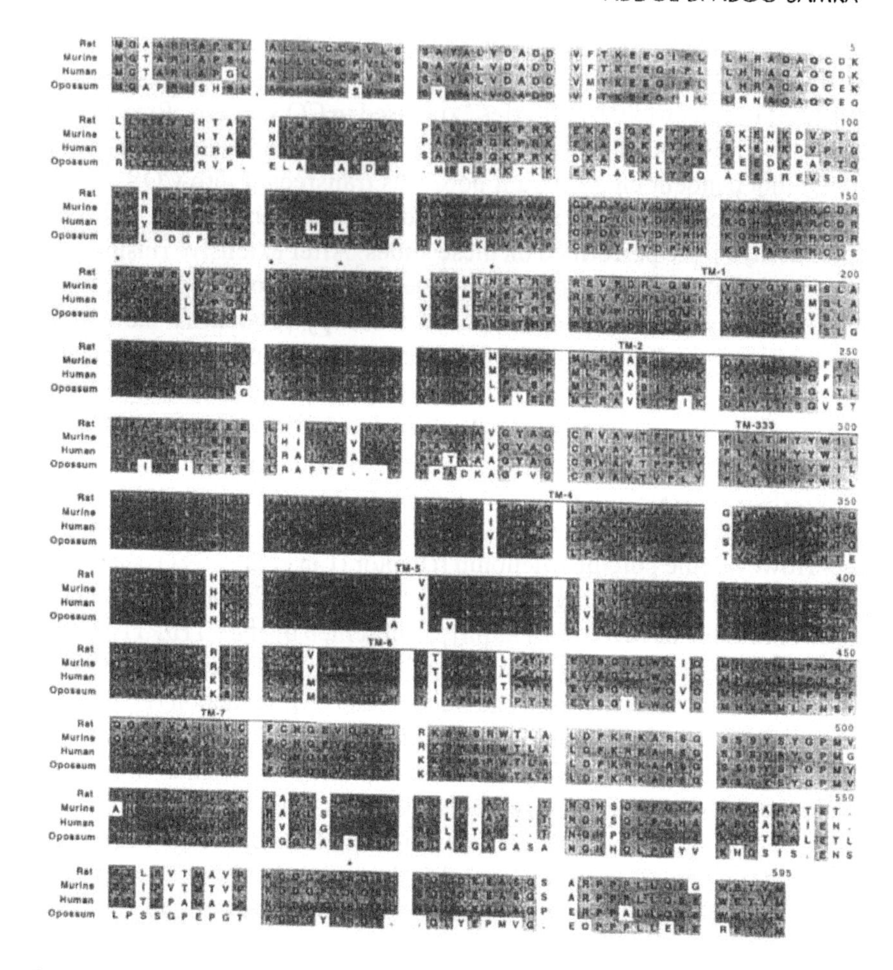

Figure 3. Sequence comparison between the rat (R), mouse (M), human (H), and opossum (O) PTH/PTHrP receptors. Identical residues in three or more species are shaded. Potential transmembrane spanning domains are labeled, potential glycosylation sites are labeled with an asterisk (*). Notice the high degree of sequence homology. There are three less conserved regions: amino acids 56 through 107 in the extracellular region, amino acids 258 through 289 in the first extracellular loop, and amino acids 544 through 595 within the carboxy-terminal tail. All three less conserved regions could be deleted without impairment of receptor expression, ligand binding, or signaling properties.

(Figure 3). The high degree of sequence conservation suggests an important role of the PTH/PTHrP receptor system in these species.

Screening by hybridization of rat kidney cDNA library using a rat PTH/PTHrP receptor probe resulted in isolation of six independent cDNA clones, all of which have the same sequence (Kong et al., 1994). No other

PTH/PTHrP receptor subtype was isolated using this technique. However, Northern blot analysis of the renal PTH/PTHrP receptor show heterogeneity with a major molecular form of 2.2 Kb and several other higher and lower size transcripts. Additionally, keratinocyte and some squamous carcinoma cell line express a large transcript that hybridizes with the PTH/PTHrP receptor cDNA probe (Orloff et al., 1995). Therefore, other receptor subtypes or alternatively spliced variants are likely to be expressed in kidney and other tissues.

B. Ligand Binding Properties of the PTH/PTHrP Receptor

The PTH/PTHrP receptor from rat, mouse, opossum, human, and *Xenopus*, expressed in COS-7 cells, binds PTH(1–34) and PTHrP(1–36) with high affinity with K_ds in the nM range (10 nM). However, the receptor from different species shows significant differences in recognition of amino-terminally truncated PTH analogue. For example, the human and opossum PTH/PTHrP receptors bind PTH(7–34) with an affinity that is about 10–100-fold higher than that displayed by the rat PTH/PTHrP receptor. These properties permitted mapping of the site conferring the high binding affinity for PTH(7–34) to the amino-terminal extracellular domain. This was achieved by constructing chimera between the rat and opossum or the rat and human PTH/PTHrP receptors. The chimeric receptors, although bound PTH(1–34) and PTHrP(1–36) with high affinity, displayed variable affinities for PTH(7–34) (Juppner et al., 1994).

Deletion mapping of the extracellular domains of the rat PTH/PTHrP receptor revealed two regions that are not essential for receptor function (Lee et al., 1994). One region corresponds to an exon encoding 48 amino acids from the extracellular region (exon E_2). Interestingly, none of the other members of the PTH receptor family contained sequences that are homologous to exon E_2. Additionally, the *Xenopus* PTH/PTHrP receptor does not contain this region (Bergwitz et al., 1994). The second region that is not essential for ligand-receptor interaction is located in the distal half of the first extracellular loop.

The conserved amino-terminal extracellular cystine residues appear essential for cell surface expression (Juppner et al., 1994). Cys to Ser mutations of these residues dramatically impaired surface expression. Single mutation of the cystine residues in the first and second extracellular loops also impaired surface expression and cAMP stimulation. However, combined Cys to Ser mutations of both residues of first and second extracellular loops did not cause additive effects. Therefore, these two residues are likely to be involved in an S-S bridge formation (Juppner et al., 1994).

Homologue scanning of the extracellular domains of the PTH/PTHrP receptor using sequences from the secretin receptor and single and cluster point mutations defined several regions in the amino-terminus and extracellular loops that are essential for high affinity ligand interaction (Lee et al., 1995). The whole picture is not yet clear, however, several determinants for ligand binding have been characterized. These include residues in the extracellular (EC) end of transmembrane domain 1, the carboxy-terminal portion of the first EC loop, the second EC loop, and the third EC loop. Mutation of these sites caused dramatic loss of ligand binding properties without affecting cell surface expression (Lee et al., 1995).

C. Signaling Properties of the PTH/PTHrP Receptor

PTH is known to stimulate cAMP accumulation in its target cells (Aurbach, 1973). PTH has also been shown to increase phosphoinositide hydrolysis (Hruska et al., 1987), raise intracellular calcium concentrations (Hruska et al., 1986), and activate protein kinase C (Abou-Samra et al., 1989b) in renal and bone cells. Therefore, it was postulated that PTH may activate several intracellular messenger systems through interaction with different receptors. The molecular cloning of the rat and opossum PTH/PTHrP receptor provided evidence that one single receptor can stimulate several effectors (Abou-Samra et al., 1992).

Stimulation of adenylate cyclase by PTH or PTHrP occurs with concentrations in the sub-nanomolar range and is detectable in cells expressing low numbers of PTH receptors (Abou-Samra et al., 1993; Bringhurst et al., 1993). Conversely, activation of phospholipase C (PLC) by PTH or PTHrP requires high PTH concentrations (10–1,000 nM) and a large number of PTH receptors (more than 100,000 receptor/cells) (Bringhurst et al., 1993). Since the plasma concentrations of PTH are in the subnanomolar range, the biologic relevance of PLC stimulation by PTH is not known. However, localized production of PTHrP in certain tissues (Weaver et al., 1995) may result in ligand concentrations that may be sufficient to stimulate PLC.

An extensive site-directed mutagenesis approach has been undertaken to delineate the regions of the receptor that couple to the G protein(s). Truncation of most of the carboxy-terminal tail of the PTH/PTHrP receptor enhanced G_s coupling (Iida-Klein et al., 1995). Some mutations in the second cytoplasmic loop uncoupled the PTH/PTHrP receptor from PLC stimulation though adenylate cyclase stimulation remained intact (Iida-Klein et al., 1997). These data suggested that G_q, or a G_q-like G protein, couples to the

PTH receptor independently from G_s and that PLC stimulation is not secondary to activation of adenylate cyclase.

D. Regulation of the PTH/PTHrP Receptor

Maintenance of extracellular calcium homeostasis involves regulation of the responsiveness of the target cells to PTH by the hormonal milieu. Multiple factors in the hormonal environment exercise both stimulatory and inhibitory control on the steady-state levels of the receptors and on their downstream intracellular signaling molecules. Hormonal regulation may vary from one target cell to another (cell-specific) and may occur as a result of exposure of the target cell to the agonist (homologous regulation) or to any other hormonal factor (heterologous regulation). Desensitization refers to a decrease in the responsiveness of the target cell to the agonist without any changes in the levels of receptor or the signaling molecules.

E. Homologous PTH Receptor Downregulation and Desensitization of Cellular Responsiveness to PTH

Long-term treatment (1–3 days) of ROS 17/2.8 (Yamamoto et al., 1988b; Abou-Samra et al., 1989a), UMR 106-01 (Abou-Samra et al., 1991), SaOS-2 (Fukayama et al., 1992, 1994), and OK (Abou-Samra et al., 1994) cells with PTH or PTHrP results in a dramatic decrease in the specific binding of iodinated PTH or PTHrP radioligands to the cell surface. Short-term desensitization of the responsiveness to PTH was well characterized in SaOS-2 cells. The cells were continuously perifused with PTH which caused an immediate rise in cAMP release in the effluent media. An 80% decrease in the responsiveness to subsequent PTH pulse occurred after the cells were perifused with PTH for 30 minutes. Cells recovered full responsiveness to PTH after perifusion with PTH-free medium for two hours (Bergwitz et al., 1994b).

Animal models with increased PTH levels provide a system to study the impact of chronic PTH elevation on the PTH receptor function. Newly hatched chicks maintained on low calcium and vitamin D-deficient diets developed hypocalcemia with secondary hyperparathyroidism. The number of the PTH receptors and maximal PTH-stimulated adenylate cyclase in renal membranes prepared from the secondary hyperparathyroid chicks were markedly decreased (Forte et al., 1982).

Chronic renal failure is associated with increased levels of circulating PTH. Decreased renal responsiveness to PTH in chronic renal failure could be due to downregulation of the PTH/PTHrP receptor. The steady-state levels of the PTH/PTHrP receptor mRNA were shown to be downregulated in rat with experimental renal failure (Tian et al., 1994; Urena et al., 1994b). Additionally, PTH-stimulated adenylate cyclase was shown to be decreased in kidney membranes prepared from animals with chronic renal failure (Urena et al., 1994b). The decrease in PTH/PTHrP receptor level was significantly correlated with the degree of renal dysfunction and occurred in parathyroidectimized rats; this suggested that chronic uremia per se is the cause of receptor downregulation in chronic renal failure (Urena et al., 1994c).

F. Heterologous Regulation of the PTH/PTHrP Receptor

Several hormonal and growth factors influence the function of the PTH/PTHrP receptor by direct effects on the receptor levels and/or by altering the levels of the downstream signaling molecules such as the G protein, adenylate cyclase, and protein kinase A.

Glucocorticoid dramatically increases the number of the PTH receptors on ROS 17/2.8 cells without changing their affinity and their effects were blocked by cycloheximide suggesting that the effects of glucocorticoid involve new receptor synthesis (Yamamoto et al., 1988a). The effects of glucocorticoids is mediated by a dramatic increase in the steady-state level of the PTH/PTHrP receptor mRNA (Urena et al., 1994a). In the other osteoblastic osteosarcoma cell lines, UMR 106-01 and SaOS, glucocorticoids have little or no effects at all on the PTH receptor density on the cell surface (unpublished data). In the renal tubular cell line, OK cells, glucocorticoid treatment decreases the number of PTH receptors and decreases the cAMP responsiveness to PTH (Kaufmann et al., 1991). The opposite effects of glucocorticoids on the PTH receptor levels on osteoblast-like and renal tubular-like cells indicate that cell-specific factors are involved in the direction of regulation of this receptor by glucocorticoids.

$1,25(OH)_2 D_3$, the active metabolite of vitamin D, has been shown to decrease the cellular responsiveness to PTH (Rizzoli and Fleisch, 1986) and to downregulate the number of the PTH/PTHrP receptors in ROS 17/2.8 cells (Titus et al., 1991). Additionally, treatment of ROS 17/2.8 with $1,25(OH)_2D_3$ decreased the PTH/PTHrP receptor immunoreactivity on the cell surface and downregulated the steady-state levels of the PTH/PTHrP receptor (Xie et al., 1994).

Growth factors and cytokines have multiple effects on osteoblast-like and renal tubular cells (Schneider et al., 1991; Takigawa et al., 1991; Hanevold et al., 1993; Law et al., 1994). In UMR 106-01 cells, tumor necrosis factor-α (TNF-α) decreased the binding of PTH(1–84) by 40%, significantly decreased PTH-stimulated tissue plasminogen activator (tPA) (Schneider et al., 1991), and decreased PTH-stimulated cAMP accumulation and PTH-stimulated increase in intracellular free calcium concentrations (Hanevold et al., 1993). TNFα and interleukin-1 (IL-1) were also shown to decrease the binding of PTH(1–34) and to reduce the cAMP responsiveness to PTH in UMR 106-01 cells (Katz et al., 1992). Treatment of OK cells with transforming growth factor-β (TGF-β) downregulates the number of the PTH/PTHrP receptors, desensitizes PTH-stimulated cAMP accumulation and decreases the levels of the PTH/PTHrP receptor transcript (Law et al., 1994). In contrast, treatment of ROS 17/2.8 cells with TGF-β increases the number of the PTH/PTHrP receptors, enhances PTH-stimulated cAMP accumulation and increases the steady-state levels of the PTH/PTHrP receptor mRNA (McCauley et al., 1994). Thus, similar to glucocorticoids, TGF-β also has opposite regulatory effects on the PTH/PTHrP receptor levels and functions in ROS 17/2.8 and OK cells.

V. THE PTH/PTHrP RECEPTOR GENE

The PTH/PTHrP receptor gene (Figure 4) is a complex gene consisting of 14 coding exons that are interrupted by 13 introns of variable length (Kong et al., 1994; McCuaig et al., 1994). The gene spans over 25 Kb and is located on the short arm of human chromosome 3 (Gelbert et al., 1994). In rat and mouse, the gene is located on the homologous counterparts of human chromosome 3, mouse chromosome 9, and rat chromosome 8, respectively (Pausova et al., 1994). Since several PTH/PTHrP receptor transcripts have been shown to occur in multiple tissues (Urena et al., 1993) it is likely that this gene undergoes alternative splicing to form transcripts of different sizes, and/or other closely related genes may exist in the genome. Recently, it has been shown that the 5' end of this gene contains at least three non-coding exons, and that these exons can be processed to form three alternatively-spliced mature transcripts (Joun et al., 1994). Additionally, these exons may serves as potential alternative promoters to initiate gene transcription from different transcription start sites.

Pseudohypoparathyroidism type 1 is a genetic disorder characterized by hypocalcemia, normal PTH secretion, and end organ resistance to PTH. Pseudohypoparathyroidism type 1 is further subdivided into type 1A, which is characterized by skeletal malformation (round face, short stature, and short fourth and fifth metacarpal bones) and defect in the G_s protein, and

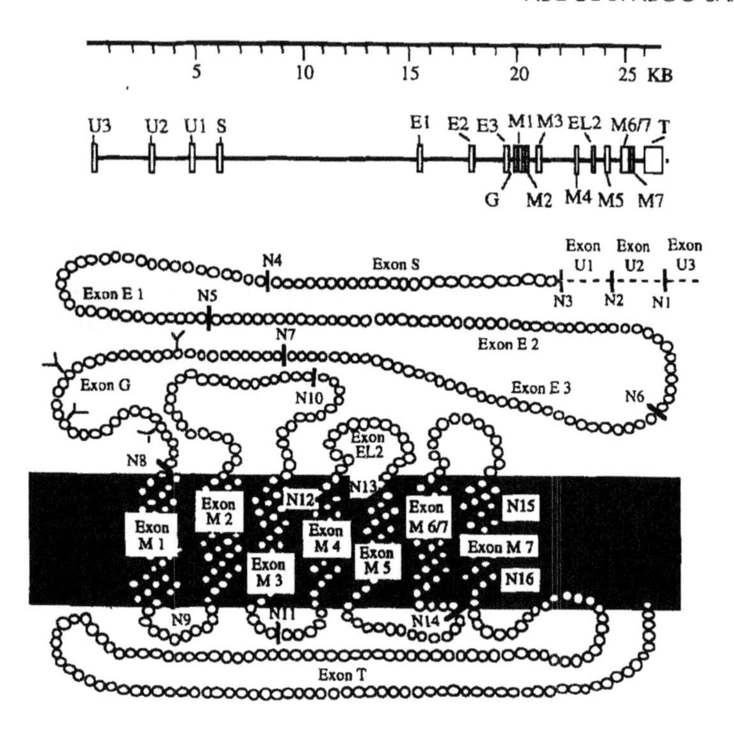

Figure 4. Organization of the coding region of the rat PTH/PTHrP receptor gene. Exon S encodes signal peptide. Exons E1, E2, E3, and G encode the amino-terminal extracellular extension of the receptor. Exon G contains all four potential extracellular glycosylation sites. Exons M1, M2, M3, M4, EL2, M5, M6/7, and M7 encode the transmembrane spanning domains of the receptor and their connecting cytoplasmic and extracellular loops. Exon T, the largest exon of this gene, encodes the cytoplasmic tail of the receptor and the 3' uncoding region. The introns (N1–N16) are numbered by their order from 5' to 3'. The position of exon and introns are shown in the top. The sizes of exons are not drawn to scale. Transcription start sites are found at the 5' end of exons U3 and U1. Exons U3, U2, and U1 encode the 5' untranslated regions in the different splice variants of this gene.

type 1B, which is characterized by a selective resistance to parathyroid hormone without skeletal malformation and with normal G_s protein. A PTH receptor defect was suggested to be the underlying cause of pseudohypoparathyroidism type 1B. However, extensive genetic analysis revealed that pseudohypoparathyroidism type 1B is not caused by mutations in the coding exons of the human parathyroid hormone (PTH)/PTHrP receptor gene (Schipani et al., 1995b). Since deletion of the PTH/PTHrP receptor gene from the mouse gnome was lethal (Lanske et al.,1996), it is unlikely to find an inactivating mutation of the PTH/PTHrP receptor in pseudohypoparathyroidism type 1B. A more distal mutation in the PTH signaling cascade, or a defect in the receptor expression levels can not be ruled out by this analysis. It has been recently reported that the levels of the PTH/PTHrP receptor transcript are decreased in fibroblast cultures obtained from patients with pseudohypothyroidism type 1B (Suarez et al., 1995).

Activating mutations within the PTH/PTHrP receptor may be the underlying cause of a rare genetic disease, Jansen's disease, that is characterized by hypercalcemia, low PTH levels, and skeletal malformation (Schipani et al., 1995a). Therefore five patients with Jansen's disease were screened. A His^{233} to Arg^{233} mutation within the second transmembrane spanning domain was found in four patients and a Thr^{410} to Pro^{410} mutation at the junction of the third cytoplasmic loop and the sixth transmembrane spanning domain was found in the fifth patient (Schipani et al., 1996). Both mutations caused constitutive activation of the receptor, i.e., raised cAMP levels when expressed in COS-7 cells without addition of PTH or PTHrP.

VI. PTH_2 RECEPTOR

A novel receptor cDNA, with 52% sequence identity to the PTH/PTHrP receptor, was cloned by PCR using primers from conserved sequences within the PTH/PTHrP receptor family (Usdin et al., 1995). Since the novel receptor was activated by PTH but not by PTHrP, it was named the PTH_2 receptor (PTH_2R). Tissue distribution of the PTH_2R transcript was limited to brain, placenta, and pancreas. The physiological role of the PTH_2R is not established yet. Since PTHrP does not bind to PTH_2R, and since PTHrP can cause hypercalcemia in cancer patients, PTH_2R is unlikely to mediate the effects of PTH on calcium homeostasis. It is possible that the PTH_2R is a receptor for another ligand that is closely related to PTH. Alternatively, it may represent a novel PTH-specific receptor in non-classical target tissues.

VII. SUMMARY

PTH is an essential hormone for the maintenance of calcium homeostasis. PTHrP subserves several biological functions in the development, differentiation, and maturation of several tissues, particularly bone. PTH is secreted by the parathyroid glands and act on target tissues at a distance, whereas PTHrP is produced by several tissues and acts locally as an autocrine/paracrine factor. The PTH/PTHrP receptor appears to mediate the hormonal effects of PTH and the autocrine/paracrine effects of PTHrP. The PTH/PTHrP receptor is a G protein-coupled receptor that spans the plasma membrane seven times, binds both PTH and PTHrP, and activates multiple intracellular signals. The PTH_2R is a PTH-specific receptor that binds selectively PTH but not PTHrP. PTH_2R is not located in classical PTH target tissues that regulate calcium homeostasis and may be a receptor for a PTH-like ligand that acts in the central nervous system, pancreas, and placenta.

The PTH endocrine system is tightly regulated by a negative feedback mechanism. An increase in extracellular calcium concentrations leads to suppression of PTH secretion. Conversely, decreased extracellular calcium concentrations lead to stimulation of PTH secretion. The negative feedback effects of calcium are mediated by a specific receptor, a G protein-coupled calcium sensor, located on the parathyroid cells. Long-term feedback control is exerted by $1,25(OH)_2D_3$. An increase in $1,25(OH)_2D_3$ level suppresses PTH secretion by inhibiting PTH gene transcription. Additionally, $1,25(OH)_2D_3$ decreases the levels of the PTH receptor on osteoblastic cells. The long-term effects of $1,25(OH)_2D_3$ on PTH gene transcription and on PTH receptor expression complements the minute-to-minute effects of extracellular calcium on PTH secretion. The PTH/PTHrP receptor is also regulated by the hormonal environment in a cell specific manner. Glucocorticoids increase its levels whereas PTH and vitamin D_3 suppress its levels in osteoblastic cells. Many other growth factor and cytokines influence the PTH responsiveness by modifying the steady-state levels of the PTH/PTHrP receptor in the target cells.

REFERENCES

Abou-Samra, A.B., Jüppner, H., Force, T., Freeman, M.W., Kong, X.F., Schipani, E., Urena, P., Richards, J., Bonventre, J.V., Potts, J.J., Kronenberg, H.M., and Segre G.V. (1992). Expression cloning of a common receptor for parathyroid hormone and parathyroid hormone-related peptide from rat osteoblastlike cells: A single receptor stimulates intracellular accumulation of both cAMP and inositol trisphosphates and increases intracellular free calcium. Proc. Natl. Acad. Sci. USA 89 (7), 2732-2736.

Abou-Samra, A.B., Jüppner, H., Khalifa, A., Karga, H., Kong, X.F., Schiffer, A.D., Xie, L.Y., and Segre, G.V. (1993). Parathyroid hormone (PTH) stimulates adrenocorticotropin release in AtT-20 cells stably expressing a common receptor for PTH and PTH-related peptide. Endocrinology 132 (2), 801-805.

Abou-Samra, A.B., Goldsmith, P., Xie, L., Jüppner, H., Spiegel, A., and Segre, G.V. (1994). Regulation of the PTH/PTHrP receptor immunoreactivity and PTH binding in opossum kidney cells by PTH and dexamethasone. Endocrinology 135, 2588-2594.

Abou-Samra, A.B., Jüppner, H., Potts, J.J., and Segre, G.V. (1989a). Inactivation of pertussis toxin-sensitive guanyl nucleotide-binding proteins increase parathyroid hormone receptors and reverse agonist-induced receptor downregulation in ROS 17/2.8 cells. Endocrinology 125 (5), 2594-2599.

Abou-Samra, A.B., Jüppner, H., Westerberg, D., Potts, Jr., J.T., and Segre, G.V. (1989b). Parathyroid hormone causes translocation of protein kinase-C from cytosol to membranes in rat osteosarcoma cells. Endocrinology 124, 1107-1113.

Abou-Samra, A.B., Uneno, S., Jüppner, H., Keutmann, H., Potts, J.J., Segre, G.V. and Nussbaum, S.R. (1989c). Nonhomologous sequences of parathyroid hormone and the parathyroid hormone-related peptide bind to a common receptor on ROS 17/2.8 cells. Endocrinology 125 (4), 2215-2217.

Abou-Samra, A.B., Zajac, J.D., Schiffer, A.D., Skurat, R., Kearns, A., Segre, G.V., and Bringhurst, F.R. (1991). Cyclic adenosine 3',5'-monophosphate cAMP-dependent and cAMP-independent regulation of parathyroid hormone receptors on UMR 106-01 osteoblastic osteosarcoma cells. Endocrinology 129 (5), 2547-2554.

Aurbach, G.D. (1973). Biosynthesis, secretion, and mechanism of action of parathyroid hormone. Trans. Am. Clin. Climatol. Assoc. 85 (0), 78-99.

Aurbach, G.D., Keutmann, H.T., Niall, H.D., Tregear, G.W., O'Riordan, J.L., Marcus, R., Marx, S.J. and Potts, J.J. (1972). Structure, synthesis, and mechanism of action of parathyroid hormone. Recent Prog. Horm. Res. 28 (353), 353-398.

Barden, J.A. and Cuthbertson, R.M. (1993). Stabilized NMR structure of human parathyroid hormone(1-34). Eur. J. Biochem. 215 (2), 315-321.

Barden, J.A. and Kemp, B.E. (1989). NMR study of a 34-residue N-terminal fragment of the parathyroid hormone-related protein secreted during humoral hypercalcemia of malignancy. Eur. J. Biochem. 184 (2), 379-394.

Barden, J.A. and Kemp, B.E. (1993). NMR solution structure of human parathyroid hormone (1-34). Biochemistry 32 (28), 7126-7132.

Barden, J.A. and Kemp, B.E. (1994). Stabilized NMR structure of the hypercalcemia of malignancy peptide PTHrP[Ala-26] (1-34) amide. Biochim. Biophys. Acta 1208 (2), 256-262.

Bergwitz, C., Klein, P., Kohno, H., Forman, S.A., Lee, K., Rubin, D., and Jüppner, H. (1997). Identification, functional characterization, and developmental expression of two nonallelic parathyroid hormone (PTH)/PTH-related peptide receptor isoforms in *Xenopus laevis* (Daudin). Endocrinology 139, 723-732.

Bergwitz, C., Abou-Samra, A.B., Hesch, R.D., and Jüppner, H. (1994b). Rapid desensitization of parathyroid hormone-dependent adenylate cyclase in perifused human osteosarcoma cells (SaOS-2). Biochim. Biophys. Acta 1222 (3), 447-456.

Bringhurst, F.R., Jüppner, H., Guo, J., Urena, P., Potts, J.J., Kronenberg, H.M., Abou-Samra, A.B., and Segre, G.V. (1993). Cloned, stably expressed parathyroid hormone (PTH)/PTH-related peptide receptors activate multiple messenger signals and biological responses in LLC-PK1 kidney cells. Endocrinology 132 (5), 2090-2098.

Bringhurst, F.R., Segre, G.V., Lampman, G.W., and Potts, J.J. (1982). Metabolism of parathyroid hormone by Kupffer cells: Analysis by reverse-phase, high-performance liquid chromatography. Biochemistry 21 (18), 4252-4258.

Bringhurst, F.R., Stern, A.M., Yotts, M., Mizrahi, N., Segre, G.V., and Potts, J.J. (1988). Peripheral metabolism of PTH: Fate of biologically active amino terminus in vivo. Am. J. Physiol. E886-E893.

Bundi, A., Andreatta, R., Rittel, W., and Wuthrich, K. (1976). Conformational studies of the synthetic fragment 1-34 of human parathyroid hormone by NMR techniques. Febs. Lett. 64 (1), 126-129.

Burtis, W.J., Brady, T.G., Orloff, J.J., Ersbak, J.B., Warrell, R.J., Olson, B.R., Wu, T.L., Mitnick, M.E., Broadus, A.E., and Stewart, A.F. (1990). Immunochemical characterization of circulating parathyroid hormone-related protein in patients with humoral hypercalcemia of cancer [see comments]. N. Engl. J. Med. 322 (16), 1106-1112.

Chen, R., Lewis, K.A., Perrin, M.H., and Vale, W.W. (1993). Expression cloning of a human corticotropin-releasing-factor receptor. Proc. Natl. Acad. Sci. USA 90 (19), 8967-8971.

Forte, L.R., Langeluttig, S.G., Poelling, R.E., and Thomas, M.L. (1982). Renal parathyroid hormone receptors in the chick: Downregulation in secondary hyperparathyroid animal models. Am. J. Physiol. 242 (3), E154-E163.

Fukayama, S., Schipani, E., Jüppner, H., Lanske, B., Kronenberg, H.M., Abou-Samra, A.B., and Bringhurst, F.R. (1994). Role of protein kinase-A in homologous downregulation of parathyroid hormone (PTH)/PTH-related peptide receptor messenger ribonucleic acid in human osteoblastlike SaOS-2 cells. Endocrinology 134 (4), 1851-1858.

Fukayama, S., Tashjian, A.J., and Bringhurst, F.R. (1992). Mechanisms of desensitization to parathyroid hormone in human osteoblastlike SaOS-2 cells. Endocrinology 131 (4), 1757-1769.

Gardella, T.J., Luck, M.D., Wilson, A.K., Keutmann, H.T., Nussbaum, S.R., Potts, J.J., and Kronenberg, H.M. (1995). Parathyroid hormone (PTH)-PTH-related peptide hybrid peptides reveal functional interactions between the 1-14 and 15-34 domains of the ligand. J. Biol. Chem. 270 (12), 6584-6588.

Gaylinn, B.D., Harrison, J.K., Zysk, J.R., Lyons, C.E., Lynch, K.R., and Thorner, M.O. (1993). Molecular cloning and expression of a human anterior pituitary receptor for growth hormone-releasing hormone. Mol. Endocrinol. 7 (1), 77-84.

Gelbert, L., Schipani, E., Juppner, H., Abou-Samra, A., Segre, G., Naylor, S., Drabkin, H., and Heath III, H. (1994). Chromosomal localization of the parathyroid hormone/parathyroid hormone-related protein receptor gene to human chromosome 3p21.1-p24.2. J. Clin. Endocrinol. Metab. 79, 1046-1048.

Goltzman, D., Callahan, E.N., Tregear, G.W., and Potts, J.J. (1976). Conversion of proparathyroid hormone to parathyroid hormone: Studies in vitro with trypsin. Biochemistry 15 (23), 5076-5082.

Habener, J.F., Kronenberg, H.M., and Potts, J.T. (1981). Biosynthesis of preproparathyroid hormone. Methods Cell Biol. 23 (51), 51-71.

Habener, J.F., Mayer, G.P., Dee, P.C., and Potts, J.T. (1976). Metabolism of amino- and carboxyl-sequence immunoreactive parathyroid hormone in the bovine: Evidence for peripheral cleavage of hormone. Metabolism 25 (4), 385-395.

Hanevold, C.D., Yamaguchi, D.T., and Jordan, S.C. (1993). Tumor necrosis factor α modulates parathyroid hormone action in UMR-106-01 osteoblastic cells. J. Bone Miner. Res. 8 (10), 1191-1200.

Hashimoto, H., Ishihara, T., Shigemoto, R., Mori, K., and Nagata, S. (1993). Molecular cloning and tissue distribution of a receptor for pituitary adenylate cyclase-activating polypeptide. Neuron 11 (2), 333-342.

Heinrich, G., Kronenberg, H.M., Potts, J.T., and Habener, J.F. (1984). Gene-encoding parathyroid hormone. Nucleotide sequence of the rat gene and deduced amino acid sequence of rat preproparathyroid hormone. J. Biol. Chem. 259 (5), 3320-3329.

Hruska, K.A., Goligorsky, M., Scoble, J., Tsutsumi, M., Westbrook, S., and Moskowitz, D. (1986). Effects of parathyroid hormone on cytosolic calcium in renal proximal tubular primary cultures. Am. J. Physiol. F188-F198.

Hruska, K.A., Moskowitz, D., Esbrit, P., Civitelli, R., Westbrook, S., and Huskey, M. (1987). Stimulation of inositol trisphosphate and diacylglycerol production in renal tubular cells by parathyroid hormone. J. Clin. Invest. 79 (1), 230-239.

Iida-Klein, A., Guo, J., Xie, L.Y., Juppner, H., Potts, J.T., Kronenberg, H.M., Bringhurst, F.R., Abou-Samra, A.B., and Segre, G.V. (1995). Truncation of the carboxyl-terminal region of the rat parathyroid hormone (PTH)/PTH-related peptide receptor enhances PTH stimulation of adenylyl cyclase but not phospholipase C.J. Biol. Chem. 270 (15), 8458-8565.

Iida-Klein, A., Gou, J., Takemura, M., Drake, M.T., Potts, J.T., Jr., Abou-Samra, A.B., Bringhurst, F.R., and Segre, G.V. (1997). Mutations in the second cytoplasmic loop of the rat parathyroid hormone (PTH)/PTH-related protein receptor result in selective loss of PTH-stimulated phospholipase C activity. J. Biol. Chem. 272, 6882-6889.

Inomata, N., Akiyama, M., Kubota, N., and Jüppner, H. (1995). Characterization of a novel parathyroid hormone (PTH) receptor with specificity for the carboxyl-terminal region of PTH-(1-84). (see comments). Endocrinology 136 (11), 4732-4740.

Ishihara, T., Shigemoto, R., Mori, K., Takahashi, K., and Nagata, S. (1992). Functional expression and tissue distribution of a novel receptor for vasoactive intestinal polypeptide. Neuron 8, 811-819.

Jelinek, L.J., Lok, S., Rosenberg, G.B., Smith, R.A., Grant, F.J., Biggs, S., Bensch, P.A., Kuijper, J.L., Sheppard, P.O., Sprecher, C.A., O'Hara, P.J., Foster, D., Walker, K.M., Chen, L.H.J., McKernan, P.A., and Kindsvogel, W. (1993). Expression cloning and signaling properties of the rat glucagon receptor. Science 259 (5101), 1614-1616.

Joun, H., Karperien, M., Lanske, B., Defize, L., Kronenberg, H., and Abou-Samra, A.B. (1994). The PTH/PTHrP receptor gene is alternatively spliced from its 5' end. J. Bone. Min. Res. 9 (Suppl.1), B221.

Jüppner, H., Abou-Samra, A.B., Freeman, M., Kong, X.F., Schipani, E., Richards, J., Kolakowski, L.J., Hock, J., Potts, J.J., and Kronenberg, H.M. (1991). A G protein-linked receptor for parathyroid hormone and parathyroid hormone-related peptide. Science 254 (5034), 1024-1026.

Jüppner, H., Abou-Samra, A.B., Uneno, S., Gu, W.X., Potts, J.J. and Segre, G.V. (1988). The parathyroid hormonelike peptide associated with humoral hypercalcemia of malignancy and parathyroid hormone bind to the same receptor on the plasma membrane of ROS 17/2.8 cells. J. Biol. Chem. 263 (18), 8557-8560.

Jüppner, H., Schipani, E., Bringhurst, F.R., McClure, I., Keutmann, H.T., Potts, J.J., Kronenberg, H. M., Abou-Samra, A.B., Segre, G.V. and Gardella, T.J. (1994). The extracellular amino-terminal region of the parathyroid hormone (PTH)/PTH-related peptide receptor determines the binding affinity for carboxyl-terminal fragments of PTH-(1-34). Endocrinology 134 (2), 879-884.

Karaplis, A.C., Luz, A., Glowacki, J., Bronson, R.T., Tybulewicz, V.L., Kronenberg, H.M., and Mulligan, R.C. (1994). Lethal skeletal dysplasia from targeted disruption of the parathyroid hormone-related peptide gene. Genes Dev. 8 (3), 277-289.

Karperien, M., Van Dijk, T.B., Hoeijmakers, T., Cremers, F., Abou-Samra, A.B., Boonstra, J., de, L.S. and Defize, L.H. (1994). Expression pattern of parathyroid hormone/parathyroid hormone-related peptide receptor mRNA in mouse postimplantation embryos indicates involvement in multiple developmental processes. Mech. Dev. 47 (1), 29-42.

Katz, M.S., Gutierrez, G.E., Mundy, G.R., Hymer, T.K., Caulfield, M.P., and McKee, R.L. (1992). Tumor necrosis factor and interleukin 1 inhibit parathyroid hormone-responsive adenylate cyclase in clonal osteoblastlike cells by downregulating parathyroid hormone receptors. J. Cell Physiol. 153 (1), 206-213.

Kaufmann, M., Muff, R., and Fischer, J.A. (1991). Effect of dexamethasone on parathyroid hormone (PTH) and PTH-related protein-regulated phosphate uptake in opossum kidney cells. Endocrinology 128 (4), 1819-1824.

Keutmann, H.T., Niall, H.D., O'Riordan, J.L., and Potts, J.J. (1975). A reinvestigation of the amino-terminal sequence of human parathyroid hormone. Biochemistry 14 (9), 1842-1847.

Khosla, S., Demay, M., Pines, M., Hurwitz, S., Potts, J.J., and Kronenberg, H.M. (1988). Nucleotide sequence of cloned cDNAs encoding chicken preproparathyroid hormone. J. Bone Min. Res. 3(6): 689-698.

Klaus, W., Dieckmann, T., Wray, V., Schomburg, D., Wingender, E., and Mayer, H. (1991). Investigation of the solution structure of the human parathyroid hormone fragment (1-34) by 1H NMR spectroscopy, distance geometry, and molecular dynamics calculations. Biochemistry 30 (28), 6936-6942.

Kong, X.F., Schipani, E., Lanske, B., Joun, H., Karperien, M., Defize, L.H., Jüppner, H., Potts, J.J., Segre, G.V., Kronenberg, H.M., and Abou-Samra, A.B. (1994). The rat, mouse, and human genes encoding the receptor for parathyroid hormone and parathyroid hormone-related peptide are highly homologous. Biochem. Biophys. Res. Commun. 201 (2), 1058.

Kronenberg, H.M., Igarashi, T., Freeman, M.W., Okazaki, T., Brand, S.J., Wiren, K.M., and Potts, J.J. (1986). Structure and expression of the human parathyroid hormone gene. Recent Prog. Horm. Res. 42 (641), 641-663.

Kronenberg, H.M., McDevitt, B.E., Majzoub, J.A., Nathans, J., Sharp, P.A., Potts, J.J., and Rich, A. (1979). Cloning and nucleotide sequence of DNA coding for bovine preproparathyroid hormone. Proc. Natl. Acad. Sci. USA 76 (10), 4981-4985.

Kronenberg, H.M., Roberts, B.E., Habener, J.F., Potts, J.J., and Rich, A. (1977). DNA complementary to parathyroid mRNA directs synthesis of preproparathyroid hormone in a linked transcription-translation system. Nature 267 (5614), 804-807.

Lanske, B., Karaplis, A.C., Lee, K., Luz, A., Vortkamp, A., Pirro, A., Karperien, M., Defize, L.H.K., Ho, C., Mulligan, R.C., Abou-Samra, A.B., Jüppner, H., Segre, G.V., and Kronenberg, H.M. (1996). PTH/PTHrP receptor in early development and Indian hedgehog-regulated bone growth. Science, 223, 663-666.

Law, F., Bonjour, J.P., and Rizzoli, R. (1994). Transforming growth factor-β: A downregulator of the parathyroid hormone-related protein receptor in renal epithelial cells. Endocrinology 134 (5), 2037-2043.

Lee, C., Gardella, T.J., Abou-Samra, A.B., Nussbaum, S.R., Segre, G.V., Potts, J.J., Kronenberg, H.M., and Jüppner, H. (1994). Role of the extracellular regions of the parathyroid hormone (PTH)/PTH-related peptide receptor in hormone binding. Endocrinology 135 (4), 1488-1495.

Lee, C., Luck, M.D., Jüppner, H., Potts, J.T., Kronenberg, H.M., and Gardella, T.J. (1995). Homolog-scanning mutagenesis of the parathyroid hormone (PTH) receptor reveals PTH-(1-34) binding determinants in the third extracellular loop. Mol. Endocrinol. 9 (10), 1269-1278.

Lin, C., Lin, S.C., Chang, C.P., and Rosenfeld, M.G. (1992). Pit-1-dependent expression of the receptor for growth hormone releasing factor mediates pituitary cell growth (See comments). Nature 360 (6406), 765-768.

Lin, H.Y., Harris, T.L., Flannery, M.S., Aruffo, A., Kaji, E.H., Gorn, A., Kolakowski, Jr., L.F., Lodish, H.F., and Goldring, S.R. (1991). Expression cloning of an adenylate cyclase-coupled calcitonin receptor. Science 254, 1022-1024.

Lovenberg, T., Liaw, C., Grigoriadis, D., Clevenger, W., Chalmers, D., DeSouza, E., and Oltersdorf, T. (1995). Cloning and characterization of a functionally-distinct corticotropin-releasing factor receptor subtype from rat brain. Proc. Natl. Acad. Sci. USA 92, 836-840.

Lutz, E.M., Sheward, W.J., West, K.M., Morrow, J.A., Fink, G., and Harmar, A.J. (1993). The VIP2 receptor: Molecular characterization of a cDNA-encoding a novel receptor for vasoactive intestinal peptide. Febs. Lett 334 (1), 3-8.

Martin, T.J., Allan, E.H., Caple, I.W., Care, A.D., Danks, J.A., Diefenbach, J.H., Ebeling, P.R., Gillespie, M.T., Hammonds, G., Heath, J.A. et al. (1989). Parathyroid hormone-related protein: Isolation, molecular cloning, and mechanism of action. Recent Prog. Horm. Res. 45 (467), 467-502.

Martin, T.J. and Ebeling, P.R. (1990). A novel parathyroid hormone-related protein: Role in pathology and physiology. Prog. Clin. Biol. Res. 332 (1), 1-37.

Marx, U.C., Austermann, S., Bayer, P., Adermann, K., Ejchart, A., Sticht, H., Walter, S., Schmid, F. X., Jaenicke, R., Forssmann, W.G. and Rüsch, P. (1995). Structure of human parathyroid hormone 1-37 in solution. J. Biol. Chem. 270 (25), 15194-15202.

McCauley, L., Beecher, C., Melton, M., Werkmeister, J., Jüppner, H., Abou-Samra, A., Segre, G., and Rosol, T. (1994). Transforming growth factor-b1 regulates steady-state PTH/PTHrP receptor mRNA levels and PTHrP binding in ROS 17/2.8 osteosarcoma cells. Mol. Cell Endocrinol. 101, 331-336.

McCuaig, K.A., Clarke, J.C., and White, J.H. (1994). Molecular cloning of the gene encoding the mouse parathyroid hormone/parathyroid hormone-related peptide receptor. Proc. Natl. Acad. Sci. USA 91 (11), 5051-5055.

Moseley, J.M., Kubota, M., Diefenbach, J.H., Wettenhall, R.E., Kemp, B.E., Suva, L.J., Rodda, C.P., Ebeling, P.R., Hudson, P.J., Zajac, J.D. et al. (1987). Parathyroid hormone-related protein purified from a human lung cancer cell line. Proc. Natl. Acad. Sci. USA 84 (14), 5048-5052.

Nussbaum, S.R., Rosenblatt, M., and Potts, J.T. (1980). Parathyroid hormone–Renal receptor interactions. Demonstration of two receptor-binding domains. J. Biol. Chem. 255 (21), 10183-10187.

Nussbaum, S.R., Zahradnik, R.J., Lavigne, J.R., Brennan, G.L., Nozawa, U.K., Kim, L.Y., Keutmann, H.T., Wang, C.A., Potts, J.T., and Segre, G.V. (1987). Highly sensitive, two-site immunoradiometric assay of parathyrin, and its clinical utility in evaluating patients with hypercalcemia. Clin. Chem. 33 (8), 1364-1367.

Orloff, J.J., Kats, Y., Urena, P., Schipani, E., Vasavada, R.C., Philbrick, W.M., Behal, A., Abou-Samra, A.B., Segre, G.V., and Juppner, H. (1995). Further evidence for a novel receptor for amino-terminal parathyroid hormone-related protein on keratinocytes and squamous carcinoma cell lines. Endocrinology 136 (7), 3016-3023.

Pausova, Z., Bourdon, J., Clayton, D., Mattei, M.G., Seldin, M.F., Janicic, N., Riviere, M., Szpirer, J., Levan, G., Szpirer, C. Goltzman, D. and Hendy, G.N. (1994). Cloning of a parathyroid hormone/parathyroid hormone-related peptide receptor (PTHR) cDNA from a rat osteosarcoma (UMR 106) cell line: Chromosomal assignment of the gene in the human, mouse, and rat genomes. Genomics 20 (1), 20-26.

Philbrick, W.M., Wysolmerski, J.J., Galbraith, S., Holt, E., Orloff, J.J., Yang, K.H., Vasavada, R.C., Weir, E.C., Broadus, A.E., and Stewart, A.F. (1996). Defining the role of parathyroid hormone-related protein in normal physiology. Physiol Rev, 76, 127-173.

Potts, J.J., Tregear, G.W., Keutmann, H.T., Niall, H.D., Sauer, R., Deftos, L.J., Dawson, B.F., Hogan, M.L., and Aurbach, G.D. (1971). Synthesis of a biologically active N-terminal tetratriacontapeptide of parathyroid hormone. Proc. Natl. Acad. Sci. USA 68 (1), 63-67.

Ray, F.R., Barden, J.A. and Kemp, B.E. (1993). NMR solution structure of the [Ala26]parathyroid-hormone-related protein(1-34) expressed in humoral hypercalcemia of malignancy. Eur. J. Biochem. 211 (1-2), 205-211.

Reagan, J.D. (1994). Expression cloning of an insect diuretic hormone receptor: A member of the calcitonin/secretin receptor family. J. Biol. Chem. 269, 9-12.

Rizzoli, R. and Fleisch, H. (1986). Heterologous desensitization by 1,25-dihydroxyvitamin D-3 of cyclic AMP response to parathyroid hormone in osteoblastlike cells and the role of the stimulatory guanine nucleotide regulatory protein. Biochim. Biophys. Acta 887 (2), 214-221.

Rosenblatt, M. (1981). Parathyroid hormone: Chemistry and structure-activity relations. Pathobiol Annu. 11 (53), 53-86.

Schipani, E., Langman, C.B., Prafitt, A.M., Jensen, G.S., Kikuchi, S., Kooh, S.W., Cole, W.G., Jüppner, H. (1996). Constitutively activated receptors for parathyroid hormone and parathyroid hormone-related peptide in Jansen's metaphyseal chondrodysplasia N. Eng. J. Med. 335, 708-714.

Schipani, E., Karga, H., Karaplis, A., Potts, J.J., Kronenberg, H., Segre, G., Abou-Samra, A.B., and Jüppner, H. (1993). Identical complementary deoxyribonucleic acids encode a human renal and bone parathyroid hormone (PTH)/PTH-related peptide receptor. Endocrinology 132 (5), 2157-2165.

Schipani, E., Kruse, K., and Jüppner, H. (1995a). A constitutively active mutant PTH-PTHrP receptor in Jansen-type metaphyseal chondrodysplasia. Science 268 (5207), 98-100.

Schipani, E., Weinstein, L.S., Bergwitz, C., Iida-Klein, A., Kong, X.-F., Stuhrmann, M., Kuse, K., Whyte, M.P., Murray, T., Schmidtke, J., Van Dop, C., Brickman, A.S., Crawford, J.D., Potts, J. T., Kronenberg, H.M., Abou-Samra, A.B., Segre, G.V., and Jüppner, H. (1995b). Pseudohypoparathyroidism type Ib is not caused by mutations in the coding exons of the human parathyroid hormone (PTH)/PTH-related peptide receptor gene. J. Clin. Endocrinol. Metab. 80, 1611-1621.

Schneider, H.G., Allan, E.H., Moseley, J.M., Martin, T.J., and Findlay, D.M. (1991). Specific downregulation of parathyroid hormone (PTH) receptors and responses to PTH by tumour necrosis factor α and retinoic acid in UMR 106-06 osteoblastlike osteosarcoma cells. Biochem. J. 280, 451-457.

Segre, G.V., Niall, H.D., Habener, J.F., and Potts, J.J. (1974). Metabolism of parathyroid hormone: Physiologic and clinical significance. Am. J. Med. 56 (6), 774-784.

Segre, G.V., Rosenblatt, M., Reiner, B.L., Mahaffey, J.E., and Potts, J.T. (1979). Characterization of parathyroid hormone receptors in canine renal cortical plasma membranes using a radioiodinated sulfur-free hormone analogue. Correlation of binding with adenylate cyclase activity. J. Biol. Chem. 254 (15), 6980-6986.

Segre, G.V., Tregear, G.W., and Potts, J.T. (1975). Development and application of sequence-specific radioimmunoassays for analysis of the metabolism of parathyroid hormone. Methods Enzymol. 37B, 38-66.

Sharp, M.E. and Marx, S.J. (1985). Radioimmunoassay for the middle region of human parathyroid hormone: Comparison of two radioiodinated synthetic peptides. Clin. Chim. Acta 145 (1), 59-68.

Shigeno, C., Hiraki, Y., Westerberg, D.P., Potts, J.T., and Segre, G.V. (1988a). Parathyroid hormone receptors are plasma membrane glycoproteins with asparagine-linked oligosaccharides. J. Biol. Chem. 263 (8), 3872-3878.

Shigeno, C., Hiraki, Y., Westerberg, D.P., Potts, J.T., and Segre, G.V. (1988b). Photoaffinity labeling of parathyroid hormone receptors in clonal rat osteosarcoma cells. J. Biol. Chem. 263 (8), 3864-3871.

Smith, D.P., Zhang, X.Y., Frolik, C.A., Harvey, A., Chandrasekhar, S., Black, E.C., and Hsiung, H.M., (1996). Structure and functional expression of a complementary DNA for porcine parathyroid hormone/parathyroid hormone-related peptide receptor. Biochim Biophys Acta, 1307, 339-347.

Soifer, N.E., Dee, K.E., Insogna, K.L., Burtis, W.J., Matovcik, L.M., Wu, T.L., Milstone, L.M., Broadus, A.E., Philbrick, W.M., and Stewart, A.F. (1992). Parathyroid hormone-related protein. Evidence for secretion of a novel mid-region fragment by three different cell types. J. Biol. Chem.. 267 (25), 18236-18243.

Spengler, D., Waeber, C., Pantaloni, C., Holsboer, F., Bockaert, J., Seeburg, P.H., and Journot, L. (1993). Differential signal transduction by five splice variants of the PACAP receptor. Nature 365 (6442), 170-175.

Stewart, A.F., Wu, T., Goumas, D., Burtis, W.J., and Broadus, A.E. (1987). N-terminal amino acid sequence of two novel tumor-derived adenylate cyclase-stimulating proteins: Identification of parathyroid hormonelike and parathyroid hormone-unlike domains. Biochem. Biophys. Res. Commun. 146 (2), 672-678.

Suarez, F., Lebrun, J.J., Lecossier, D., Escoubet, B., Coureau, C., and Silve, C. (1995). Expression and modulation of the parathyroid hormone (PTH)/PTH-related peptide receptor messenger ribonucleic acid in skin fibroblasts from patients with type Ib pseudohypoparathyroidism. J. Clin. Endocrinol. Metab. 80 (3), 965-970.

Suva, L.J., Winslow, G.A., Wettenhall, R.E., Hammonds, R.G., Moseley, J.M., Diefenbach, J.H., Rodda, C.P., Kemp, B.E., Rodriguez, H., Chen, E.Y., Hudson, P.J., Martin, T.J., and Wood, W.I. (1987). A parathyroid hormone-related protein implicated in malignant hypercalcemia: Cloning and expression. Science 237 (4817), 893-896.

Takigawa, M., Kinoshita, A., Enomoto, M., Asada, A., and Suzuki, F. (1991). Effects of various growth and differentiation factors on expression of parathyroid hormone receptors on rabbit costal chondrocytes in culture. Endocrinology 129 (2), 868-876.

Thorens, B. (1992). Expression cloning of the pancreatic β cell receptor for the gluco-incretin hormone glucagonlike peptide 1. Proc. Natl. Acad. Sci. USA 89 (18), 8641-8645.

Tian, J., Smogorzewski, M., Kedes, L., and Massry, S.G. (1994). PTH-PTHrP receptor mRNA is downregulated in chronic renal failure. Am. J. Nephrol. 14 (1), 41-46.

Titus, L., Jackson, E., Nanes, M.S., Rubin, J.E., and Catherwood, B.D. (1991). 1,25-dihydroxyvitamin D reduces parathyroid hormone receptor number in ROS 17/2.8 cells and prevents the glucocorticoid-induced increase in these receptors: Relationship to adenylate cyclase activation. J. Bone. Miner. Res. 6 (6), 631-637.

Urena, P., Iida, K.A., Kong, X.F., Jüppner, H., Kronenberg, H.M., Abou-Samra, A.B., and Segre, G.V. (1994a). Regulation of parathyroid hormone (PTH)/PTH-related peptide receptor messenger ribonucleic acid by glucocorticoids and PTH in ROS 17/2.8 and OK cells. Endocrinology 134 (1), 451-456.

Urena, P., Kong, X.F., Abou-Samra, A.B., Jüppner, H., Kronenberg, H.M., Potts, J.J., and Segre, G.V. (1993). Parathyroid hormone (PTH)/PTH-related peptide receptor messenger ribonucleic acids are widely distributed in rat tissues. Endocrinology 133 (2), 617-623.

Urena, P., Kubrusly, M., Mannstadt, M., Hruby, M., Trinh, M.M., Silve, C., Lacour, B., Abou-Samra, A.B., Segre, G.V., and Drueke, T. (1994b). The renal PTH/PTHrP receptor is downregulated in rats with chronic renal failure. Kidney Int. 45 (2), 605-611.

Urena, P., Mannstadt, M., Hurby, M., Ferriera, A., Segre, G., and Drueke, T. (1994c). Parathyroidectomy does not prevent renal PTH/PTHrP receptor downregulation in uremic rats. J. Bone. Min. Res. 9 (Suppl. 1), B465.

Usdin, T.B., Gruber, C., and Bonner, T.I. (1995). Identification and functional expression of a receptor selectively recognizing parathyroid hormone, the PTH2 receptor. J. Biol. Chem. 270 (26), 15455-15458.

Vasicek, T.J., McDevitt, B.E., Freeman, M.W., Fennick, B.J., Hendy, G.N., Potts, J.J., Rich, A., and Kronenberg, H.M. (1983). Nucleotide sequence of the human parathyroid hormone gene. Proc. Natl. Acad. Sci. USA 80 (8), 2127-2131.

Weaver, D.R., Deeds, J.D., Lee, K., and Segre, G.V. (1995). Localization of parathyroid hormone-related peptide (PTHrP) and PTH/PTHrP receptor mRNAs in rat brain. Brain Res. Mol. Brain Res. 28 (2), 296-310.

Weir, E.C., Philbrick, W.M., Amling, M., Neff, L.A., Baron, R., and Broadus, A.E. Targeted expression of the parthyroid hormone-related peptide in chondrocytes causes chondrodysplasia and delayed endochondral bone formation. Proc. Natl. Acad. Sciences USA 93, 10240-10245.

Wray, V., Federau, T., Gronwald, W., Mayer, H., Schomburg, D., Tegge, W., and Wingender, E. (1994). The structure of human parathyroid hormone from a study of fragments in solution using 1H NMR spectroscopy and its biological implications. Biochemistry 33 (7), 1684-1693.

Xie, L.Y., Leung, A., Segre, G.V., and Abou-Samra, A.B. (1994). Downregulation of the PTH/PTHrP receptor by vitamin D3 in the osteoblastlike ROS 17/2.8 cells. Am. J. Physiol. 270, E654-E660.

Yamamoto, I., Potts, J.J., and Segre, G.V. (1988a). Glucocorticoids increase parathyroid hormone receptors in rat osteoblastic osteosarcoma cells (ROS 17/2). J. Bone Min. Res. 3(6), 707-712.

Yamamoto, I., Shigeno, C., Potts, J.J., and Segre, G.V. (1988b). Characterization and agonist-induced downregulation of parathyroid hormone receptors in clonal rat osteosarcoma cells. Endocrinology 122 (4), 1208-1217.

Yang, K.H., dePapp, A.E., Soifer, N.E., Dreyer, B.E., Wu, T.L., Porter, S.E., Bellantoni, M., Burtis, W.J., Insogna, K.L., Broadus, A.E., Philbrick, W.M., and Stewart, A.F. (1994). Parathyroid hormone-related protein: Evidence for isoform- and tissue-specific posttranslational processing. Biochemistry 33 (23), 7460-7469.

THE STRUCTURE AND MOLECULAR BIOLOGY OF THE CALCITONIN RECEPTOR

Steven R. Goldring

I. Introduction 188
II. Characterization and Cloning of Calcitonin Receptors 189
A. Cloning of the Porcine Renal Calcitonin Receptor 189
B. Structural Features of the Porcine Calcitonin Receptor and Relationship to the Calcitonin Receptor Family 190
III. Calcitonin Receptor Isoforms 193
A. Human Calcitonin Receptor Isoforms 193
B. Rat and Murine Calcitonin Receptor Isoforms 195
IV. Calcitonin Receptor Gene Organization and Chromosomal Localization 197
A. Calcitonin Receptor Gene Structure 197
B. Chromosomal Localization of the Human and Murine Calcitonin Receptor Genes 198
C. Receptor Subtypes 199
V. Functional Properties of the Cloned Calcitonin Receptors 200

Advances in Organ Biology
Volume 5A, pages 187-211.

ISBN: 0-7623-0390-5

A. Signal Pathway Coupling . . . 200
B. Ligand Cross Reactivity . . . 201
C. Receptor Isoforms . . . 201
VI. Regulation of Calcitonin Receptors . . . 203
VII. Summary . . . 205

I. INTRODUCTION

Calcitonin (CT) is a 32 amino acid peptide identified originally as a hypocalcemic factor (Copp et al., 1962; Hirsch et al., 1964). In mammals, CT is produced by the parafollicular cells of the thyroid gland in response to elevations in extracellular calcium. Its hypocalcemic activity is related to its capacity to inhibit osteoclast-mediated bone resorption and to enhance renal calcium excretion (Friedman and Raisz, 1965; Raisz and Niemann, 1967; Warshawsky et al., 1980). These effects are mediated by high affinity CT receptors expressed on osteoclasts and a subset of renal tubular cells (Warshawsky et al., 1980; Nicholson et al., 1986).

Although characterized originally based on its activity as a hypocalcemic factor, CT has an unexpectedly diverse spectrum of biological and pharmacological activities, including effects on many different organ systems not directly involved in the regulation of mineral ion homeostasis. These include effects on the central nervous, gastrointestinal, immune, and cardiovascular systems (Azria, 1989). The application of autoradiographic and radioligand binding techniques with iodinated CT have provided insights into the unexpected diversity of these activities. High affinity CT receptors have been identified in multiple different tissues, including localized regions of the central nervous system (Fischer et al., 1981; Goltzman, 1985), placenta (Nicholson et al., 1988), ovary (Azria, 1989), testis (Chausmer et al., 1980), spermatozoa (Silvestroni et al., 1987) and lymphocytes (Marx et al., 1974), as well as certain malignant cell lines (Findlay et al., 1980, 1981; Evain et al., 1981; Binet et al., 1985; Upchurch et al., 1986; Gattei et al., 1991). Whether the CT receptors in these additional tissue sites play a physiologically relevant functional role has not been established; however, these activities have been exploited for pharmacological manipulation. For example, CT has been used clinically to produce a form of centrally-mediated analgesia (Azria, 1989).

In addition to the presence of CT receptors in the central nervous system, there is also evidence that this hormone is produced locally within the brain (Fischer et al., 1981; Sexton et al., 1993). CT has also been identified in the central nervous system of primitive organisms such as the chordate, *ciona intestinalis,* and in the brain of the cyclostome, myxine (Azria, 1989) suggesting that it may have evolved originally as a hormone with principal functions as a neurotransmitter. This activity may have existed prior to its role in regulating mineral ion homeostasis. There is also evidence that CT may function as a regulatory hormone in development. For example, in *Xenopus* embryos, addition of CT to the ambient water of the developing eggs produces larvae with multiple defects in oral-facial architecture (Burgess, 1982, 1985). Recent studies by Gorn et al. (1995a) have suggested a role for the products of the CT gene in early vertebrate embryogenesis. These investigators observed that in zebra fish embryos, overexpression of procalcitonin in the two cell stage of development results in a variable axis duplication (Gorn et al., 1995a). More definitive insights into the potential functional role of CT in development will likely be gained by the mutation or deletion of this gene and/or its receptor with the techniques of recombinant DNA technology in transgenic animals utilizing homologous recombination in embryonic stem cells.

II. CHARACTERIZATION AND CLONING OF CALCITONIN RECEPTORS

A. Cloning of the Porcine Renal Calcitonin Receptor

Radioligand binding studies using iodinated CT provided the initial insights into the characteristics of the CT receptor. These results demonstrated that the receptors were functionally heterogeneous, particularly in the central nervous system (Fischer et al., 1981; Goltzman, 1985; Nakamuta et al., 1990; Sexton, 1991) where two distinct CT receptor subtypes were identified on the basis of their differential binding affinities for radioiodinated analogues of salmon CT. Additional support for the existence of CT receptor subtypes has been provided by the mapping of CT binding and calcitonin gene-related peptide (CGRP) binding sites in brain tissues. In most regions of the brain, the binding patterns of CT and CGRP are distinct. In certain restricted areas, however, there is high affinity binding of both ligands (Sexton, 1991). These findings are consistent with the existence of a unique CT-like receptor which has been termed C3 by Sexton et al. (1991). The recent cloning and characterization of CT receptors has provided insights in the structural and molecular basis for the apparent heterogeneity of CT receptors, as will be discussed below.

The initial insights into the structure of the CT receptor were provided by covalent crosslinking studies employing a photoactivated CT derivative

(Moseley et al., 1982, 1986). This approach identified a single binding component with a predicted molecular weight in the range of 80 kDa but did not yield sufficient protein for characterization of the actual amino acid sequence of the receptor. This information was provided by the cloning of a porcine CT receptor cDNA by Lin et al. (1991a,b) using a mammalian expression system in COS cells. To accomplish this, a size-fractionated cDNA library was prepared from the LLC-PK_1 cell line in the mammalian expression vector pcDNA-1 (Invitrogen). These porcine renal epithelial cells express large numbers of high-affinity CT receptors (Goldring et al., 1978) and provided an optimal source of cells expressing abundant CT receptor RNA. The library was screened using iodinated salmon CT and emulsion autoradiography and, with this approach, two positive clones encoding the same open reading frame were identified. Subsequently, the larger of the two clones was more extensively evaluated.

The evidence that the cloned cDNA encoded an authentic CT receptor was provided by examination of the binding kinetics and functional properties of the expressed protein after transfection of the cDNA in COS cells. These results confirmed the identity of the cDNA as an authentic CT receptor. To investigate the capacity of the expressed receptor protein to transduce second messenger responses after treatment with CT, cell lines stably expressing the porcine CT receptor cDNA were prepared. Previous studies had indicated that the CT receptor could couple to multiple signaling pathways, including those associated with adenylate cyclase and phospholipase C (Chakraborty et al., 1991). Incubation of the CT receptor-expressing cell lines with salmon CT induced a concentration-dependent increase in cAMP levels and parallel increases in cytosolic free $[Ca^{2+}]$ ($[Ca^{2+}]_i$) and inositol phosphate production (Force et al., 1992). These findings are consistent with association of the expressed cloned CT receptor cDNA with G proteins coupled to two independent signaling pathways, one linked to adenylate cyclase and the other to phospholipase C.

B. Structural Features of the Porcine Calcitonin Receptor and Relationship to the Calcitonin Receptor Family

Analysis of the predicted amino acid sequence of the CT receptor cDNA cloned from the LLC-PK_1 cells, revealed that it contained seven hydrophobic segments that could form transmembrane spanning α-helices (Figure 1). This pattern of structural organization, as well as the functional coupling of the receptor through G protein-coupled signaling pathways, suggested that the CT receptor belonged to the so-called super-family of G protein-

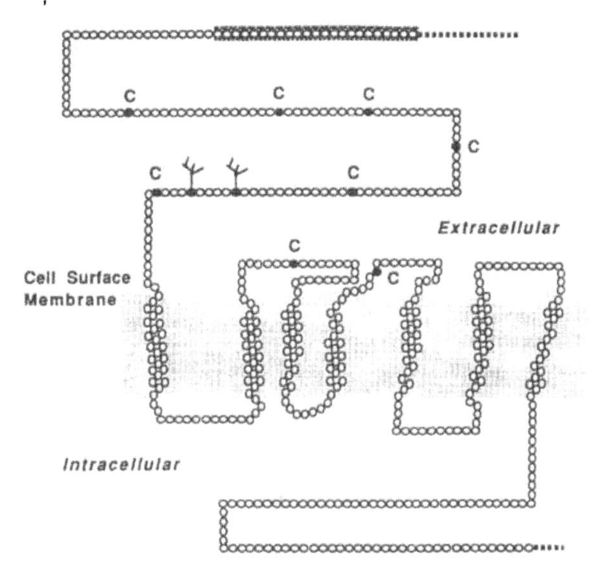

Figure 1. The cloned porcine CT receptor has seven hydrophobic regions that are predicted to form transmembrane spanning helices. A hydrophobic putative leader sequence at the amino-terminus is designated by the shaded area. Conserved cysteines are depicted by the closed circles. The shaded circles identify the potential N-linked glycosylation sites.

coupled receptors with seven transmembrane spanning helices (Strosberg, 1987, 1991; Dohlman et al., 1987, 1991; Strader et al., 1995). Surprisingly, when the predicted amino acid sequence of the cloned CT receptor was compared to the sequence of the other members of this super-family, it exhibited less than 12% identity, suggesting that it belonged to a distinct subfamily of G protein-coupled receptors with seven transmembrane spanning helices. Subsequently, the receptors for a variety of other peptide hormones have been shown to exhibit significant amino acid identity and structural homology to the CT receptor based on their similarity in amino acid sequence and commonalities in gene structure.

Examination of the structural features of the receptors that belong to the CT receptor family reveals that in addition to the presence of similarities in their amino acid sequences they share certain common structural and functional motifs. For example, they all contain an extended N-terminal domain with conserved cysteines and several potential N-linked glycosylation sites (Figure 1). In addition, the amino acid sequences comprising the last three membrane-spanning domains and the initial portion of the carboxy-terminal tail are more highly conserved. With respect to functional properties, in addition to coupling to adenylate cyclase through interaction with G proteins, many members of this family also are coupled to signaling pathways associated with phospholipase C. The members of this receptor family include the receptors for parathyroid hormone (PTH) parathyroid hormone related peptide (PTHrP) (Abou-Samra et al., 1992; Jüppner et al., 1991), corticotropin-releasing factor (Chen et al., 1993) and, in addition, the receptors for the glucagon family of peptides: glucagon (Jelinek et al., 1993), secretin (Ishihara et al., 1991), vasoactive intestinal peptide (Ishihara et al., 1992), glucagon-like peptide 1 (Thorens, 1992), growth hormone-releasing hormone (Mayo, 1992), and pituitary adenylate cyclase activating peptide (Pisegna and Wank, 1993). The most recent addition to this family is the so-called insect diuretic hormone receptor from adult *Manduca sexta* which stimulates fluid secretion and cAMP synthesis in the malphighian tubules (Reagan, 1994). The peptide that activates this receptor belongs to the corticotropin-releasing factor peptide family (Chen et al., 1993).

The mechanisms underlying the evolution and diversification of the receptors for the members of the CT receptor family are not known, however, the selection pressures for these events may have been provided when life forms moved from the oceans to the land environment. The ligands for these receptors, in addition to other activities, are involved in the regulation of ion transport in the gastrointestinal, renal, and skeletal systems. In the terrestrial habitats, in which calcium was not immediately available from the aqueous surroundings, new hormonal systems were necessary for rigorously regulating the levels of extracellular calcium and other ions and these hormones, and their receptors may have evolved to accommodate these unique demands.

Particularly surprising is the relationship of the receptors for PTH/PTHrP and CT. Although both ligands are peptide hormones, they are products of unrelated genes and exhibit contrasting biological activities (Rosenblatt et al., 1989). PTH is the principal hormone responsible for the regulation of extracellular calcium levels. It increases extracellular calcium levels by increasing osteoclastic bone resorption and decreasing renal calcium clearance. In contrast, CT decreases extracellular calcium levels by inhibiting osteoclastic bone resorption and enhancing renal calcium clearance. Most data indicate that the effects of PTH on osteoclasts are indirect and not mediated by receptors on osteoclasts or their precursors, but rather through interactions of PTH with receptors on osteoblast-lineage cells that, in turn, release products that are responsible for the recruitment and activation of osteoclasts (Raisz, 1988; Suda et al., 1992).

Results from studies in our laboratory would support this hypothesis. Using *in situ* hybridization techniques with ^{35}S-labeled restriction fragments prepared from the cloned human PTH/PTHrP and CT receptors, we have examined human bone tissues for the presence of mRNA encoding the cloned PTH and CT receptors (Harada et al., 1994). PTH receptor mRNA was abundantly expressed in bone marrow stromal cells adjacent to regions of osteoclastic bone resorption and in osteoblasts, but not in osteoclasts. CT receptor mRNA was detected exclusively in osteoclasts on bone surfaces.

III. CALCITONIN RECEPTOR ISOFORMS

A. Human Calcitonin Receptor Isoforms

The initial indication of the existence of CT receptor isoforms was provided by the characterization of a CT receptor cDNA that was cloned from a human small cell ovarian carcinoma cell line, BIN-67 by Gorn et al. (1992b). Analysis of the structure of this clone predicted a polypeptide of 490 amino acids which demonstrated many of the features characteristic of the cloned porcine CT receptor, including seven hydrophobic domains, an attenuated third intracellular loop, and a long amino-terminal presumed exocytoplasmic stretch with a hydrophobic N-terminal putative signal sequence (Figure 2). Cysteines in the N-terminal region and the first and second extracellular loops, as well as several of the N-linked glycosylation sites, were also conserved. The major area of divergence between the porcine and human receptors was in the first intracellular loop, where the human CT receptor contained a consecutive 48 nucleotide cassette that encoded a 16 amino acid insert that was not present in the porcine CT receptor. The presence of this inserted peptide sequence suggested that the human CT receptor cDNA could represent a transcript splice variant of the CT receptor.

Subsequently, Kuestner et al. (1994) cloned a human CT receptor cDNA from T47D mammary carcinoma cells that helped to more firmly establish the existence of CT receptor isoforms. Analysis of the predicted structural features of this cDNA revealed that, similar to the porcine CT receptor, it lacked the 16 amino acid insert in the first intracellular loop. The identification of two additional CT receptor isoforms cloned from human giant cell tumor of bone confirmed the presence of additional CT receptor isoforms (Gorn et al., 1995b). Both clones differed structurally from the ovarian and

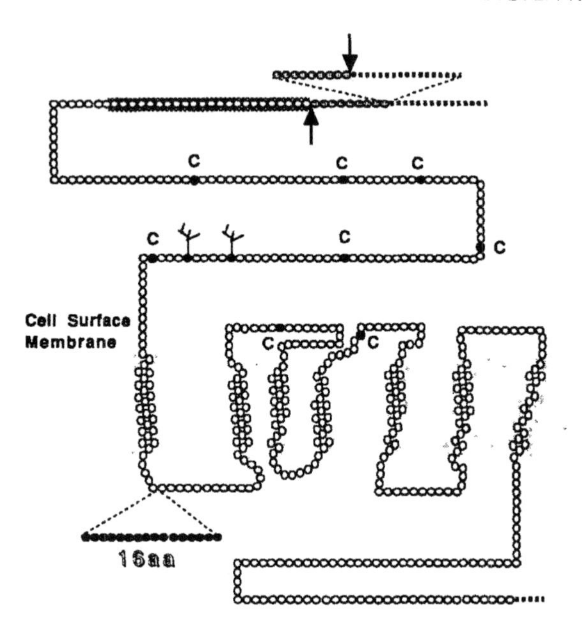

Figure 2. The cloned human ovarian CT receptor has seven hydrophobic regions that are predicted to form transmembrane spanning helices. The presence of a consecutive 48 nucleotide sequence that encodes a 16 amino acid insert in the first intracellular loop distinguishes the human CT receptor cDNA from the porcine CT receptor. In addition, the human ovarian CT receptor contains a 71 nucleotide insert at the amino-terminus that includes a potential translation initiation site. Arrows depict the potential translation initiation sites. Conserved cysteines are depicted by the closed circles.

breast carcinoma CT receptor cDNAs (Figure 3). The first clone (designated GC-10) lacked a 71 bp segment in the 5'-region that was present in the ovarian and breast carcinoma cDNAs. It was, however, otherwise identical to the ovarian clone in the more 3'-regions of the open reading frame, including the presence of the 48 bp insert in the putative first intracellular loop. The second human CT receptor cDNA from giant cell tumor (designated GC-2) lacked the 71 bp 5'-insert, but also lacked the 48 nucleotides that encoded the insert in the first intracellular loop. More recently, Moore et al. (1995) identified CT receptor cDNAs that encode similar isoforms and also detected a less common form with an inserted sequence in the first intracellular loop that contains an inframe stop codon that would lead to premature termination of the receptor at the carboxy-terminal portion of the first transmembrane domain. An additional isoform has been cloned from human mammary carcinoma MCF-7 cells (Albrandt et al., 1995) which has a truncation of the first 47 amino acids of the amino-terminal extracellular do-

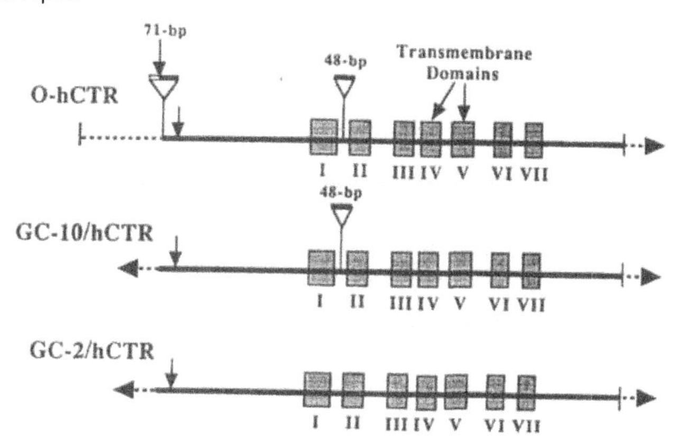

Figure 3. Comparison of the predicted structural features of cloned human CT receptor cDNAs. The CT receptor cloned from the ovarian carcinoma cell line, BIN-67, contains two inserts. The first insert consists of a 71 bp cassette localized to the 5'-end of the cDNA. It contains a potential translation initiation consensus designated by the arrow. This clone also contains a 48 bp insert in the first intracellular loop between the first and second transmembrane spanning helices. The two CT receptor cDNAs cloned from giant cell tumor of bone, GC-10 and GC-2, lack the 5' insert and are identical with the exception of the presence or absence of the 48 bp insert in the first intracellular loop.

main and, similar to the GC-2 clone, lacks the 48 bp insert in the first intracellular loop.

B. Rat and Murine Calcitonin Receptor Isoforms

The identification of additional cDNAs encoding rat and murine CT receptor cDNAs has firmly established the existence of CT receptor isoforms in different species, and analysis of the structure of the CT receptor gene confirms that, similar to the human CT receptor isoforms, they represent splice variants of a single gene generated by alternative RNA processing. Characterization of two CT receptors cloned from rat (Albrandt et al., 1993; Sexton et al., 1993) and murine (Yamin et al., 1994) brain cDNA libraries has provided evidence for the existence of a novel CT receptor isoform designated Clb. The most common and widely distributed CT receptor in rat and murine species, designated Cla, is similar in structure to the porcine and GC-2 human CT receptors in that it lacks the 48 bp insert in the first intracellular loop. The Clb isoform is identical to C1a except that it contains a 111 bp insert that is localized to the predicted first extracellular loop between the second and third transmembrane spanning helices (Figure 4).

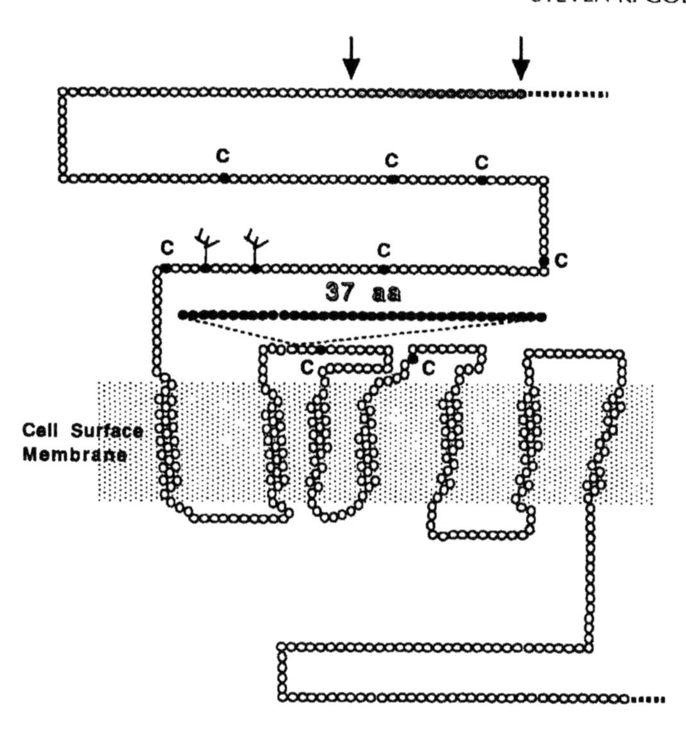

Figure 4. Predicted structural features of the cloned murine brain CT receptor cDNA (similar to the rat Clb isoform). The presence of a 111 bp cassette encoding a 37 amino acid insert in the first extracellular loop characterizes this receptor isoform and distinguishes it from the Cla receptor isoform. The closed circles indicate the conserved cysteines and the arrows depict the potential translation initiation consensus sequences.

Examination of the tissue distribution of the CT receptor isoforms using reverse transcriptase-polymerase chain reaction (RT-PCR) with RNA prepared from different tissue sites reveals that the Cla isoform is the most abundant form of the receptor and is more widely distributed than the Clb or other isoforms of the receptor. More extensive analyses are needed to specifically identify the pattern of tissue and cellular distribution of these less common isoforms of the CT receptor.

Results obtained with RT-PCR have established the widespread distribution of CT receptor mRNA. These findings have been confirmed by Northern analysis using 32p-labeled restriction fragments prepared from the CT receptor. With this approach, transcripts have been identified in brain, kidney, spinal cord, bone marrow, stomach, ovary, testis, skeletal muscle, and uterus. The widespread distribution of the CT receptor mRNA is consistent with the results of radioligand-binding studies that indicate that the receptor is expressed in diverse organs, including many in which the functional role of CT is not known.

IV. CALCITONIN RECEPTOR GENE ORGANIZATION AND CHROMOSOMAL LOCALIZATION

A. Calcitonin Receptor Gene Structure

The most extensive characterization of the structure of the CT receptor gene has been reported by Zolnierowicz et al. 1994) who characterized the porcine calcitonin receptor gene. Their analyses confirmed that the gene spanned approximately 70 kilobases and exhibited a complex organization consisting of multiple exons many of which were separated by lengthy introns. In preliminary studies, we have confirmed that the structure of the murine CT receptor is of comparable length and exhibits a similar structural organization (Yamin et al., 1993).

Zolnierowicz et al. 1994) identified at least 14 distinct exons in the porcine CT receptor gene. Although their analyses did not define all of the distinct inserts that characterize the thus far identified CT receptor isoforms, sequencing of the intron/exon boundaries revealed that the 3'-end of intron 7 had two potential splice sites separated by 48 nucleotides that fit the consensus sequence for acceptor sites. They used RT-PCR to amplify RNA from LLC-PK_1 cells and were able to identify in the reverse transcribed product the presence of this 48 nucleotide cassette which conformed to the insert present in the CT receptor cDNAs cloned from the human ovarian carcinoma cell line and human giant cell tumor of bone (GC-10). Subsequently, Nussenzveig et al. (1995) characterized the organization of the human CT receptor gene in this region and demonstrated that, in contrast to the porcine gene in which the two acceptor sites that included the 48 bp cassette are present in exon 8, in the human gene there is a distinct 48 nucleotide exon that is separated from the upstream exon 7 by over 6 kilobases and from the downstream exon 8 by over 1 kilobase. Splicing of this exon provides the mechanism for generating the two isoforms of the human receptor. These results suggest that, although the overall complex structural organization of the CT receptor is conserved across species, the specific structural features may vary. Of particular interest is the observation that the CT receptor isoform with the 48 bp insert is present in at least two different species. This provides evidence that this, and perhaps other CT receptor isoforms, may have some relevant physiological role, and that this function may provide the selective pressure for their evolutionary conservation.

B. Chromosomal Localization of the Human and Murine Calcitonin Receptor Genes

The technique of *in situ* hybridization with the ovarian CT receptor cDNA has been used to probe spread metaphase chromosomes in order to identify the chromosomal location of the human CT receptor (Gorn et al., 1995b). Results confirmed the presence of grains that localized to chromosome 7, confined predominantly to 7q22. A secondary peak was observed on chromosome 6. To more definitively establish the localization of the human CT receptor gene, human-hamster somatic cell hybrids that contained either human chromosome 6 or 7 were amplified by PCR using primers from the human CT receptor cDNA that corresponded to two exons identified in the mouse CT receptor (Yamin et al., 1994), and only the DNA from the hybrid that contained chromosome 7 yielded a product that hybridized with the CT receptor probe. These results are consistent with the presence of a single human CT receptor gene that resides on chromosome 7 at band q22. Nussenzveig et al. (1995) have used a similar strategy and have also identified a single human CT receptor gene that they localized to chromosome 7.

Further evidence for the existence of a single CT receptor gene is provided by the results of studies in mice in which interspecific back-cross analysis was used to map the murine CT receptor gene (Yamin et al., 1995). These results indicate that there is a single CT receptor gene that localized to the proximal region of mouse chromosome 6 linked to three previously characterized genes, *Met, Ptn,* and *Tcrb.* The CT receptor gene maps to a region of the composite map that contains one mutation, sightless (Sig). Animals homozygous for this mutation die at birth and have pronounced hydrocephaly and hindfoot abnormalities. A relationship between these developmental abnormalities and the presence of the CT receptor gene in this region of the mouse genome has not been established. It is of interest, however, that the addition of CT to developing *Xenopus* eggs results in the production of larvae with abnormalities of the central nervous system and oral-facial bony architecture, suggesting a possible role for CT in early development.

Characterization of the structural organization of the genes for several members of the CT receptor family, including the genes for growth hormone releasing hormone (Lin et al., 1993), PTH/PTHrP (Kong et al., 1994), pituitary adenylate cyclase activating peptide (Spengler et al., 1993), and corticotropin releasing factor (Chen et al., 1993), reveals that, similar to the CT receptor gene, they are large genes consisting of multiple exons and introns. Two or more receptor isoforms, generated by alternative splicing, have been identified for several of these genes. Analysis of the tissue distribution of these isoforms indicates that they are distributed in a cell-specific pattern, thus providing a unique mechanism for producing differential tissue responses to the respective ligands.

Analysis of the structural features of the genes for other members of G protein coupled receptors with seven transmembrane spanning helices reveals that there is precedence for the existence of isoforms. For example, alternative processing generates two forms of the D2 dopamine receptor (Neve et al., 1991) and the glutamate operated channel (Monyer et al., 1991). In contrast, for other members of this receptor superfamily, the heterogeneity in receptor structure is related to the existence of multiple distinct genes that encode the individual receptor subtypes. These include, for example, the adrenergic, muscarinic, seratonergic, vasopressin, and angiotensin H receptors (Strosberg, 1991). The similarity in the structural organization of the CT receptor family supports the hypothesis of an evolutionary relationship among the individual members of this family, and indicates their possible evolution from a common ancestral gene.

C. Receptor Subtypes

In addition to the presence of multiple CT receptor isoforms, there is precedence for the existence of CT receptor subtypes that are closely related to the CT receptor but are products of different genes. For example, Fluhmann et al. (1995) have identified a cDNA in a cDNA library of human cerebellum that encodes a receptorlike structure that is 56% identical in amino acid sequence to the human CT receptor. It is 91% identical to a CT receptorlike sequence in rat pulmonary blood vessels (Njuki et al., 1993). After expression, this receptor fails to bind CT or transduce second messenger responses when treated with calcitonin gene related peptide (CGRIP) or amylin, and thus far its ligand has not been identified. The pattern of expression of this cDNA which is localized to the brain, lung, heart and kidney differs from that of the CT receptor. Although it should appropriately be classified as a member of the CT receptor family, the ligand for this receptor has not yet been identified. In contrast, a subtype of the PTH receptor has been identified by Usdin et al. (1995). This receptor, designated PTH2, binds PTH with high affinity but does not interact with PTHrP. Thus, it appears to function exclusively as a receptor for PTH. Its level of expression in bone and kidney is low and, therefore, is not likely to mediate the major effects of PTH on calcium and phosphate metabolism. The PTH2 receptor is encoded by a gene that is distinct from the gene that encodes the PTH/PTHrP receptor and, thus, these two receptor cDNAs should more appropriately be considered as receptor subtypes rather than isoforms. Although only a single gene encoding a CT receptor has been identified, it is possible that, similar to the PTH receptor, additional members of the CT receptor family, i.e., subtypes, will be identified. The presence of receptor subtypes with differential ligand specificities and tissue distribution could provide an additional mechanism for producing tissue and cell-specific agonist-induced responses to CT.

V. FUNCTIONAL PROPERTIES OF THE CLONED CALCITONIN RECEPTORS

A. Signal Pathway Coupling

Previous studies have established that the members of the CT receptor family are coupled to adenylate cyclase, presumably through interactions with the G_s transducer molecule. Additional studies have suggested that many of these receptors, including the CT receptor, are also coupled to activation of phospholipase C which induces the breakdown of membrane phosphoinositol lipids to yield inositol 1,4,5-trisphosphate and diacylglycerol. These two second messenger molecules, in turn, stimulate Ca^{2+} release into the cytoplasm from intracellular pools, and activate the serine/threonine protein kinase C. The capacity of the members of the CT receptor family to activate both adenylate cyclase and phospholipase C is not unique to this family, since receptors for many other members of the G protein seven transmembrane superfamily also exhibit this functional property including, for example, the thyroid-stimulating hormone (TSH) receptor (Nagayama and Rapoport, 1992), the m2-muscarinic (Hosey, 1992), and the α adrenergic receptors (Lomasney et al., 1991).

The availability of the cloned CT receptor has permitted definitive demonstration that this receptor can couple to multiple independent signaling pathways, presumably by interaction with G proteins that are coupled to distinct second messenger responses (Chabre et al., 1992; Force et al., 1992; Teti et al., 1995). Several different approaches have been used to define the mechanisms by which these effects are mediated. One approach has involved the construction of chimeric CT/insulin growth factor II (IGF II) receptors in which nucleotides encoding the putative G-protein interacting sequences of the porcine CT receptor were subcloned into the intracellular G-protein binding region of a mutagenized IGF-II receptor resulting in replacement of the endogenous IGF-II G protein binding domain. Two distinct segments of the porcine CT receptor that have the capacity to interact with G proteins and transduce distinct second messenger responses were identified. One segment is localized to the third intracellular loop between the fifth and sixth transmembrane spanning helices, and the other is in the proximal portion of the carboxy-terminal intracellular domain.

B. Ligand Cross Reactivity

Previous studies had suggested that CGRP, which is a potent vasodilator, and amylin, a hormone that elevates blood glucose levels (two peptide hormones that are structurally related to CT) could bind with low affinity to the CT receptor and produce biological and pharmacological effects (Azria, 1989; Wimalawansa, 1990). These investigations had not established, however, whether the effects of these ligands were mediated directly through interaction with the CT receptor. The availability of the cloned CT receptor has provided a model for addressing this question. Employing cell lines stably expressing the CT receptor, it was possible to demonstrate that amylin or CGRP induce a concentration-dependent increase in cAMP levels via interaction with the cloned CT receptor (Force et al., 1992). The E.C. 50s for CGRP and amylin were almost two orders of magnitude higher compared to the binding affinity for CT, a result consistent with the observed potency of these ligands *in vivo*. Secretin, which exhibits minimal structural relationship to these ligands, but whose receptor is related to the CT receptor, also increases cAMP levels in cells stably expressing the CT receptor. Of interest, incubation of the cloned CT receptor with CGRP or amylin does not affect $[Ca2+]_i$ or inositol phosphate levels (Force et al., 1992).

C. Receptor Isoforms

Additional insights into the relationship between certain structural features of the CT receptor and several of its distinctive functional properties have been derived from the identification and characterization of the properties of multiple different receptor isoforms. Analysis of isoforms that differ from each other in their structural organization in specific regions of the receptor have led to the identification of two domains of the receptor that appear to be important in the determination of ligand binding specificity and/or signal transduction. The first domain is localized to the intracellular loop between the first and second transmembrane spanning helices; the second site is in the extracellular loop between the second and third transmembrane helices.

Insights into the functional importance of the intracellular region of the receptor between the first and second transmembrane spanning helices is provided by comparison of the properties of cloned human CT receptor cDNAs with or without a 48 bp cassette that encodes a 16 amino acid insert that localizes to this predicted intracellular region. Ligand binding and signaling properties of the isoforms differ markedly (Nussenzveig et al., 1994; Gorn et al., 1995b; Moore et al., 1995). The cDNA that encodes a CT receptor that contains the 16 amino acid insert in this region exhibits higher binding affinity for salmon and human CT compared to the cloned receptor without the insert. Because the insert is present in a region of the CT receptor that is predicted to be located intracellularly, it is unlikely that it interacts directly with peptide hormones such as CT, amylin, or CGRP. Rather, it is more likely that the presence of this amino acid insertion somehow alters the conformation of the receptor in such a way as to enhance the capacity of the extracellular domains of the receptor to interact with CT and related ligands.

In contrast to the effects of the 16 amino acid insert to enhance ligand binding affinity, its presence markedly attenuates the capacity of the expressed receptor to transduce second messenger responses via activation of adenylate cyclase or phopholipase C (Nussenzveig et al., 1994; Gorn et al., 1995b; Moore et al., 1995). This effect could be related to the capacity of this intracellular insertion to interfere with binding of the receptor to G proteins that transduce the second messenger responses. The amino acids in this insert do not demonstrate the characteristic motif of a G protein binding moiety (Okamoto et al., 1990; Okamoto and Nishimoto, 1991, 1992) and it is possible, therefore, that the presence of the insert modifies the receptor conformation in such a way as to impair coupling of other intracellular domains to G-proteins.

Insights into the functional importance of the extracellular regions of the receptor between the second and third transmembrane spanning helices is provided by comparison of the binding kinetics and patterns of signal transduction of CT receptor cDNAs cloned from rat and mouse brain. The rat Clb receptor (Sexton et al., 1993; Albrandt et al., 1995) and the mouse brain CT receptor cloned by Yamin et al. (1994) contain a 111 bp sequence that encodes a 37 amino acid insert that is predicted to localize to the first extracellular loop between the second and third transmembrane domains. The Cla receptor cloned from rat brain lacks this insert and is similar in structure to the porcine renal and human receptor cDNAs cloned from the T47D breast carcinoma cell line and from giant cell tumor of bone designated GC-2. Cells expressing the Cla receptor bind human and salmon CT with high affinity and increase cAMP levels after treatment with these ligands. In contrast, cells transfected with the murine brain or Clb cDNAs do not bind human CT and show minimal increases in cAMP levels after incubation with rat or human CT. The failure of the rat Clb isoform of the CT receptor to bind or transduce a second messenger response when exposed to rat CT is particularly surprising since this ligand would be expected to represent the principal endogenous ligand for this receptor. These findings suggest that the Clb isoform of the CT receptor may function as a receptor for some as yet unidentified ligand (Sexton et al., 1993). The existence of a salmon CT-like peptide has been described in the human brain (Fischer et al., 1981), and it is possible that this (or a related peptide) could function as the natural ligand for the Clb CT receptor.

An additional approach used to define the structural basis for the functional properties of the CT receptor has involved the construction of chimeric receptors and the use of receptor site directed mutagenesis. These studies have helped to localize the ligand binding domains of many of the members of the G-protein-coupled super-family of hormone receptors (Strosberg, 1987, 1991; Strader et al., 1989, 1995). In our studies, we have used chimeric receptors containing regions of the CT and PTH/PTHrP receptors and, in addition, have prepared chimeric ligands from portions of the CT and PTH ligands. This approach has helped to provide initial insights into the structural basis for ligand binding specificities and signal transduction (Jüppner et al., 1993; Bergwitz et al., 1995). Results indicate that the ligand binding specificity of the receptors for CT or PTH/PTHrP is determined by the extracellular portions of the receptors as well as certain membrane-embedded regions. The amino-terminal, extracellular domain of the receptors modulates the capacity of the receptors to couple to signal pathways and transduce second messenger responses, presumably by effects on binding affinity for the respective agonist.

VI. REGULATION OF CALCITONIN RECEPTORS

A significant problem with the clinical efficacy of CT in treating disorders of skeletal remodeling relates to the tendency of prolonged or continuous treatment with CT to induce a state of refractoriness to the effects of CT. This condition has been termed the "escape" or "plateau" phenomenon. Although CT effectively inhibits osteoclast-mediated bone resorption after acute administration, continuous exposure to this ligand under certain conditions can result in the development of a state of refractoriness in which bone resorption increases despite the continued presence of CT. This phenomenon was first observed in bone organ cultures (Werner et al., 1972; Tashjian et al., 1978), but also occurs *in vivo*.

Previous studies have suggested that the loss of responsiveness to CT may be multifactorial. In some subjects receiving salmon or eel CT, the development of refractoriness has been attributed to the presence of antibodies that interfere with the action of CT. In this situation, switching to the use of human CT restores biological and clinical efficacy. However, in others, it has been suggested that the loss of responsiveness to CT may be related to downregulation of CT receptors on osteoclasts and/or the possible recruitment of osteoclasts that lack CT receptors (Tashjian et al., 1978; Krieger et al., 1982; Nicholson et al., 1987).

The availability of cloned CT receptors has provided useful reagents for more rigorously defining the factors and conditions responsible for regulation of CT receptor expression, particularly with respect to the elucidation of the molecular mechanisms responsible for the phenomenon of escape. Several *in vitro* models have been employed in these studies (Lee et al., 1995; Takahashi et al., 1995; Wada et al., 1995; Ikegame et al., 1996). Based on observations in marrow culture systems in which osteoclast-like cells have been induced, the expression of the CT receptor appears to occur late in the sequence of osteoclast differentiation associated with the process of multinucleation and acquisition of the capacity to resorb bone (Takahashi et al., 1988; Hattersley and Chambers, 1989; Suda et al., 1992). These findings have been confirmed in studies employing RNA samples from murine (Lee et al., 1995; Wada et al., 1995) as well as human (Takahashi et al., 1995) bone marrow cultures that have been induced to form osteoclast-like multinucleated cells.

Results of studies employing isolated osteoclast-like cells from human giant cell tumor of bone or osteoclast-like cells generated in bone marrow culture systems indicate that continuous treatment with CT results in a rapid loss of CT mRNA and down-regulation of CT receptor binding activity (Lee et al., 1995; Takahashi et al., 1995; Wada et al., 1995). Removal of CT from the culture media results in a slow return of CT receptor message accompanied by restoration of CT binding activity. The pattern of CT receptor mRNA expression after CT treatment differs in nonosteoclast lineage cells. In these cells, CT treatment only partially decreases steady-state CT receptor mRNA levels, although CT binding activity is lost. These results indicate that the regulation of CT receptor transcription and mRNA processing may differ in osteoclasts and cells of nonosteoclast lineage.

VII. SUMMARY

The cloning of the CT receptor has helped to provide insights into the molecular basis for extreme diversity and pleiotropy of the *in vivo* activities of CT. These effects can be attributed to the widespread distribution of CT receptors, including tissues not directly involved in regulation of mineral ion homeostasis. In addition, the CT receptor exists in the form of multiple structurally and functionally distinct receptor isoforms. These isoforms exhibit differential ligand binding specificities and vary in their pattern of coupling to signal transduction pathways and second messenger responses. The individual isoforms are expressed in a tissue- and cell-specific fashion and this provides a unique system for producing organ-specific responses to CT and related ligands. The physiological role of many of these isoforms has not been established, but they could be important in mediating the effects of CT in tissues not directly involved in calcium and phosphorus metabolism. The availability of the reagents derived from the cloning of the CT receptor gene will also permit further elucidation of the molecular mechanisms responsible for the regulation of the CT receptor during osteoclast development. This should provide important insights into the mechanisms underlying the escape phenomenon and help to define the possible role of CT in development.

REFERENCES

Abou-Samra, A.B., Jüppner, H., Force, T., Freeman, M.W., Kong, X.F., Schipani, E., Urena, P., Richard, J., Bonventre, J.V., Potts, Jr., J.T., Kronenberg, H.M., and Sege, G.V. (1992). Expression cloning of a PTH/PTHrp receptor from rat osteoblastlike cells: A single receptor stimulates intracellular accumulation of both cAMP and inositol triphosphates and increases intracellular free calcium. Proc. Natl. Acad. Sci. USA 89, 2732-2736.

Albrandt, K., Brady, E.M.G., Moore, C.X., Mull, E., Sierzega, M.E., and Beaumont, K. (1995). Molecular cloning and functional expression of a third isoform of the human calcitonin receptor and partial characterization of the calcitonin receptor gene. 136, 5377-5384.

Albrandt, K., Mull, E., Brady, E.M.G., Herich, J., Moore, C.X., and Beaumont, K. (1993). Molecular cloning of two receptors from rat brain with high affinity for salmon calcitonin. FEBS Lett. 325, 225-232.

Azria, M. (Ed.) (1989). The Calcitonins: Physiology and Pharmacology. Karger, Basel, Switzerland

Bergwitz, C., Gardella, T.J., Flannery, M.R., Potts, J.T., Jr., Goldring, S.R., and Jüppner, H. (1995). Activation of PTH-CT receptor chimeras by hybrid ligands provides evidence for a common functional architecture of ligand-receptor pairs. J. Bone Min. Res. 10, S141.

Binet, E., Laurent, P., and Evain-Brion, D. (1985). F-9 embryonal carcinoma cells autocrine system: Correlation between immunoreactive calcitonin secretion and calcitonin receptor number. J. Cell Physiol. 124, 288-292.

Burgess, A.M.C. (1982). The developmental effect of calcitonin on the interocular distance in early *Xenopus* embryos. J. Anat. 135, 745-751.

Burgess, A.M.C. (1985). The effect of calcitonin on the prechordal mesoderm, neural plate, and neural crest *of Xenopus* embryos. J. Anat. 140, 49-55.

Chabre, O., Conklin, B.R., Lin, H.Y., Lodish, H.F., Wilson, E., Ives, H.E., Catanzariti, L., Hemmings, B.A., and Boume, H.R. (1992). A recombinant calcitonin receptor independently stimulates cAMP and Ca^{2+}/inositol phosphate signaling pathways. Mol. Endocrinol. 6, 551-556.

Chakraborty, M., Chatterjee, D., Kellokumpu, S., Rasmussen, H., and Baron, R. (1991). Cell cycle-dependent coupling of the calcitonin receptor to different G proteins. Science 251, 1078-1082.

Chausmer, A., Stuart, C., and Stevens, M. (1980). Identification of testicular cell plasma membrane receptors for calcitonin. J. Lab. Clin. Med. 96, 933-938.

Chen, R., Lewis, K.A., Perrin, M.H., and Vale, W.W. (1993). Expression cloning of a human corticotropin-releasing factor receptor. Proc. Natl. Acad. Sci. USA 90, 8967-8971.

Copp, D.H., Cameron, E.C., Cheney, B.A., Davidson, G.F., and Henze, K.G. (1962). Evidence for calcitonin-a new hormone from the parathyroid that lowers blood calcium. Endocrinology 70, 638-649.

Dohlman, H.G., Bouvier, M., Benovic, J.L., Caron, M.G., and Lefkowitz, R.J. (1987). The multiple membrane spanning topography of the β-2 adrenergic receptor. J. Biol. Chem. 262, 14282-14288.

Dohlman, H.G., Thomer, J., Caron, M.G., and Lefkowitz, R.J. (1991). Model systems for the study of seven-transmembrane-segment receptors. Annu. Rev. Biochem. 60, 653-688.

Evain, D., Binet, E., and Anderson, W.B. (1981). Alterations in calcitonin and parathyroid hormone responsiveness to adenylate cyclase in F-9 cells treated with retinoic acid and dibutryl cyclic AMP. J. Cell Physiol. 109, 453-459.

Findlay, D.M., DeLuise, M., Michelangeli, V.P., Ellison, M., and Martin, T.J. (1980). Properties of a calcitonin receptor and adenylate cyclase in BEN cells, a human cancer cell line. Cancer Res. 40, 1311-1317.

Findlay, D.M., Michelangeli, V.P., Moseley, J.M., and Martin, T.J. (1981). Calcitonin binding and deregulation by two cultured human breast cancer cell lines (MCF7 and T47D). J. Biochem. 196, 513-520.

Fischer, J.A., Tobler, P.H., Kaufmann, M., Born, W., Henke, H., Cooper, P.E., Sagar, S.M., and Martin, J.B. (1981). Calcitonin: Regional distribution of the hormone and its binding sites in the human brain and pituitary. Proc. Natl. Acad. Sci. USA. 78, 7801-7805.

Fluhmann, B., Muff, R., Hunziker, W., Fisher, J.A., and Born, W. (1995). A human orphan calcitonin receptorlike structure. Biochem. Biophys. Res. Commun. 206, 341-347.

Force, T., Bonventre, J.V., Flannery, M.R., Gorn, A.H., Yamin, M., and Goldring, S.R. (1992). A cloned porcine renal calcitonin receptor couples to adenylyl cyclase and phospholipase C. Am. J. Phys. 262, F1110-F1115.

Friedman, J. and Raisz, L.G. (1965). Thyrocalcitonin: Inhibitor of bone resorption in tissue culture. Science 150, 1465-1467.

Gattei, V., Bemabei, P.A., Pinto, A., Bezzini, R., Ringressi, A., Fonnigli, L., Tanini, A., Attadia, V., and Brandi, M.L. (1991). Phorbol ester-induced osteoclastlike differentiation of a novel human leukemic cell line (FLG 29.1). J. Cell Biol. 116, 437-447.

Goldring, S.R., Dayer, J.-M., Ausiello, D.A., and Krane, S.M. (1978). A cell strain cultured from porcine kidney increase cyclic AMP content upon exposure to calcitonin or vasopressin. Biochem. Biophys. Res. Commun. 83, 434-440.

Goltzman, D. (1985). Interaction of calcitonin and calcitonin gene-related peptide at receptor sites in target tissues. Science 227, 1343-1345.

Gorn, A.H., Lin, H.Y., Yamin, M., Auron, P.E., Flannery, M.R., Tapp, D.R., Manning, C.A., Lodish, H.F., Krane, S.M., and Goldring, S.R. (1992). The cloning, characterization, and expression of a human calcitonin receptor from an ovarian carcinoma cell line. J. Clin. Invest. 90, 1726-1735.

Gorn, A.H., Rudolph, S.M., and Fishman, M.C. (1995a). Procalcitonin may encode a developmental signal for axial patterning in zebrfish embryos. J. Bone Min. Res. 10, S156.

Gorn, A.H., Rudolph, S.M., Flannery, M.R., Morton, C., Weremowicz, S., Wang, J.-T., Krane, S.M., and Goldring, S.R. (1995b). Expression of two human skeletal calcitonin receptor isoforms cloned from a giant cell tumor of bone. J. Clin. Invest. 95, 2680-2691.

Harada, Y., Wang, J.T., Gorn, A.H., Gravallese, E.M., Thornhill, T.S., Jasty, M., Harris, W.H., Juppner, H., Goldring, S.R. (1994). Identification of the cell types responsible for bone resorption in rheumatoid arthritis. Arthritis Rheum. 37 (Suppl. 9), S211.

Hattersley, G. and Chambers, T.J. (1989). Generation of osteoclastic function in mouse bone marrow cultures: Multinuclearity and tartrate-resistant acid phosphatase are unreliable markers for osteoclastic differentiation. Endocrinology 124, 1989.

Hirsch, P.F., Voelkel, E.F., and Munson, P.L. (1964). Thyrocalcitonin: Hypocalcemic, hypophosphatemic principle of the thyroid gland. Science 146, 412-413.

Hosey, M.M. (1992). Diversity of structure, signaling, and regulation within the family of muscarinic cholinergic receptors. FASEB J. 6, 845-852.

Ikegame, M., Rakopoulos, M., Martin, T.J., Moseley, J.M., and Findlay, D.M. (1996). Effects of continuous calcitonin treatment on osteoclastlike cell development and calcitonin receptor expression in mouse marrow cultures. J. Bone. Min. Res. 11, 456-465.

Ishihara, T., Nakamura, S., Kaziro, Y., Takahashi, T., Takahashi, K., and Nagata, S. (1991). Molecular cloning and expression of a cDNA encoding the secretin receptor. EMBO J. 10, 1635-1641.

Ishihara, T., Shigemoto, R., Mori, K., Takahashi, K., and Nagata, S. (1992). Functional expression and tissue distribution of a novel receptor for vasoactive intestinal polypeptide. Neuron 8, 811-819.

Jelinek, L.J., Lok, S., Rosenberg, G.B., Smith, R.A., Grant, F.J., Biggs, S., Bensch, P.A., Kuijper, J.L., Sheppard, P.O., Sprecher, C.A., O'Hara, P.J., Foster, D., Walker, K.M.,

Chen, L. H.J., Mckeman, P.A., and Kindsvogel, W. (1993). Expression cloning and signalling properties of the rat glucagon receptor. Science 259, 1614-1616.

Jüppner, H., Abou-Samra, A., Freeman, M., Kong, X.F., Schipani, E., Richards, J., Kolakowski, Jr., L.F., Hock, J., Potts, Jr., J.T., Kronenberg, H.M., and Sege, G.V. (1991). A G protein-linked receptor for parathyroid hormone and parathyroid hormone-related peptide. Science 254, 1024-1026.

Jüppner, H., Flannery, M.S., McClure, I., Abou-Samra, A.B., Gardella, T.J., and Goldring, S.R. (1993). Chimeras between the receptors for calcitonin (CTR) and parathyroid hormone (PTH)/PTH-related peptide (PTHRP) define functionally important domains. J. Bone Min. Res. 8, S183.

Kong, X.-F., Schipani, E., Lanske, B., Joun, H., Karperien, M., Defize, L.H.K., Juppner, H., Potts, Jr, J.T., Kronenberg, H.M., and Abou-Samra, A.B. (1994). The rat, mouse, and human genes encoding the receptor for parathyroid hormone and parathyroid hormone-related peptide are highly homologous. Biochem. Biophys. Res. Commun. 200, 1290-1299.

Krieger, N.S., Feldman, R.S., and Tashjian, Jr., A.H. (1982). Parathyroid hormone and calcitonin interactions in bone: Irradiation-induced inhibition of escape in vitro. Calcif. Tissue Int. 34, 197-203.

Kuestner, R.E., Elrod, R.D., Grant, F.J., Hagen, F.S., Kuijper, J.L., Matthewes, S.L., O'Hara, P.J., Sheppard, P.O., Stroop, S.D., Thompson, D.L., Whitmore, T.E., Findlay, D.M., Houssami, S., Sexton, P.M., and Moore, E.E. (1994). Cloning and characterization of an abundant subtype of the human calcitonin receptor. Molecular Pharmacology 46, 246-255.

Lee, S.K., Goldring, S.R., and Lorenzo, J. (1995). Expression of the calcitonin receptor in bone marrow cell cultures and in bone: A specific marker of the differentiated osteoclast that is regulated by calcitonin. Endocrinology 136, 4572-4581.

Lin, H.Y., Harris, T.L., Flannery, M.S., Aruffo, A., Kaji, E.H., Gorn, A., Kolakowski, L.F., Jr., Lodish, H.F., and Goldring, S.R. (1991a). Expression cloning of an adenylate cyclase-coupled calcitonin receptor. Science 254, 1022-1024.

Lin, H.Y., Harris, T.L., Flannery, M.S., Aruffo, A., Kaji, E.H., Gorn, A., Kolakowski, Jr., L.F., Yamin, M., Lodish, H.F., and Goldring, S.R. (1991b). Expression cloning and characterization of a porcine renal calcitonin receptor. Trans. Assoc. Am. Phys. CIV, 265-272.

Lin, S.-C., Lin, C.R., Gukovsky, I., Lusis, A.J., and Rosenfeld, M.G. (1993). Molecular basis of the little mouse phenotype and implications for cell-specific growth. Nature 364, 208-213.

Lomasney, J.W., Cotecchia, S., Leftkowitz, R.J., and Caron, M.G. (1991). Molecular biology of α-adrenergic receptors: Implications for receptor classification and for structure-function relationships. Biochem. Biophysica Acta 1095, 127-139.

Marx, S.J., Aurbach, G.D., Gavin, J.R., and Buell, D.W. (1974). Calcitonin receptors on cultured human lymphocytes. J. Biol. Chem. 249, 6812-6816.

Mayo, K.E. (1992). Molecular cloning and expression of a pituitary-specific receptor for growth hormone-releasing hormone. Mol. Endocrinol. 6, 1734-1744.

Monyer, H., Seeburg, P.H., and Wisden, W. (1991). Glutamate-operated channels: Developmentally early and mature forms arise by alternative splicing. Neuron 6, 799-810.

Moore, E.E., Kuestner, R.E., Stroop, S.D., Grant, F.J., Matthewes, S.L., Brady, C.L., Sexton, P.M., and Findlay, D.M. (1995). Functionally different isoforms of the human calcitonin receptor result from alternative splicing of the gene transcript. Mol. Endocrinology 9, 959-968.

Moseley, J.M., Findlay, D.M., Martin, T.J., and Gorman, J.J. (1982). Covalent crosslinking a photoactive derivative of calcitonin to human breast cancer cell receptors. J. Biol. 257, 5846-5851.

Moseley, J.M., Smith, P., and Martin, T.J. (1986). Identification of the calcitonin receptor by chemical crosslinking and photoaffinity labeling in human cancer cell lines. J. Bone Min. Res. 1, 293-297.

Nagayama, Y. and Rapoport, B. (1992). The thyrotropic receptor 25 years after its discovery: New insights after its molecular cloning. Mol. Endocrinol. 6, 145-156.

Nakamuta, H., Orlowski, R.C., and Epand, R.M. (1990). Evidence for calcitonin receptor heterogeneity: Binding studies with nonhelical analogs. Endocrinol. 127, 163-169.

Neve, K.A., Neve, R.L., Fidel, S., Janowsky, A., and Higgins, G.A. (1991). Increased abundance of alternatively spliced forms of D2 dopamine receptor MRNA after denervation. Proc. Natl. Acad. Sci. USA 88, 2802-2806.

Nicholson, G.C., D'Santos, C.S., Evans, T., Moseley, J.M., Kemp, B.E., Michelangeli, V.P., and Martin, T.J. (1988). Human placental calcitonin receptors. Biochem. J. 250, 877-882.

Nicholson, G.C., Moseley, J.M., Sexton, P.M., Mendelsohn, F.A.O., and Martin, T.J. (1986). Abundant calcitonin receptors in isolated rat osteoclasts. J. Clin. Invest. 78, 355-360.

Nicholson, G.C., Moseley, J.M., Yates, J.P., and Martin, T.J. (1987). Control of cyclic Adenosine 3',5'-monophosphate production in osteoclasts: Calcitonin-induced persistent activation and homologous desensitization of adenylate cyclase. Endocrinol 120, 1902-1908.

Njuki, F., Nicholl, C.G., Howard, A., Mak, J.C.W., Barnes, P.J., Girgis, S.I., and Legon, S. (1993). A new calcitonin-receptor-like sequence in rat pulmonary blood vessels. Clinical Science 85, 385-388.

Nussenzveig, D.R., Mathew, S., and Gershengom, M.C. (1995). Alternative splicing of a 48-nucleotide exon generates two isoforms of the human calcitonin receptor. Endocrinology, 2047-2051.

Nussenzveig, D.R., Thaw, C.N., and Gershengom (1994). Inhibition of inositol phosphate second messenger formation by intracellular loop one of a calcitonin receptor. J. Biol. Chem. 269, 28123-28129.

Okamoto, T., Katada, T., Murayama, Y., Ui, M., Etsuro, O. and Nishimoto, I. (1990). A simple structure encodes G protein-activating function of the IGF R/mannose 6-phosphate receptor. Cell 62, 709-717.

Okamoto, T. and Nishimoto, I. (1991). Analysis of stimulation–G protein subunit coupling by using active insulinlike growth factor II receptor peptide. Proc. Natl. Acad. Sci. USA. 88, 8020-8023.

Okamoto, T. and Nishimoto, I. (1992). Detection of G protein-activator regions in M4 subtype muscarinic, cholinergic and α-adrenergic receptors based upon characteristics in primary structure. J. Biol. Chem. 267, 8342-8346.

Pisegna, J.R. and Wank, S.A. (1993). Molecular cloning and functional expression of the pituitary adenylate cyclase-activating peptide type I receptor. Proc. Nat. Acad. Sci. 90, 6345-6349.

Raisz, L.G. (1988). Local and systemic factors in the pathogenesis of osteoporosis. N. Engl. J. Med. 318, 818-828.

Raisz, L.G. and Niemann, I. (1967). Early effects of parathyroid hormone and thyrocalcitonin on bone in organ culture. Nature 214, 486-487.

Reagan, J.D. (1994). Expression cloning of an insect diuretic hormone receptor. J. Biol. Chem. 269, 1-4.

Rosenblatt, M., Kronenberg, H.M., and Potts, Jr., J.T., (1989). Parathyroid hormone physiology, chemistry, biosynthesis, secretion, metabolism, and mode of action. In: Endocrinology. (DeGroot, Ed.), pp. 848-891. W.B. Saunders, Philadelphia.

Sexton, P.M. (1991). Central nervous system binding sites for calcitonin and calcitonin gene-related peptide. Molecular Neurobiology 5, 251-273.

Sexton, P.M., Houssami, S., Hilton, J.M., O'Keeffe, L.M., Center, R.J., Gillespie, M.T., Darcy, P., and Findlay, D.M. (1993). Identification of brain isoforms of the rat calcitonin receptor. Mol. Endocrinol. 7, 815-821.

Silvestroni, L., Menditto, A., Frajese, G., and Gnessi, L. (1987). Identification of calcitonin receptors in human spermatozoa. J. Clin. Endocrinol. Metab. 65, 742-746.

Spengler, D., Waeber, C., Pantaloni, C., Holdboer, F., Bockaert, J., Seeburg, P.H., and Joumot, L. (1993). Differential signal transduction by five splice variants of the PACAP receptor. Nature 365, 170-175.

Strader, C.D., Fong, T.M., Graziano, M.P., and Tota, M.R. (1995). The family of G-protein coupled receptors. FASEB J. 9, 745-754.

Strader, C.D., Sigal, I.S., and Dixon, R.A.F. (1989). Structural basis of β-adrenergic receptor function. FASEB 3, 1825-1832.

Strosberg, A.D. (Ed.) (1987). The Molecular Biology of Receptors. Techniques and Applications of Receptor Research., Ellis Horwood.

Strosberg, A.D. (1991). Structure/function relationship of proteins belonging to the family of receptors coupled to GTP-binding proteins. Eur. J. Biochem. 196, 1-10.

Suda, T., Takahashi, N., and Martin, T.J. (1992). Modulation of osteoclast differentiation. Endocr. Rev. 13, 66-80.

Takahashi, E., Goldring, S.R., Katz, M., Hilsenbeck, S., Williams, R., and Roodman, G.D. (1995). Downregulation of calcitonin receptor mRNA expression by calcitonin during human osteoclastlike cell differentiation. J. Clin. Invest. 95, 167-171.

Takahashi, N., Akatsu, T., Sasaki, T., Nicholson, G.C., Moseley, J.M., Martin, T.J., and Suda, T. (1988). Induction of calcitonin receptors by 1 α 25-dihydroxyvitamin D3 in osteoclastlike multinucleated cells formed from mouse bone marrow cells. Endocrinology 123, 1504.

Tashjian, A.M., Wright, D.R., Ivey, J.L., and Pont, A. (1978). Calcitonin binding sites in bone: Relationships to biological response and escape. Recent Progress in Hormone Res. 34, 285-334.

Teti, A., Paniccia, R., and Goldring, S.R. (1995). Calcitonin increases cytosolic free calcium concentrations via capacitative calcium influx. J. Biol. Chem. 270, 16666-16670.

Thorens, B. (1992). Expression cloning of the pancreatic β cell receptor for the gluco-incretin hormone glucagonlike peptide 1. Proc. Natl. Acad. Sci. USA 85, 8641-8645.

Upchurch, K.S., Parker, L.M., Scully, R.E., and Krane, S.M. (1986). Differential cyclic AMP responses to calcitonin among human ovarian carcinoma cell lines: A calcitonin-responsive line derived from a rare tumor type. J. Bone Min. Res. 1, 299-304.

Usdin, T.B., Gruber, C., and Bonner, T.I. (1995). Identification and functional expression of a receptor selectively recognizing parathyroid hormone, the PTH2 receptor. J. Biol. Chem. 270, 15455-15458.

Wada, S., Martin, T.J., and Findlay, D.M. (1995). Homologous regulation of the calcitonin receptor in mouse osteoclastlike cells and human breast cancer T47D cells. Endocrinology 136, 2611-2621.

Warshawsky, H., Goltzman, D., Rouleau, M.F., and Bergeron, J.J.M. (1980). Direct in vivo demonstration by radioautography of specific binding sites for calcitonin in skeletal and renal tissues of the rat. J. Cell Biol. 85, 682-694.

Werner, J.A., Gorton, S.J., and Raisz, L.G. (1972). Escape from inhibition of resorption in cultures of fetal bone treated with calcitonin and parathyroid hormone. Endocrinology 90, 752-759.

Wimalawansa, S.J. (1990). Calcitonin: Molecular biology, physiology, pathophysiology, and its therapeutic uses. In: Advances in Bone Regulatory Factors: Morphology, Biochemistry, Physiology, and Phartncology. (A. Pecile and B. Bernard, Ed.), pp. 121-160. Plenum Press, England.

Yamin, M., Flannery, M.R., Tapp, D.R., Gorn, A.H., Krane, S.M., and Goldring, S.R. (1993).

Analysis of a unique murine brain calcitonin receptor (CTR) cDNA and preliminary characterization of the murine CTR gene; evidence for the existence of functionally distinct isoforms of the CTR. J. Bone. Min. Res. 8, S129.

Yamin, M., Gorn, A.H., Flannery, M.R., Jenkins, N.A., Gilbert, D.J., Copeland, N.G., Tapp, D.R., Krane, S.M., and Goldring, S.R. (1994). Cloning and characterization of a mouse brain calcitonin receptor complementary deoxyribonucleic acid and mapping of the calcitonin receptor gene. Endocrinology 135, 2635-2643.

Zolnierowicz, S., Cron, P., Solinas-Toldo, S., Fries, R., Lin, H.Y., and Hemmings, B.A. (1994). Isolation, characterization, and chromosomal localization of the porcine calcitonin receptor gene. J. Biol. Chem. 269, 19530-19538.

THE VITAMIN D RECEPTOR: DISCOVERY, STRUCTURE, AND FUNCTION

J. Wesley Pike

I. Introduction 214
II. The Discovery of the Vitamin D Receptor 214
III. Biochemical Properties and Organization of the Vitamin D Receptor 216
IV. The Structural Gene for the Vitamin D Receptor 217
A. Molecular Cloning 217
B. Member of the Intracellular Receptor Superfamily of Genes 219
C. Vitamin D Receptor Domains 220
V. Functional Analysis of the Vitamin D Receptor 225
A. The Osteocalcin Gene as a Model for the Mechanism of Action of Vitamin D . . 225
B. DNA Binding *In Vitro* 227
C. Polarity of DNA Binding 229
D. Transactivation by the Vitamin D Receptor 230
E. Role of $1,25(OH)_2D_3$ in Vitamin D Receptor Activation 231
VI. Concluding Remarks 232
VII. Summary 233

Advances in Organ Biology
Volume 5A, pages 213-241.

ISBN: 0-7623-0390-5

I. INTRODUCTION

Steroid, thyroid, and vitamin D hormones are known to exert profound regulatory control over complex gene networks. Many, if not all, of these actions occur at the level of the cellular genome (O'Malley et al., 1969; Beato, 1989). The products of these modulated genes control processes essential to cellular growth and differentiation as well as to extracellular homeostasis. The actions of these signals are mediated by unique intracellular receptors (Evans, 1988; O'Malley, 1990; Beato et al., 1995; Mangelsdorf and Evans, 1995; Mangelsdorf et al., 1995). The presence of these signaling receptor molecules in cells and tissues represents a principal although not exclusive determinant of response to a particular hormone. These soluble transducers of hormonal and environmental signals are members of one of the largest gene regulating families of latent transcription factors that acquire unique regulatory capacity upon activation by their respective cognate ligands. Certain members of this class of regulators, however, do not appear to require ligands and represent modulators controlled by the temporal expression of their chromosomal genes or through other activating pathways. These receptors are currently termed orphan receptors as it is formally possible that unknown or nontraditional ligands will be discovered that are capable of their activation. While hormone interaction with the classic receptors that bind ligand has been well characterized, the events that follow association of ligand with its receptor remain less well understood. Significant advances have been made during the past decade, however, in elucidating key events associated with both activation and repression by this receptor family.

II. THE DISCOVERY OF THE VITAMIN D RECEPTOR

While several lines of evidence had suggested that vitamin D or an active metabolite of the vitamin might function to regulate the expression of genes, the first successful studies that hinted at the existence of a binding protein or vitamin D receptor (VDR) were carried out in 1969 by Haussler and Norman (1969). Following these pioneering studies in target intestinal tissues of the chicken, more definitive evidence for the VDR began to emerge. In studies by Brumbaugh and Haussler (1974) and Lawson and Wilson (1974), the protein nature of the receptor was established through proteolytic digestion studies and equilibrium sedimentation analysis. Further evaluation of the binding properties of the receptor by Brumbaugh and Haussler (1975) suggested an affinity of the protein for labeled 1,25-dihydroxyvitamin D_3 (1,25$(OH)_2D_3$) to be in the low nanomolar range. Finally, *in vitro* experiments carried out in 1975 enabled the conclusion that the cytosol-derived VDR which displayed a sedimentation coefficient (S) of about 3.5 could bind to chromatin fractions in the presence of the hormonal ligand (Brumbaugh and Haussler, 1975). These studies collectively provided definitive support for the existence of the VDR and prompted the suggestion that a cytoplasmic VDR translocated to the nucleus as seen in Figure 1, pathway A.

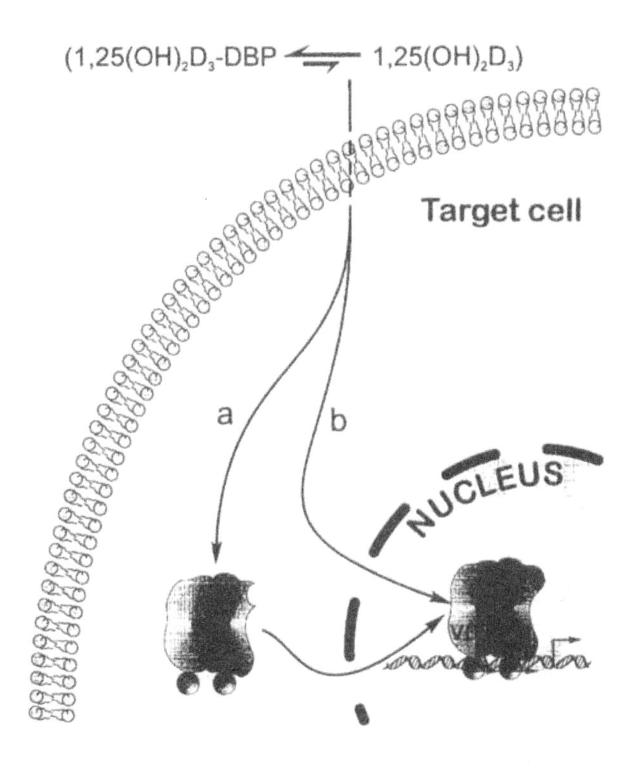

Figure 1. Model for the molecular mechanism of action of the vitamin D hormone. 1,25$(OH)_2D_3$ dissociates from serum vitamin D binding protein (DBP), enters the cell by diffusion, and interacts with the VDR. Activation by the ligand leads to the interaction of the VDR with responsive genes and the modulation of gene expression. VDR-MU or VDR modulatory unit is comprised of one VDR molecule and an associated protein which is exemplified by but not restricted to a retinoid X receptor isoform. (**A**) An early model wherein the VDR is shown located in the cytoplasm, undergoes cytoplasmic to nuclear translocation upon ligand activation, and eventually binds to the regulatory region of a modulated gene. (**B**) A current view of the location of the VDR wherein the receptor is located in the nucleus and following ligand activation bound to the regulatory region of a vitamin D modulated gene.

III. BIOCHEMICAL PROPERTIES AND ORGANIZATION OF THE VITAMIN D RECEPTOR

The VDR exists in relatively low abundance in target tissues and cultured cells, an abundance consistent with the fact that it is a potent transcription regulatory molecule (Pike and Haussler, 1979; Haussler et al., 1981). Estimates of receptor abundance range from under 500 VDR molecules/cell to over 25,000 copies of VDR/cell (2 to 100 fmoles/mg protein) depending upon the cell type or cell line examined, and up to a pmole/mg of protein in certain tissue extracts. These estimates of abundance are based upon the capacity of extracts to bind $1,25(OH)_2D_3$ and thus presumably reflect active functional receptor. The existence of this wide range in VDR abundance suggests that those cells with higher VDR content may be more highly responsive to the $1,25(OH)_2D_3$ hormone than those with lower levels of expression. While the latter concept is intuitive and some evidence for this postulate exists, it is important to note that numerous other factors also play an important role in individual cellular responsivity to the hormone. These factors include cellular capacity to internalize and subsequently metabolize $1,25(OH)_2D_3$; differential activation events that may modulate VDR activity in a cell- or tissue- specific manner (phosphorylation of VDR?); the nature, availability, and concentration of numerous partner proteins that are required for gene activation; and finally the accessibility and regulable nature of specific genes. These as well as additional events contribute significantly to the sensitivity and biological responsivity of a particular cell to $1,25(OH)_2D_3$.

Both physical and functional properties of the VDR emerged immediately following the discovery of the protein. With the exception of molecular mass, no evidence emerged to suggest that the VDR differed significantly in biochemical properties from cell to cell or from species to species. Sedimentation analysis revealed a protein of 3 to 3.7S that exhibited an elongated shape. Gel filtration estimates of the protein ranged from 50,000 to 70,000 Da depending upon species. Perhaps the most important biochemical and functional property of the VDR was its capacity to bind $1,25(OH)_2D_3$ with both high affinity and selectivity (Kream et al., 1977; Wecksler et al., 1978; Mellon and DeLuca, 1979; Wecksler and Norman, 1980). In that regard, numerous experiments were performed that led to the determination of an equilibrium dissociation constant of 10^{-10}M for its natural ligand $1,25(OH)_2D_3$. VDR also binds $1,25(OH)_2D_3$ precursors as well as other metabolites of vitamin D with substantially lower affinity (Kream et al., 1977; Wecksler et al., 1978). The contribution of both the 25-hydroxyl and the 1α-hydroxyl groups on the $1,25(OH)_2D_3$ molecule in specific high affinity binding to VDR has been studied extensively (Wecksler et al., 1978).

While numerous additional properties of the receptor emerged in the late 1970s, the discovery that the VDR exhibited DNA-binding capabilities consistent with its role as a nuclear transcription factor represented a considerable advance (Pike and Haussler, 1979). A much more precise understanding of the properties of VDR DNA-binding emerged following the identification of specific DNA binding sites (vitamin D response elements, VDREs) located adjacent to promoter for vitamin D-inducible genes (see below). Nevertheless, the finding that the VDR bound to nonspecific DNA not only set the stage for ensuing studies aimed at a preliminary understanding of the structural organization and function of the VDR, but provided the initial means whereby the VDR could be isolated in quantities of sufficient purity to generate valuable immunological reagents. These reagents were ultimately useful in further characterization of the receptor and in the molecular cloning of its structural gene. Two important observations on the DNA-binding properties of the VDR were made that are believed to reflect the role of $1,25(OH)_2D_3$ in the receptor activation process (Hunziker et al., 1983; Pike and Haussler, 1983). First, VDR was capable of binding DNA in the absence of ligand, an observation that suggested that, unlike the latent DNA binding properties of the sex steroid receptors, the VDR was fully capable of binding to DNA in the absence of $1,25(OH)_2D_3$. Second, the "affinity" of the receptor for DNA was quantitatively increased following complex formation with $1,25(OH)_2D_3$. This latter property implied that the structure of the VDR or perhaps the composition of the active receptor was transformed in the presence of the hormonal ligand. More recent studies to be described later in this chapter more precisely define the nature of the effects of ligand on the VDR and its DNA binding capabilities.

IV. THE STRUCTURAL GENE FOR THE VITAMIN D RECEPTOR

A. Molecular Cloning

The anti-VDR monoclonal antibody 9A7 was utilized by McDonnell et al. (1987) to screen randomly primed chicken intestinal cDNA expression libraries prepared in a viral expression system. A single cDNA clone was selected that produced a protein that exhibited immunological crossreactivity not only with the screening probe but with an additional anti-VDR antibody as well. The DNA sequence of this cDNA clone and several additional clones recovered through cross hybridization revealed them to contain a sequence that exhibited a high degree of homology to a domain located in the genes for glucocorticoid, estrogen, and progesterone receptors. This domain was initially believed to be related to that found in the transcription factor TFIIIA and was hypothesized to be responsible for receptor DNA binding. An important repeating module within this region that occurred twice in the domain for the receptors but multiple times in TFIIIA was a zinc-coordinated DNA-binding finger structure. It is now known that there exists minimal structural relatedness between the zinc fingers of TFIIIA and the DNA-binding domain of the receptors (Berg, 1989), although it is clear that this region is responsible for DNA binding (see below). The presence of this domain and its reactivity to the 9A7 antibody (known to interact adjacent to the DNA binding domain of the VDR) (Pike et al., 1988) led to initial confidence that these initial cDNAs represented a portion of the transcript encoding the VDR. Subsequent hybridization-selected *in vitro* translation techniques using these clones substantiated the authenticity of the cDNA clones (McDonnell et al., 1987). The recovery of these cDNAs constitute the molecular cloning of the VDR. More importantly, they provided the first direct evidence of a structural relationship between the VDR and other bona fide members of the steroid receptor family of genes (McDonnell et al., 1988).

The recovery of the first cDNA for the VDR from the chicken enabled subsequent recovery of full-length VDR cDNA transcripts from human (Baker et al., 1988) and rat (Pike et al., 1988) tissue sources. The rat intestinal VDR was also cloned independently using monoclonal antibody selection by Burmester et al. (1988). Subsequently, cDNA sequences for the VDR have been reported from mouse (Kamei et al., 1995), Japanese quail (Elaroussi et al., 1992), and from *Xenopus* (Li et al., 1996). Recovery of a cDNA transcript from the human HL-60 cell line (Goto et al., 1992) revealed virtual identity to that of the original human VDR cDNA cloned by Baker et al. (1988). This provided important evidence that the VDR involved in cellular differentiation was not different from that involved in calcium metabolism. A comparison of the sequences of the VDR from the above reported cDNAs has revealed that, in addition to several domains of homology with other members of the nuclear receptor family, the VDR is also highly conserved across tissue sources and species. The overall homology of rat, mouse, and avian receptors to that of human VDR is 79, 86, and 66%, respectively. However, within specific domains such as the DNA binding domain this homology rises to above 95%. One substantial difference, noted early based upon receptor protein size, is the variability in the number of amino acid residues located amino-terminal to the DNA binding domain. This region varies from 21 amino acids in the human VDR (the smallest of the VDRs) to approximately 57 amino acids in the chicken protein (McDonnell et al., 1987; Baker et al., 1988). Additional inserts within the hinge region are also evident in the rat (Burmester et al., 1988). Two initiation sites also appear to be present in the human cDNA sequence; the second start site, however, lies only three codons downstream of the first (Baker et al., 1988). Whether both are used to produce two proteins of almost equivalent mass in the human (424 and 427 amino acids) is unknown. The molecular cloning of VDR transcripts thus confirmed a number of initial observations made at the protein level. The molecular cloning also confirmed and extended the original hypothesis that the VDR was a member of the steroid receptor family and enabled significant structure/function analyses to be conducted.

B. Member of the Intracellular Receptor Superfamily of Genes

The cloning of glucocorticoid (Hollenberg et al., 1985) and estrogen (Green et al., 1986) receptors in 1985 and 1986 represented the first of a long series of successful efforts to clone each of the known intracellular receptor genes. Over 150 members of this intracellular receptor gene family now exist (Mangelsdorf and Evans, 1995). The size of this particular gene family eclipses that of any other currently known transcription factor group. It suggests that the common structural motifs within this family that include DNA binding domains paired with activity regulating domains under the control of chemically diverse small signaling molecules have been highly successful evolutionarily. These hormonal ligand-activated transcription factors control an incredibly wide range of biological processes that include fundamental growth and differentiation functions in the developing animal as well as a wide range of physiological and homeostatic functions in the adult. As can be seen in Figure 2, the genes are derived from both vertebrates as well as invertebrates. In cases such as that for the retinoic acid (RAR) and retinoid X (RXR) receptors, multiple genes exists. The expression of these genes is tissue specific, suggesting that certain receptor subtypes play more significant roles in the biology of the tissue than others. Finally, as stated earlier, while many receptors are activated by ligands, some of which are hormonal in nature while others are intracellularly derived, the vast majority of the members of this family are regulators that do not appear to be activated by ligands.

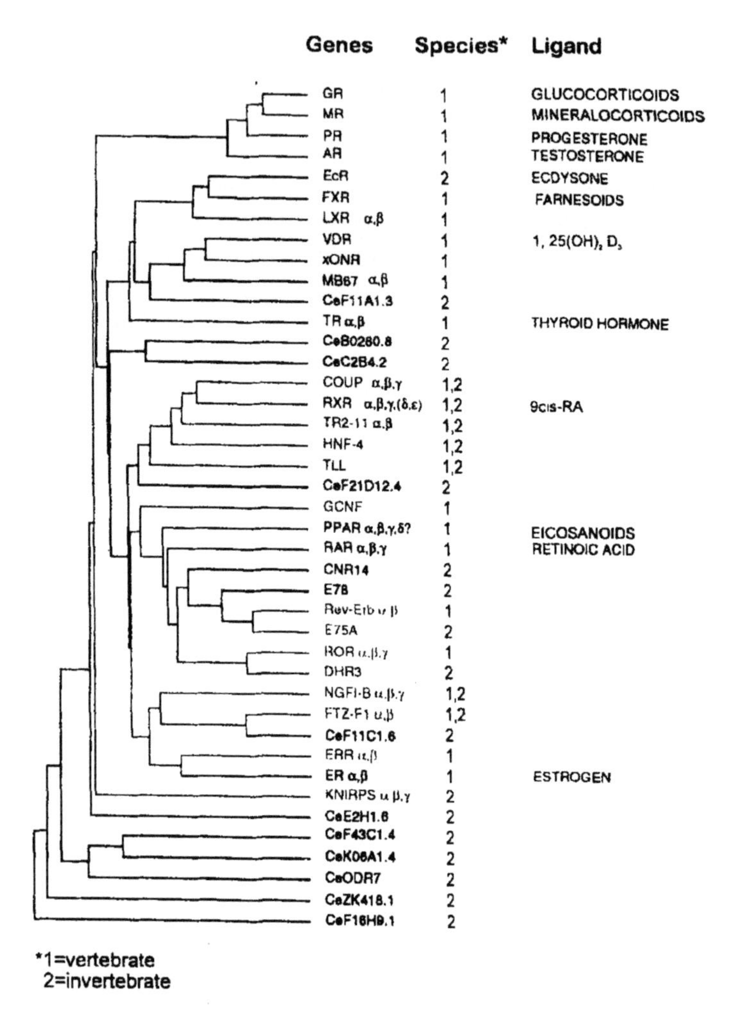

Figure 2. Members of the nuclear receptor superfamily of genes. The figure illustrates the relationship between cloned nuclear receptor family members based upon multiple sequence alignments. This figure represents a modification of that found in Mangelsdorf et al., (1995), wherein further details can be obtained.

C. Vitamin D Receptor Domains

The cloning of the estrogen receptor in 1986 led to the general designation of receptor segments as A, B, C, D, E, and F domains (Green et al., 1986). As illustrated in Figure 3A, segment A/B includes residues amino-terminal to the DNA-binding domain. The C region comprises the highly conserved DNA binding domain. The hinge region which lies between the C domain and the ligand binding domain is designated the D domain. Finally, the carboxy-terminal region that contains the ligand-binding domain in the activated receptors is termed the E or E/F domain. Three regions of homology among members of the nuclear receptor family exist within the E region. The F domain is not conserved and exhibits extensive variability; the VDR appears not to contain the F domain segment. Figure 3B depicts the domain structure of mammalian VDRs. As can be seen, the A/B domain is highly abbreviated relative to other members of the nuclear receptor family, particularly those for the sex and adrenal steroids. The C region that comprises the DNA binding domain of the VDR represents the most highly conserved domain across all the nuclear receptors (McDonnell et al., 1989). This domain is the hallmark of the nuclear receptor family. The D domain within the VDR appears to link in a highly flexible fashion the DNA-

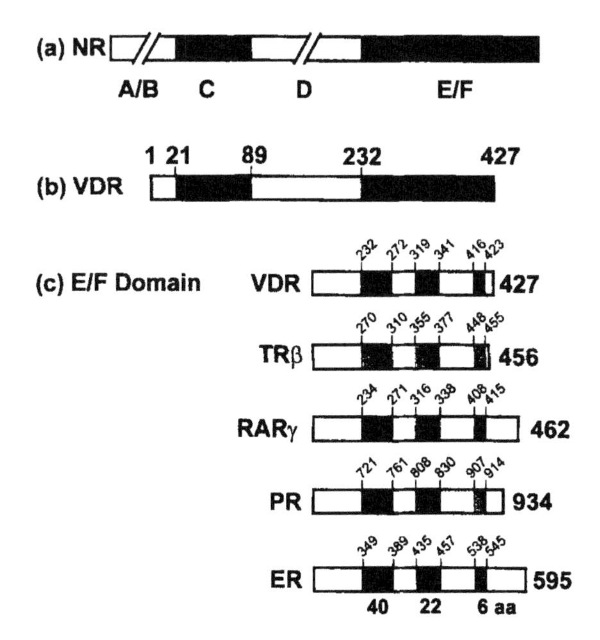

Figure 3. Functional domain structure of the nuclear receptor superfamily. (**A**) The nuclear receptors (NR) are separated into five regions designated A/B, C, D, and E/F. (**B**) The residue boundaries of corresponding regions within the VDR. (**C**) Three regions of sequence homology and residue boundaries within the E/F domain of the nuclear receptor family. Thyroid receptor b, TRb; retinoid receptor a, RARa; progesterone receptor, PR; and estrogen receptor, ER. Functions associated with these regions include: transactivation (A/B), DNA binding (C), flexible hinge (D), dimerization, ligand binding, transactivation, and repression (E/F).

binding and hormone-binding domains (McDonnell et al., 1989). Conservation within this segment is low among the VDRs from different species and is not conserved either in length or sequence with that of other members of the nuclear receptor family (Baker et al., 1988; Burmester et al., 1988). Finally, the E/F domain of the VDR contains the $1,25(OH)_2D_3$-binding function of the protein (McDonnell et al., 1989). In addition to ligand-binding, this domain serves as a highly complex protein/protein interface for a series of additional proteins of varied function (MacDonald et al., 1995; Jin and Pike, 1996; Jin et al., 1996). These features highlight the role of the VDR as a recruitment center for other transcription factors that contribute to the VDR's DNA binding function as well as its transcriptional-regulating functions. As these biological activities are consistent with that of all the members of the receptor gene family, it is not surprising that this extended region contains several subdomains documented in Figure 3C that exhibit moderate conservation across the entire transcription factor gene family.

VDR domain C encodes the DNA binding domain. This domain consists of two similar modules each comprised of a zinc-coordinated finger structure. Each zinc atom is tetrahedrally coordinated through four highly conserved cysteine residues and serves to stabilize the finger structure itself. As stated earlier, these finger modules are structurally unrelated to the zinc fingers found in TFIIIA wherein the zinc atom is coordinated through two cysteines and two histidines (Berg, 1988, 1989). While the two zinc modules of the VDR appear to be highly related structurally, they are not equivalent topologically due to the chirality of the residues in each module that coordinate the zinc atom (Berg, 1988). More importantly, the function of each of these modules in DNA binding is known to be substantially different. Thus, although it is possible that the evolution of two exons that encode these modules began from a common ancestral gene through duplication and then diverged as a result of different selective pressures, the more likely possibility is that the two modules evolved independently. Whereas the amino-terminal module functions to direct specific DNA-binding in the major grove of the DNA binding site, the carboxy-terminal module serves as a dimerization interface for interaction with a partner protein (Mader et al., 1989; Umesono and Evans, 1989). In the case of the VDR, at least one of these protein partners is RXR (see below). As the three-dimensional structure of the DNA binding domain of several of the receptors has been determined through both nuclear magnetic resonance spectroscopy as well as through x-ray crystallography, our understanding of the structural organization of these modules as well as the mechanisms by which they function to interact with DNA is beginning to significantly advance (Hard et al., 1990; Schwabe et al., 1990, 1993; Luisi et al., 1991; Lee et al., 1993; Rastinjead et al., 1995).

The E/F region of the VDR represents a multifunctional domain that exerts absolute regulatory control on the DNA binding as well as transcriptional modifying properties of the VDR. The switch that converts this latent transcription factor into an active gene regulator is 1,25$(OH)_2D_3$. Indeed, 1,25$(OH)_2D_3$ binding is hypothesized to induce conformational changes in the ligand-binding domain of the VDR, much like that of all other small molecule hormones in this class. It is these conformational changes that presumably are responsible for the reduction in proteolytic sensitivity observed in early studies by Allegretto and Pike (1985) and Allegretto et al., (1987). A more sensitive version of the proteolytic digestion assay has been recently developed. Application of this assay to an analysis of VDR structural domains has confirmed that VDR binding to 1,25$(OH)_2D_3$ results in the appearance of a proteolytically resistant 34 kDa polypeptide which is largely comprised of the E/F domain of the VDR (Peleg et al., 1995). Whether ligand-induced conformational changes are restricted to the E/F region of the VDR is unknown. Conversion to the active form following hormone binding results in increased formation of dimers that comprise the fundamental DNA binding subunit structure of the VDR as well as exposure of additional regions of the molecule which ultimately allow contact with the core transcriptional machinery. Much is known regarding the former; little is currently known regarding the latter. In addition, it is likely that other protein surfaces are affected that play a direct role in modifying both negatively as well as positively the activity of the receptor in perhaps cell-specific and gene promoter-specific ways. The complexity of the ligand-regulated domain coupled to the mechanistic similarities by which this family of receptors modify gene expression lead to a prediction that several regions of homology should exist. As observed in Figure 3C, at least three regions of the VDR E/F domain exhibit significant sequence similarity within the E/F domain of other nuclear receptors; these regions are in fact conserved among all family members (Wang et al., 1989). Functional mapping studies have suggested that amino acids in the first two amino terminally located regions of homology are essential for dimer formation by the VDR (Nakajima et al., 1994; Jin et al., 1996). The E/F region has been structurally elucidated through determination of the three-dimensional structure of the ligand binding domains of RXRα (Bourguet et al., 1995), RARγ (Renaud et al., 1995), and TRα1 (Wagner et al., 1995). The latter two receptors were crystallized in the presence of ligand (holodomains) whereas the RXRα structure was determined in the absence of ligand (apodomain). Twelve α-helices (H1–H12) arranged as an antiparallel α-helical sandwich comprise the bulk of the structure of each of the receptors. It is likely that the VDR will be arranged in a structurally similar although not identical manner. These three-dimensional structures support H9 and H10 as essential for the formation of dimers of RAR, RXR, and TR. While functional studies support the essentially of these helical sequences in RAR, RXR, and TR interactions, however, H9 and H10 may not be insufficient for VDR dimerization. This conclusion is supported by a more complete evaluation of the dimerization properties of the carboxy-terminal E/F region of the VDR (Jin et al., 1996). These results suggest that while significant insights will be gained through structural modeling of the VDR based upon other members of the nuclear receptor family, true insights will require direct structural determination of the VDR.

An additional function inherent to the E/F region of the VDR is an activation function termed AF2. This function lies within the smallest most carboxy-terminal third homology domain of the receptors and virtually at the carboxy-terminus of the VDR (Danielian et al., 1992; Whitfield et al., 1995) (see Figure 3c). The core of this function appears to be associated with H12, although it is clear from activity studies involving mutagenesis that additional components including those in the amino-terminal homology domain of the E/F regions also play a role (Jin et al., 1996). It is important to note here, however, that neither study rules out the possibility of protein destabilization as a principal mechanism for the loss of transcriptional function. Nevertheless, H12 is clearly repositioned back upon the hydrophobic core of the E domain in the crystal structure. The interaction between H12 and other α-helices may account for loss of transcriptional capacity following mutagenesis of residues well upstream of H12. Determination of the three-dimensional structure of the VDR will no doubt answer these important questions.

Binding of $1,25(OH)_2D_3$ substantially alters the conformation of the E/F region of the VDR. This hypothesis was suggested by very early studies demonstrating a decrease in the lability of the VDR (McCain et al., 1978), more recent studies which suggest that the presence of $1,25(OH)_2D_3$ stabilizes the VDR (Sone et al., 1990; Santiso-Mere et al., 1993), and through the demonstration that hormone binding increases the resistance of the receptor to proteolytic degradation (Peleg et al., 1995). Despite these indirect observations, the actual structural rearrangement that occurs upon ligand occupancy can only be inferred based upon the rearrangements that occur in presently crystallized holoreceptors. Likewise, the nature of the ligand binding pocket of the VDR remains undefined. Loss of function studies demonstrate that mutagenesis of a number of amino acids throughout the entire E/F region can produce an alteration in $1,25(OH)_2D_3$ binding. This suggests that the three-dimensional binding pocket is comprised of many segments spanning the entire carboxy-terminal domain. Since the function of $1,25(OH)_2D_3$ is to act as a small molecular switch capable of receptor activation, it should not be surprising that the binding of ligand should induce sweeping conformational changes in the E/F domain. The exact positioning of the $1,25(OH)_2D_3$ molecule within the ligand pocket awaits the solution structure of the VDR. It should be anticipated, however, that at least some of the residue contact sites that serve to stabilize the natural hormone within the pocket will not be identical to those that stabilize the binding of lower affinity vitamin D metabolites such as $24R,25(OH)_2D_3$ or synthetic analogues such as $1\alpha,25(OH)_2$-16-ene-23 yne-D_3, and 20-epi- $1,25(OH)_2D_3$ (Norman, 1995). The theoretical result of occupancy of the VDR by ligands other than $1,25(OH)_2D_3$ is a spectrum of receptor conformations potentially capable of unique and perhaps selective biological actions. The potential for this to occur may account, at least in part, for the interesting and selective biological actions of an array of new metabolites and analogues (Bikle, 1994; Brown et al., 1994). While theoretical for the VDR, the concept of ligand-induced conformational specificity that results in unique biological actions is now well established for several of the sex steroids (Tzukerman et al., 1994; McDonnell et al., 1995). The discovery of ligands with unique properties has resulted in a broad array of therapeutic opportunities.

V. FUNCTIONAL ANALYSIS OF THE VITAMIN D RECEPTOR

A. The Osteocalcin Gene as a Model for the Mechanism of Action of Vitamin D

Osteocalcin is a small abundant noncollagenous bone protein whose exact function remains unclear. Genetic ablation of the osteocalcin gene suggests that this osteoblast-specific protein contributes to the density and structural integrity of bone (Ducy et al., 1996). Despite the uncertainty surrounding the function of osteocalcin, a broad number of cytokines, growth factors, and systemic hormones control its expression. One of the most potent regulators of osteocalcin production is $1,25(OH)_2D_3$ (Price and Baukol, 1980; Pan and Price, 1986). This fact together with the cloning of the human osteocalcin gene and its promoter in 1987 (Celeste et al., 1986) provided a unique blend of opportunities for researchers to study the molecular determinants through which $1,25(OH)_2D_3$ modulated the expression of this gene. Initial investigations demonstrated that the activity of $1,25(OH)_2D_3$ on the osteocalcin promoter was direct (Kerner et al., 1989; McDonnell et al., 1989; Morrison et al., 1989). Thus, introduction of a plasmid containing a large upstream fragment of the human osteocalcin promoter (fused to the reporter gene chloramphenicol acetyltransferase) into osteoblast-like osteosarcoma cells revealed that the activity of the chimeric gene was sensitive to $1,25(OH)_2D_3$. An abbreviated version of this upstream sequence containing less than several hundred base pairs exhibited only basal activity (Yoon et al., 1988). Kerner et al. (1989) initially localized the *cis*-acting element to a region approximately 500 bp upstream of the transcriptional start site. This study and an additional one by Ozono et al. (1990) helped define the first VDRE as a directly repeated hexanucletotide sequence separated by three base pairs. Parallel studies using the rat osteocalcin gene promoter led to a similar conclusion regarding the organizational motif of the VDRE (Lian et al., 1989; Demay et al., 1990; Terpening et al., 1991). These studies collectively provided the first insight regarding a specific DNA sequence that mediated vitamin D-inducible action.

Since these experiments were carried out, several additional genes have been explored for their sensitivity to $1,25(OH)_2D_3$, including the mouse osteopontin gene (Noda et al., 1990), mouse calbindin D-28K (Gill and Christakos, 1993), rat calbindin D-9K (Darwish and DeLuca, 1992), the rat (Ohyama et al., 1994; Zierold et al., 1995), human (Chen and DeLuca, 1995) 25-hydroxyvitamin D_3-24-hydroxylase genes (two apparent VDREs), and the human p21 gene (Liu et al., 1996). A list of the sequences that were shown to mediate vitamin D action as well as their locations within the promoters are documented in Table 1. It is clear from inspection of these sequences that a "typical" VDRE is comprised of two hexad repeats separated by a three base pair spacer. Whereas the sequence of the spacer appears not to be conserved, the general consensus hexad is AGGTCA or preferentially GGTTCA. Considerable variability in these hexad sequences is apparent, however, particularly in the downstream halfsite. The above efforts to define *cis*-acting elements that mediate vitamin D action contributed in part to the current view of hormone response elements (Umesono et al., 1991). The overall nature of these response elements permit classification of the DNA binding sites and thus the nuclear receptor family into three categories: palindromic halfsites which interact with sex steroid receptors; directly repeated halfsites that interact with the small receptors represented by the VDR, retinoic acid, and thyroid receptors; and single halfsites that

Table 1. Location and Sequence of Positive Natural Vitamin D Response Elements

Gene	*Location*	*Nucleotide*	*Sequence*
Rat osteocalcin	-460/-446	GGGTGA	atg AGGACA
Human osteocalcin	-499/-485	GGGTGA	acg GGGGCA
Mouse osteopontin	-757/-743	GGTTCA	cga GGTTCA
Rat calbindin-D9K	-489/-475	GGGTGT	cgg AAGCCC
Mouse calbindin-D28K	-198/-183	GGGGGA	tgtg AGGAGA
Rat 24 hydroxylase	-150/-136 (proximal)	AGGTGA	gtg AGGGCG
	-258/-244 (distal)	GGTTCA	gcg GGTGCG
Human 24 hydroxylase	-169-/155 (proximal)	AGGTGA	gcg AGGGCG
	-291/-277 (distal)	AGTTCA	ccg GGTGTG
Rat pit 1	-67/-52	AGTTCA	gcga AGTTCA
Human p21	-779/-765	AGGGAG	att GGTTCA

mediate the actions of monomeric receptors such as NGFI-B (Mangelsdorf et al., 1995). As will be discussed below, whether a receptor functions on a repeated halfsite as a homodimer or heterodimer allows further categorization of the nuclear receptor family. The reader is referred to references (Evans, 1988; Beato, 1989; O'Malley, 1990; Beato et al., 1995; Mangelsdorf and Evans, 1995; Mangelsdorf et al., 1995) for a complete review of the nature of these DNA binding sites.

B. DNA binding *In Vitro*

Two domains within the VDR are required for high affinity DNA binding *in vitro*, the D domain (DNA-binding domain per se) and the E region (carboxy-terminal ligand-binding domain). This conclusion is based upon an extensive battery of mutations that have been introduced into the VDR by numerous investigators. Thus, point mutations that lead to amino acid changes in the zinc finger modules (Hughes et al., 1988, 1991; Sone et al., 1989, 1991a; Freedman and Towers, 1991; Towers et al., 1993; Nishikawa et al., 1994; Lemon and Freedman, 1996) as well as mutations that alter or delete residues across a majority of the carboxy-terminus (McDonnell et al.,1989; Nakajima et al., 1994; Whitfield et al., 1995; Jin et al., 1996) can block DNA binding. Interestingly, the molecular basis for abrogation of DNA binding by mutations in the two regions is different. In the first case, certain alterations in the DNA binding domain directly prevent interaction with DNA. In the second instance, carboxy-terminal mutations block the ability of the receptor to form dimers that are in turn capable of DNA binding of high affinity, selectivity, and cooperativity (Nakajima et al., 1994; Jin et al., 1996).

The requirement that the VDR must form a dimer in order to interact with DNA was suggested by the repeated halfsite nature of VDREs. The surprising finding, however, was that the VDR bound to DNA not as a dimer but rather as a heterodimer (see Figure 4). Liao et al. (1990) and Sone et al. (1990, 1991a,b) observed that while the VDR derived from mammalian cell extracts was fully capable of binding to DNA *in vitro*, the production of the VDR through either *in vitro* transcription/translation reactions or through recombinant means from nonmammalian cell sources such as yeast failed to produce a DNA binding-competent VDR. The addition of mammalian cell extract to yeast extract-derived VDRs, however, led to recovery of VDR DNA binding, suggesting the necessity for a DNA-binding facilitator. The requirement for this factor, which could be found in a variety of tissues and cell sources (Sone et al., 1991b), was confirmed by others (MacDonald et al., 1991; Ross et al., 1992). This factor(s) was termed nuclear accessory factor (NAF). NAFs were simultaneously discovered for other nuclear receptors including the thyroid receptor (TR) and RAR. In 1991 and 1992, Yu et al. (1991), Leid et al. (1992), Zhang, et al. (1992), and Kliewer et al.

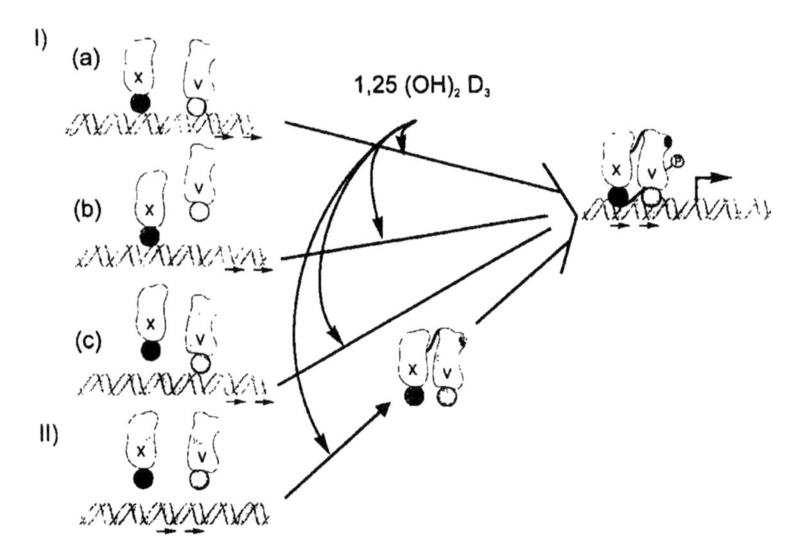

Figure 4. A role for $1,25(OH)_2D_3$ in the formation of the functional VDR heterodimer. Monomers of VDR (v) and a partner protein such as RXR (x) are either associated with DNA (I) or free in the nucleus (II). $1,25(OH)_2D_3$ binding to the VDR increase the latter's affinity for its partner leading to the formation of dimers that associate with specific binding sites within vitamin D responsive gene promoters. Arrows indicate DNA sequence halfsites for VDR binding in a responsive gene promoter and P represents a phosphorylation site on hVDR serine 208.

(1992) discovered that a previously cloned member of the nuclear receptor family, namely the RXRs were the likely protein partners for the above nuclear receptors as well as for VDR. Thus, NAF emerged as the cumulative partnering activity of the RXRα, RXRβ, and/or RXRγ subtypes expressed in a given cell. Considerable efforts over the past several years have gone into defining the nature, function, and complexity of receptor heterodimer DNA binding activity and the reader is referred to reviews which deal with this complex issue (Evans, 1988; Mangelsdorf and Evans, 1995). It is now widely believed, however, that the two protein partners contribute to the activation process despite the fact that only the signaling partner (in this case the VDR) is activated by ligand. True dimerization between the two proteins has been confirmed through definition of the VDR dimerization domain (Nakajima et al., 1994; Jin et al., 1996). Jin et al. (1996) utilized an extensive series of internal deletions of the VDR to define two regions essential for interaction with RXR. These regions coincide with the two moderately conserved subregions of homology located in the E/F domain of the VDR. Interestingly, while a recent study by Perlmann et al. (1996) suggests that a small region of 40 amino acids lying within the second E/F homology domain (corresponding to H9 and H10 of the crystal structure of the RXR ligand binding domain) is sufficient for RXR to form dimers with the retinoic acid and the thyroid hormone receptors, this same region is not sufficient to permit formation of RXR/VDR heterodimers. This finding suggests that the domains responsible for interaction between RXR and certain of the other signaling partners may be different. The deletion analysis described by Jin et al. (1996), however, was unable to distinquish between structural abnormalities induced by the deletion and the actual dimerization domain itself. Thus, elucidation of the three dimensional structure of VDR/RXR ligand-binding domain dimers will still be required to precisely define the interaction between the two proteins. The existence of a single permissive partner protein that functions as a master or at least a central regulator for several endocrine systems suggests that substantial cross-talk may exist among these endocrine systems.

C. Polarity of DNA Binding

Both the asymmetric nature of natural VDREs (direct repeats) coupled with the heteromeric nature of the receptor activation unit (VDR/RXR) indicates that the two receptor subunits must bind to the VDRE with a defined polarity. Studies by Jin et al. (1996) and Towers et al. (1993) addressed this question in detail. Through the use of chimeric receptors and chimeric response elements, it is now clear that RXR binds to the upstream 5' half element and the VDR binds to the downstream 3' half element as in Figure 4. This organization is consistent with the relative polarity noted for both RXR/thyroid receptor (Perlmann et al., 1993) and RXR/retinoic acid receptor (Kurokawa et al., 1993) heterodimers bound to their respective response elements.

D. Transactivation by the Vitamin D Receptor

Early studies that introduced the human osteocalcin promoter into cell lines containing the VDR enabled evaluation of the elements which act in *cis* to mediate vitamin D action. Nevertheless, it was the capacity to introduce both the osteocalcin promoter and the VDR into a cell line initially devoid of endogenous VDR that established an absolute requirement for that protein in vitamin D action. Studies by McDonnell et al. (1989) first demonstrated that while the human osteocalcin gene promoter was unresponsive to $1,25(OH)_2D_3$ when introduced into a VDR negative cell line (CV-1), introduction of an expression vector for the VDR permitted the recovery of vitamin D response. This biological assay enabled further examination of the transcriptional activity of the VDR, particularly as it related to DNA-binding and dimerization functions. Indeed, each of the mutations within the VDR that were examined for DNA binding and dimerization *in vitro* led to transcriptionally inactive receptors in intact cells. These studies therefore confirm the crucial nature of both of the above functional activities of the VDR in transactivation. This assay was also essential in establishing the inactive nature of many of the mutant VDRs that were identified in the human syndrome of hereditary $1,25(OH)_2D_3$ resistant rickets (Hughes et al., 1988, 1991; Sone et al., 1989).

The transactivation assay described above was essential in characterizing many features of the VDR. However, the ubiquitous expression of endogenous NAF (RXR) in mammalian cell lines precluded an experiment designed to unequivocally demonstrate a requirement for NAF in VDR function in intact cells. As a result of this, Jin et al. (1996) utilized yeast to recreate the VDR transcriptional response unit and to test for the requirement of RXRs in VDR-induced transcription. While eukaryotic in nature, yeast do not express either VDR or the RXR genes. Yeast therefore represent a potential opportunity to evaluate a requirement for RXR by introducing each of the receptors into this cell background through recombinant means. While VDR exhibited little capacity to activate a chimeric gene promoter which contained a fused osteocalcin VDRE sequence upstream of the yeast Cyc-1 promoter, the introduction of an RXR expression vector dramatically stimulated the VDR's ability to activate transcription. This study, as well as additional efforts which examined the role of the RXR ligand 9-*cis* retinoic acid (MacDonald et al., 1993; Lemon and Freedman, 1996) on VDR function, lend additional support to the notion that RXR also represents an essential partner for the VDR in intact cells.

The participation of the VDR in transcriptional activation likely requires additional regulatory proteins. Unique transactivation domains have been identified within many eukaryotic transcription factors, and the nuclear receptor superfamily is no exception. These domains represent interactive surfaces on the receptor or protein that facilitate the protein/protein interactions necessary for contact with the core promoter machinery of a gene. Indeed, mutations that compromise the transcriptional regulating capacity of the VDR without abrogating either DNA-binding or dimerization have been defined (Whitfield et al., 1995; Jin et al., 1996). These mutations lie at the extreme carboxy-terminus of the VDR (Whitfield et al., 1995) and may be analogous to the activation function II (AFII) domain of other members of the receptor family (Danielian et al., 1992). Activation is not restricted to this region, however. Jin et al. (1996) utilized the power of yeast selection system to define single mutations within the VDR that selectively compromise transactivation. These mutations lie within the two most amino-terminal homology domains found in the E region of the VDR. How these mutations impact the proposed AFII region of the VDR remain unknown. A protein(s) that interacts with the AFII domain of the VDR and presumably mediates contact with the core promoter has yet to be discovered. The additional observation that TFIIB, a core promoter transcription factor that can associate with the VDR (MacDonald et al., 1995; Blanco et al., 1995), suggests that the biochemical mechanisms by which the VDR contributes regulatory inputs into the basal transcriptional apparatus will be multi-faceted and complex.

E. Role of 1,25$(OH)_2D_3$ in Vitamin D Receptor Activation

Most of the nuclear receptors which have well defined biological roles are activated by ligands, the vast majority of which are true endocrine hormones. Despite the knowledge that this activation event occurs, an understanding of the mechanisms by which an otherwise latent transcription factor is converted to an active form is only beginning to emerge. Ligand-induced receptor conformational changes provide the theoretical basis, however, for activation and likely drive many if not all the downstream events inherent to the acquisition of transcriptional activity. A key event precipitated by activating ligands for the sex steroid receptor subfamily is the dissociation of an inhibitory complex comprised of various heat shock proteins (Pratt et al., 1988). This permits homodimer formation by the receptor and subsequent DNA binding (Beato, 1989). Other events such as induction of phosphorylation, interaction with other transcription factors, and contact with the core transcriptional machinery also probably occur as a result of structural changes in the protein. The VDR, in contrast, is not found associated with heat shock proteins. The discovery that the VDR requires a dimerization partner (NAF) for VDRE binding and that this interaction is a ligand-modulated phenomenon led to the hypothesis that at least one role for $1,25(OH)_2D_3$ might be to promote an increase in affinity of the VDR for its partner. Sone et al. (1991) provided initial support for this hypothesis by demonstrating that in the absence of DNA the affinity of the VDR for NAF increased ninefold in response to $1,25(OH)_2D_3$. This increase in affinity of the VDR for NAF almost certainly reflects a conformational change in the VDR induced by $1,25(OH)_2D_3$ (discussed earlier in this chapter). The effect of ligand on dimerization has been confirmed between VDR and RXR partners through surface plasmon resonance techniques (Cheskis and Freedman, 1996). These experiments collectively support a fundamental role for $1,25(OH)_2D_3$ in promoting formation of an active VDR/RXR heterodimer as illustrated in Figure 4. Whether the receptors are bound to DNA or free in the nucleoplasm prior to heterodimer formation is unknown as are other downstream events that are required for eventual gene activation. Interestingly, the VDR appears not to require ligand for activation in yeast (Jin and Pike, 1996; Jin et al., 1996). This observation hints at the existence in mammalian cells of an inhibitor, analogous to those that regulate the steroid receptors, or a transcriptional repressor which might be released upon ligand binding. It remains for future studies to define additional important events that are initiated through binding of $1,25(OH)_2D_3$ to VDR.

VI. CONCLUDING REMARKS

The pace of exploration into the actions of vitamin D has accelerated enormously over the past decade largely as a result of the molecular cloning of the VDR in 1987, but also as a result of the cloning and availability of vitamin D target genes. As described herein, we have gained considerable insight into the structure of the VDR and its compartmentalization into definable functional domains. The availability of recombinant clones has allowed investigation of the interaction of the VDR with vitamin D inducible gene promoters and definition of VDREs. Further studies have revealed that the VDR requires a protein partner for DNA binding in the form of RXR, a central regulator of several additional nuclear receptors. While additional research will be necessary, a preliminary understanding of the role of receptor and its ligand in the regulation of transcription has begun to emerge. These insights are currently being utilized to understand the tissue-selective mechanisms of action of a new generation of vitamin D analogues under consideration as therapeutics for a broad range of indications that include skin and immunologic diseases and cancer. The cloning of the VDR enabled the recovery of its chromosomal gene. Investigations into the nature of the hereditary human syndrome of $1,25(OH)_2D_3$ resistance revealed the underlying cause to be point mutations in the VDR. This discovery together with genetic ablations studies of the VDR in mice (Yoshizawa et al., 1996) that mimic hereditary resistance to $1,25(OH)_2D_3$ confirm the central role of the VDR in the regulation of mineral metabolism.

For the future, it is likely that the three-dimensional structure of the VDR will be determined, revealing its organization, the nature of the ligand binding site, and the changes that are induced in the protein upon ligand binding. It is likely that new proteins which play a role in the vitamin D activation pathway by facilitating the receptor's interaction with the core gene promoter elements will be identified and cloned. It is at this level that a better understanding of the selective actions of vitamin D analogues will emerge. Perhaps more important than the molecular details of vitamin D action is the likelihood that we will achieve a better understanding of how vitamin D directly and indirectly controls the expression of broad networks of genes which in turn coordinate complex cellular and tissue activities.

VII. SUMMARY

This chapter describes research over the past two decades which has defined the fundamental genomic mechanism of action of the hormonal form of vitamin D. This research revealed that the actions of vitamin D are mediated by a nuclear receptor protein that binds 1,25-dihydroxyvitamin D with high affinity which in turn functions to modulate gene expression. Following its discovery in 1969 and subsequent characterization during the 1970s, the vitamin D receptor was molecularly cloned in 1986. This event resulted in an exponential increase in research focused upon the molecular actions of vitamin D, and resulted in a substantial increase in our understanding of how the hormone functions at the molecular level. The vitamin D receptor was shown to belong to the steroid receptor family of genes, and molecular dissection of the protein genetically has revealed a complex protein composed of multiple domains and functions. These functions include the capacity to bind $1,25(OH)_2D_3$, interact with partner proteins, associate specifically with unique DNA sequences upstream of vitamin D regulated genes, and activate transcription through contacts with the general transcriptional apparatus which assembles at initiator sequences. It is likely that the next decade will reveal the three-dimensional structure of the vitamin D receptor as well as additional molecular details that define precisely how the receptor activates or in some cases selectively inhibits transcription. Genetic studies which further describe the human syndrome wherein the vitamin D receptor is inactivated as well as the genetic ablation of the vitamin D receptor gene in mice should prove helpful in understanding the role of the vitamin D receptor in the numerous biological functions currently ascribed to the vitamin D hormone.

REFERENCES

Allegretto, E.A. and Pike, J.W. (1985). Trypsin cleavage of chick 1,25-dihydroxyvitamin D_3 receptors. Generation of discrete polypeptides, which retain hormone but are unreactive to DNA and monoclonal antibody. J. Biol. Chem. 260, 10139-10145.

Allegretto, E.A., Pike, J.W., and Haussler, M.R. (1987). Immunological detection of unique proteolytic fragments of the chick 1,25-dihydroxyvitamin D_3 receptor. Distinct 20-kDa DNA binding and 45-kDa hormone-binding species. J. Biol. Chem. 262, 1312-1319.

Baker, A.R., McDonnell, D.P., Hughes, M.R., Crisp, T.M., Mangelsdorf, D.J., Haussler, M.R., Shine, J., Pike, J.W., and O'Malley, B.W. (1988). Molecular cloning and expression of human vitamin D receptor complementary DNA: Structural homology with thyroid hormone receptor. Proc. Natl. Acad. Sci.USA 85, 3294-3298.

Beato, M. (1989) Gene regulation by steroid hormones. Cell 56, 335-344.

Beato, M., Herrliche, P., and Schutz, G. (1995) Steroid hormone receptors: Many actors in search of a plot. Cell 83, 851-857.

Berg, J.M. (1988) Proposed structure for the zinc-binding domains from transcription factor IIIA and related proteins. Proc. Natl. Acad. Sci. USA 85, 99-102.

Berg, J.M. (1989 DNA binding specificity and steroid receptors. Cell 57, 1065-1068.

Bikle, D.D. 1994 Clinical Counterpoint: Vitamin D: New actions, new analogs, new therapeutic potential. Endocrine Rev. 13, 765-788.

Blanco, J.C.G., Wang, I.-M., Tsai, S.Y., Tsai, M.-J., O'Malley, B.W., Jurutka, P.W., Haussler, and M.R., Ozato, K. (1995). Transcription factor TFIIB and the vitamin D receptor cooperatively activate ligand-dependent transcription. Proc. Natl. Acad. Sci. USA 92, 1535-1539.

Bourguet, W., Ruff, D., Chambon, P., Gronemeyer, H., and Moras, D. (1995). Crystal structure of the ligand binding domain of the human nuclear receptor RXRa. Nature 375, 377-382.

Brown, A.J. Dusso, A., and Slatopolsky, E. (1994). Selective vitamin D analogs and their therapeutic application. Sem. Nephrol. 14, 156-174.

Brumbaugh, P.F. and Haussler, M.R. (1974). 1,25-Dihydroxycholecalciferol receptors in intestine. Temperature-dependent transfer of the hormone to chromatin via a specific receptor. J. Biol. Chem. 249, 1258-1262.

Brumbaugh, P.F. and Haussler, M.R. (1975). Specific binding of 1,25-dihydroxycholecalciferol to nuclear components of chick intestine. J. Biol. Chem. 250, 1588-1594.

Burmester, J.K., Maeda, N., and DeLuca, H.F. (1988). Isolation and expression of rat 1,25-dihydroxyvitamin D receptor cDNA. Proc. Natl. Acad. Sci. USA 85, 1005-1009.

Celeste, A.J., Rosen, V. Buecker, J.L., Kriz, R., Wang, E.A., and Wozney, J.M. (1986). Isolation of the human gene for bone gla protein utilizing mouse and rat cDNA clones. EMBO J. 5, 1885-1890.

Chen, K.S., and DeLuca, H.F. (1995). Cloning of the human 1,25-dihydroxyvitamin D_3-24-hydroxylase gene promoter and identification of two vitamin D responsive elements. Biochim. Biophys. Acta 1263, 1-9.

Cheskis, B. and Freedman, L.P. (1996). Modulation of nuclear receptor interactions by ligands: kinetic analysis using surface plasmon resonance. Biochemistry 35, 3309-3318.

Danielian, P.S., White, R., Lees, J.A., and Parker, M.G. (1992). Identification of a conserved region required for hormone-dependent transcriptional activation by steroid hormone receptors. EMBO J. 11, 1025-1033.

Darwish, H.M. and DeLuca, H.F. (1992 Identification of a 1,25-dihydroxyvitamin D_3 response element in the 5' flanking region of the rat calbindin *D-9K* gene. Proc. Natl. Acad. Sci. USA 89, 603-607.

Demay, M.B., Gerardi, J.M., DeLuca, H.F., and Kronenberg, H.M. (1990). DNA sequences in the rat *osteocalcin* gene that bind the 1,25-dihydroxyvitamin D receptor and confer responsiveness to 1,25-dihydroxyvitamin D_3. Proc. Natl. Acad. Sci. USA 87, 369-373.

Ducy, P., Desbois, C., Boyce, B., Pinero, G., Story, B., Dunstan, C., Smith, E., Bonadio, J., Goldstein, S., Gundberg, C., Bradley, A., and Karsenty, G. (1996). Increased bone formation in osteocalcin-deficient mice. Nature 382, 448-452.

Elaroussi, M.A., Prahl, J.M., and DeLuca, H.F. (1992). The avian vitamin D receptors: Primary structures and their origins. Proc. Natl. Acad. Sci. USA 91, 11596-11600.

Evans, R.M. (1988). The steroid and thyroid hormone receptor superfamily. Science 240, 889-895.

Freedman, L.P. and Towers, T. (1991). DNA binding properties of the vitamin D receptor zinc fingers region. Mol. Endocrinol. 5, 1815-1826.

Gill, R.K. and Christakos, S. (1993). Identification of sequence elements in mouse *calbindin-D28K* gene that confer 1,25-dihydroxyvitamin D_3 and butyrate-inducible responses. Proc. Natl. Acad. Sci. USA 90, 2984-2988.

Goto, H., Chen, K.S., Prahl, J.M., and DeLuca, H.F. (1992). A single receptor identical to that for intestinal/T47D cells mediates the action of 1,25-dihydroxyvitamin D_3 in HL-60 cells. Biochim. Biophys Acta 1132, 103-108.

Green, S., Walter, P. Kumar, V., Krust, A., Bornert, J.M. Argos, P., and Chambon, P. (1986) Human oestrogen receptor cDNA: Sequence expression and homology to v-erbA. Nature 320, 134-139.

Gronemeyer, H. (1993). Transcriptional activation by nuclear receptors. J. Receptor Res. 13, 667-691.

Hard, T., Kellenbach, E., Boelens, R., Maler, B.A., Dahlman, K., Freedman, L.P., Carlstedt-Duke, J.,, Yamamoto K.R., Gustafsson, J.A., and Kaptein, R. (1990). Solution structure of the glucocorticoid receptor DNA binding domain. Science 249, 157-160.

Haussler, M.R. and Norman, A.W (1969). Chromosomal receptor for a vitamin D metabolite. Proc. Natl. Acad. Sci. USA 62, 155-162.

Haussler, M.R., Pike, J.W., Chandler, J.S., Manolagas, S.C., and Deftos, L.J. (1981). Molecular actions of 1,25-dihydroxyvitamin D_3: New cultured cell models. Annals N.Y. Acad. Sci. 372, 502-517.

Hollenberg, S.M., Weinberger, C., Ong, E.S., Cerelli, G., Oro, A.E., Lebo, R., Thompson, E.B., Rosenfeld, M.G., and Evans, R.M. (1985). Primary structure and expression of a functional human glucocorticoid receptor cDNA. Nature 318, 635-641.

Hughes, M.R., Malloy, P.J. Kieback, D.G., Kesterson, R.A., Pike, J.W., Feldman, D., and O'Malley, B.W. (1988). Point mutations in the human vitamin D receptor gene associated with hypocalcemic rickets. Science 242, 1702-1705.

Hughes, M.R., Malloy, P.J., O'Malley, B.W., Pike, J.W., and Feldman, D. (1991). Genetic defects of the 1,25-dihydroxyvitamin D receptor. J. Recept Res. 11, 699-716.

Hunziker, W., Walters, M.R., Bishop, J.E., and Norman, A.W. (1983). Unoccupied and in vitro and in vivo occupied 1,25-dihydroxyvitamin D_3 intestinal receptors. Multiple biochemical forms and evidence for transformations. J. Biol. Chem. 258, 8642-8648.

Jin, C.H., Kerner, S.A., Hong, M.H., and Pike, J.W. (1996). Transcriptional activation and dimerization functions of the human vitamin D receptor. Mol. Endocrinol. 10, 945-957.

Jin, C.H. and Pike, J.W. (1996). Human vitamin D receptor dependent transactivation in *Saccharomyces cerevisiae* requires retinoid X receptor. Mol. Endocrinol. 10, 196-205.

Kamei, Y., Kawada, T., Fukuwatari, T., Ono, T., Kato, S., and Sugimoto, E. (1995). Cloning and sequence of the gene encoding the mouse vitamin D receptor. Gene 152, 281-282.

Kerner, S.A., Scott, R.A., and Pike, J.W. (1989). Sequence elements in the human osteocalcin gene confer basal activation and inducible response to hormonal vitamin D_3. Proc. Natl. Acad. Sci. USA 86, 4455-4459.

Kleiwer, S.A., Umesono, K., Mangelsdorf, D.J., and Evans, R.M. (1992). Retinoid X receptor interacts with nuclear receptors in retinoic acid, thyroid, and vitamin D signaling. Nature (London) 355, 446-449.

Kream, B.E., Jose, J.L., and DeLuca, H.F. (1977). The chick intestinal cytosol binding protein for 1,25-dihydroxyvitamin D_3: A study of analog binding. Arch. Biochem. Biophys. 179, 462-468.

Kurokawa, B., Yu, V.C., Naar, A., Kyakumoto, S., Han, Z., Silverman, S., Rosenfeld, M.G., and Glass, C.K. (1993). Differential orientation of the DNA-binding domain and carboxy terminal dimerization interface regulate binding site selection by nuclear receptor heterodimers. Genes Dev. 7, 1423-14335.

Lawson, D.E.M. and Wilson, P.W. (1974). Intranuclear localization and receptor proteins for 1,25-dihydroxycholecalciferol in chick intestine. Biochem. J. 144, 573-583.

Lee, M.S., Kliewer, S.A., Provencal, J., Wright, P.E., and Evans, R.M. (1993). Structure of the retinoid X receptor a DNA binding domain: A helix required for homodimeric DNA binding. Science 260, 1117-1121.

Leid, M., Kastner, P., Lyons, R., Nakshatro, H., Saunders, M., Zacharewski, T., Chen, J.-Y. Staub, A., Garnier, J.-M. Mader, S., and Chambon, P. (1992). Purification, cloning, and

RXR identity of the Hela cell factor with which RAR or TR heterodimerizes to bind target sequences efficiently. Cell 68, 377-395.

Lemon, B. and Freedman, L.P. (1996). Selective effects of ligands on vitamin D_3 receptor and retinoid X receptor mediated gene activation in vivo. Mol. Cell. Biol. 16. 1006-1016.

Li, Y.C., Bergwitz, C., Juppner, H., and Demay, M.B. (1996). Cloning and characterization of the vitamin D receptor from *Xenopus laevis*. J. Bone Min. Res. 11, (Supple. 1), S161.

Lian, J.B., Stewart, C., Puchacz, E., Mackowiak, S., Shalhoub, V., Colart, D., Sambetti, G., and Stein, G. (1989). Structure of the rat osteocalcin gene and regulation of vitamin D-dependent expression. Proc. Natl. Acad. Sci. USA 86, 1143-1147.

Liao, J. Ozono, K., Sone, T., McDonnell, D.P., and Pike, J.W. (1990). Vitamin D receptor interaction with specific DNA requires a nuclear protein and 1,25-dihydroxyvitamin D_3. Proc. Natl. Acad. Sci. USA 87, 9751-9755.

Liu, M., Lee, M.-H., Cohen, M., Freedman, L.P. (1996). Transcriptional activation of the *p21* gene by vitamin D_3 leads to the differentiation of the myelomonocytic cell line U937. Genes Dev. 10, 142-153.

Luisi, B.F., Xu, W., Otwinowski, Z., Freedman, L.P., Yamamoto, K.R., and Sigler, P.B. (1991). Crystallographic analysis of the interaction of the glucocorticoid receptor with DNA. Nature 352, 497-505.

McCain, T.A., Haussler, M.R., Hughes, M.R., and Okrent, D. (1978). Partial purification of the chick 1,25-dihydroxyvitamin D receptor. FEBS Lett. 86, 65-70.

McDonnell, D.P., Clemm, D.L., Hermann, T., Goldman, M.E., and Pike, J.W. (1995). Analysis of estrogen receptor function in vitro reveals three distinct classes of anti-estrogens. Mol. Endocrinol. 9, 659-669.

McDonnell, D.P., Mangelsdorf, D.J., Pike, J.W., Haussler, M.R., and O'Malley, B.W. (1987). Molecular cloning of complementary DNA encoding the avian receptor for vitamin D. Science, 1214-1217.

McDonnell, D.P., Pike, J.W., and O'Malley, B.W. (1988). Vitamin D receptor: A primitive steroid receptor related to the thyroid hormone receptor. J. Steroid Biochem. 30, 41-47.

McDonnell, D.P., Scott, R.A., Kerner, R.A., O'Malley, B.W., and Pike, J.W. (1989). Functional domains of the human vitamin D receptor regulate osteocalcin gene expression. Mol. Endorinol. 3, 635-644.

MacDonald, P.N., Dowd, D.R., Nakajima, S., Galligan, M.A., Reeder, M.C., Haussler, C.A., Ozato, K., and Haussler, M.R. (1993). Retinoid X receptors stimulate and 9-*cis* retinoic acid inhibits 1,25-dihydroxyvitamin D_3-activated expression of the rat osteocalcin gene. Mol. Cell. Biol. 13, 5907-5917.

MacDonald, P.N., Haussler, C.A., Terpening, C.M., Galligan, M.A., Reeder, M.C., Whitfield, G.K., and Haussler, M.R. (1991). Baculovirus-mediated expression of the human vitamin D receptor: Functional characterization, vitamin D response element interactions, and evidence for a receptor auxiliary factor. J. Biol. Chem. 266, 18808-18813.

MacDonald, P.N., Sherman, D.R., Dowd, D.R., Jefcoat, Jr., S.C., and DeLisle, R.K. (1995). The vitamin D receptor interacts with general transcription factor IIB. J. Biol. Chem. 270, 4748-4752.

Mader, S., Kumar, V., deVereneuil, H., and Chambon, P. (1989). Three amino acids of the oestrogen receptor are essential to its ability to distinguish an oestrogen from a glucocorticoid responsive receptor. Nature 338, 271-274.

Mangelsdorf, D.J. and Evans, R.M. (1995). The RXR heterodimer and orphan receptors. Cell 83, 841-850.

Mangelsdorf, D.J., Thummel, C., Beato, M., Herrliche, P., Schutz, G., Umesono, K., Blumberg, B., Kastner, P., Mark, M., Chambon, P., and Evans, R.M. (1995). The nuclear receptor superfamily: The second decade. Cell 83, 835-839.

Mellon, W.S. and DeLuca, H.F. (1979) An equilibrium and kinetic study of 1,25-dihydroxyvitamin D_3 binding to chicken intestinal cytosol employing high specific activity 1,25-dihydroxy[3H-26,27]-vitamin D_3. Arch. Biochem. Biophys. 197, 90-95.

Morrison, N.A., Shine, J., Fragonas, J.-C., Verkest, V., McMenemy, M.L., and Eisman, J.A. (1989). 1,25-Dihydroxyvitamin D_3-responsive element and glucocorticoid repression in the osteocalcin gene. Science 246, 1158-1161.

Nakajima, S., Hsieh, J.C., MacDonald, P.N., Galligan, M.A., Haussler, C.A., Whitfield, G.K., and Haussler, M.R. (1994). The C-terminal region of the vitamin D receptor is essential to form a complex with a receptor auxilliary factor required for high affinity binding to the vitamin D responsive element. Mol. Endocrinol. 8, 159-172.

Nishikawa, J., Kitaura, M., Matsumoto, M., Imagawa, M., and Nishihara, T. (1994). Difference and similarity of DNA sequence recognized by VDR homodimer and VDR/RXR heterodimer. Nucl. Acid Res. 22, 2902-2907.

Noda, M., Vogel, R.L., Craig, A.M., Prahl, J., DeLuca, H.F., and Denhardt, D. (1990). Identification of a DNA sequence resonsible for binding of the 1,25-dihydroxyvitamin D_3 receptor and 1,25-dihydroxyvitamin D_3 enhancement of mouse secreted phosphoprotein 1 (Supp-l or osteopontin) gene expression. Proc. Natl. Acad. Sci. USA 87, 9995-9999.

Norman, A.W. (1995). The vitamin D endocrine system: Manipulation of structure-function relationships to provide opportunities for development of new cancer chemopreventive and immunosuppressive agents. J. Cell Biochem. 22, S218-S225.

Ohyama, Y. Ozono, K., Uchida, M., Shinki, T., Kato, S., Suda, T., Yamamoto, O., Noshiro, M., and Kato, Y. (1994). Identification of a vitamin D-responsive element in the 5'-flanking region of the rat 25-hydroxyvitamin D_3 24 hydroxlyase gene. J. Biol. Chem. 269, 10545-10550.

O'Malley, B.W., McGuire, W.L., Kohler, P.O., and Korenman, S.G. (1969). Studies on the mechanism of steroid hormone regulation of specific proteins. Recent Prog. Hormone. Res. 25, 105-160.

O'Malley, B.W. (1990). The steroid receptor superfamily: More excitement predicted for the future. Mol. Endocrinol. 4, 363-344.

Ozono, K., Liao, J., Scott, R.A., Kerner, S.A., and Pike, J.W., (1990). The vitamin D responsive element in the human osteocalcion gene: Association with a nuclear protooncogene enhancer. J. Biol. Chem. 265, 21881-21888.

Pan, L.C., and Price, P.A. (1986). 1,25-Dihydroxyvitamin D_3 stimulates transcription of bone gla protein. J. Bone Min. Res. 1, (Supple 1) A20.

Peleg, S., Sastry, M., Collins, E.D., Bishop, J.E., and Norman, A.W. (1995). Distinct conformational changes induced by 20-epi analogues of 1,25-dihydroxyvitamin D_3 are with enhanced activation of the vitamin D receptor. J. Biol. Chem. 270, 10551-10558.

Perlmann, T., Rangarajan, P.N., Umesono, K., and Evans, R.M. (1993). Determinants for selective RAR and TR recognition of direct repeat HREs. Genes Dev. 7, 1411-1422.

Perlmann, T., Umesono, K., Rangarajan, P.N., Forman, B,M., and Evans, R.M. (1996). Two distinct dimerization interfaces differentially modulate target gene specificity of nuclear hormone receptors. Mol. Endocrinol. 10, 958-966.

Pike, J.W. and Haussler, M.R. (1979). Purification of chicken intestinal receptor for 1,25-dihydroxyvitamin D_3. Proc. Natl. Acad. Sci. USA 76, 5488-5494.

Pike, J.W. and Haussler, M.R. (1983). Association of 1,25-dihydroxyvitamin D_3 with cultured 3T6 mouse fibroblast. J. Biol. Chem. 258, 8554-8560.

Pike, J.W., Kesterson, R.A., Scott, R.A., Kerner, S.A., McDonnell, D.P., and O'Malley, B.W. (1988). Vitamin D receptors: Molecular structure of the protein and its chromosomal gene. In: Vitamin D: Molecular, Cellular and Clilnical Endocrinology. (Norman, A.W., Schaefer, K., Grigoleit, H.-G., and Herrath, D.v., Eds.), pp. 215-224. Walter de Gruyter, Berlin.

Pratt, W., Jolly, D.J., Pratt, D.V., Hollenberg, S.M., Giguere, V., Cadepon, F.M., Schweizer-Groyer, G., Cartelli, M.G., Evans, R.M., and Baulieu, E.E. (1988). A region in the steroid-binding domain determines formation of the non-DNA-binding glucocorticoid receptor complex. J. Biol. Chem. 263, 267-273.

Price, P.A. and Baukol, S.A. (1980). 1,25-Dihydroxyvitamin D_3 increases synthesis of the vitamin D-dependent bone proteins by osteosarcopma cells. J. Biol. Chem. 255, 11660-11663.

Rastinejad, F., Perlmann, T., Evans, R.M., and Sigler, P.B. (1995). Structural determinants of nuclear receptor assembly on DNA direct repeats. Nature 375, 203-211.

Renaud, J.-P., Natacha, R., Ruff, M., Vivat, V., Chambon, P., Gronemyer, H., and Moras, D. (1995). Crystal structure of the RARg ligand binding domain bound to all-*trans* retinoic acid. Nature 378, 681-689.

Ross, T.K., Moss, V.E., Prahl, J.M., and DeLuca, H.F. (1992). A nuclear protein essential for binding of rat 1,25-dihydroxyvitamin D_3 receptor to its response elements. Proc. Natl. Acad. Sci. USA 89, 256-260.

Santiso-Mere, D., Sone, T., Hilliard, G.M., Pike, J.W., and McDonnell, D.P. (1993). Positive regulation of the vitamin D receptor by its cognate ligand in heterologous expression systems. Mol. Endocrinol. 7, 833-839.

Schwabe, J.W.R., Chapman, L., Finch, J.T. and Rhodes, D. (1993). The crystal structure of the oestrogen receptor DNA binding domain bound to DNA: How receptors discriminate between their response elements. Cell 75, 567-578.

Schwabe, J.W.R., Neuhaus, D. and Rhodes, D. (1990). Solution structure of the DNA binding domain of the oestrogen receptor. Nature 348, 458-461.

Sone, T., Kerner, S.A., and Pike, J.W. (1991a). Vitamin D receptor interaction with specific DNA. Association as a 1,25-dihydroxyvitamin D_3-modulated heterodimer. J. Biol. Chem. 266, 23296-23305.

Sone, T., McDonnell, D.P., O'Malley, B.W., and Pike, J.W. (1990). Expression of the human vitamin D receptor in *Saccharomyces cerevisiae*: Purification properties and generation of polyclonal antibodies. J. Biol. Chem. 265, 21997-22003.

Sone, T., Ozono, K., and Pike, J.W. (1991b). A 55-kilodalton accessory factor facilitates vitamin D receptor DNA binding. Mol. Endocrinol. 5, 1578-1586.

Sone, T., Scott, R., Hughes, M., Malloy, P., Feldman, D., O'Malley, B.W., and Pike, J.W. (1989). Mutant vitamin D receptors which confer hereditary resistance to 1,25-dihydroxyvitamin D_3 in humans are transcriptionally inactive in vitro. J. Biol. Chem. 264, 20230-20234.

Terpening, C.M., Haussler, M.R., Jurutka, P.W., Galligan, M.A., Komm, B.S., and Haussler, M.R. (1991). The vitamin D responsive element in the rat bone gla protein gene is an imperfect direct repeat that cooperates with other *cis* elements in 1,25-dihydroxyvitamin D_3-mediated transcriptional activation. Mol. Endocrinol. 5, 373-385.

Towers, T., Luisi, B.L., Asianov, A., and Freedman, L.P. (1993). DNA target selectivity by the vitamin D receptor: Mechanism for dimer binding to an asymmetric repeat element. Proc. Natl. Acad. Sci. USA 90, 6310-6314.

Tzukerman, M.T., Esty, A., Santiso-Mere, D., Danielian, P., Parker, M.G., Stein, R.B., Pike, J.W., and McDonnell, D.P. (1994). Human estrogen receptor transcriptional capacity is determined by both cellular and promoter context and mediated by two functionally distinct intramolecular regions. Mol. Endorinol. 8, 21-30.

Umesono, K. and Evans, R.M. (1989). Determinants of target gene specificity for steroid/thyroid hormone receptors. Cell 57, 1139-1146.

Umesono, K., Murikami, K.K., Thompson, C.C., and Evans, R.M. (1991). Direct repeats as selective response elements for the thyroid hormone, retinoic acid, and vitamin D_3 receptors. Cell 65, 1255-1266.

Wang, L.H., Tsai, S.Y., Cook, R.G., Beattie, W.G., Tsai, M.-J., and O'Malley, B.W (1989). COUP transcription factor is a member of the steroid receptor superfamily. Nature 340, 163-166.

Wagner, R.L., Apriletti, J.W., McGrath, M.E., West, B.L., Baxter, J.D., and Fletterick, R.J. (1995). A structural role for hormone in the thyroid hormone receptor. Nature 378, 690-697.

Wecksler, W.R., and Norman, A.W. (1980). A kinetic and equilibrium binding study of 1,25-dihydroxyvitamin D_3 with its cytosol receptor from chick intestinal mucosa. J. Biol. Chem. 255, 3571-3574.

Wecksler, W.R. Okamura, W.H., and Norman, A.W. (1978). Quantitative assessment of the structural requirements for the interaction of 1,25-dihydroxyvitamin D_3 with its chick intestinal mucosa receptor system. J. Steroid Biochem. 9, 929-937.

Whitfield, G.K., Hsieh, J.C., Nakajima, S., MacDonald, P.N., Thompson, P.D., Jurutka, P.W., Haussler, C.A., and Haussler, M.R. (1995). A highly conserved region in the hormone-binding domain of the human vitamin D receptor contains residues vital for heterodimerization with retinoid X receptor and for transcriptional activation. Mol. Endocrinol. 9, 1166-1179.

Yoon, K., Rutledge, S.J.C., Buenaga, R.F., and Rodan, G.A. (1988). Characterization of the rat osteocalcin gene: Stimulation of promoter activity by 1,25-dihydroxyvitamin D_3. Biochemistry 27, 8521-8526.

Yoshizawa, T., Handa, Y., Uematsu, Y., Sekine, K., Takeda, S., Yoshihara, Y., Kawakami, T., Sato, H., Alioka, K., Tanimoto, K., Fukamizu, A., Masushige, S., Matsumoto, T., and Kato, S. (1996). Disruption of the vitamin D receptor in the mouse. J. Bone Min. Res. 11 (Suppl. 1), S62.

Yu, V., Delsert, C., Andersen, B., Holloway, J.M., Devary, O.V., Naar, A.M., Kim, S.Y., Boutin, J.-M., Glass, C.K., and Rosenfeld, M.G. (1991). RXRb: A coregulator that enhances binding of retinoic acid, thyroid hormone, and vitamin D receptors to their cognate response elements. Cell 67, 1251-1266.

Zhang, X.-K. , Hoffman, B., Tran, P.B., Graupner, G., and Pfahl, M. (1992). Retinoid X receptor is an auxilliary protein for thyroid hormone and retinoic acid receptors. Nature (London) 355, 441-446.

Zierold, C., Darwish, H.M., and DeLuca, H.F. (1995). Two vitamin D response elements function in the rat 1,25-dihydroxyvitamin D_3-24 hydroxylase promoter. J. Biol. Chem. 269, 1675-1678.

MOLECULAR PHYSIOLOGY OF AVIAN BONE

Christopher G. Dacke

I. Introduction 244
II. Plasma Calcium Regulation 244
III. Avian Bone Morphology 246
A. Structure and Formation of Medullary Bone 246
B. Function of Medullary Bone 248
C. Medullary Bone Osteoclasts 250
D. Assessment of Avian Osteoclast Motility 252
E. Osteoclastic Tartrate-Resistant Acid Phosphatase 253
IV. Parathyroid Hormone, Related Peptides, and Avian Bone 255
A. Skeletal Actions 257
B. Osteoclast Activation and the Role of PTH 261
C. Parathyroid Hormone-Related Peptide 262
V. Calcitonin, Related Peptides, and Avian Bone 263
A. Calcitonin 263
B. Calcitonin Gene-Related Peptide and Amylin 265
VI. Is There a Calcium Receptor in Avian Bone? 267
VII. Are There Estrogen Receptors in Avian Bone? 271
VIII. The Avian Vitamin D System 272

Advances in Organ Biology
Volume 5A, pages 243-285.

ISBN: 0-7623-0390-5

IX. Prostaglandins and Avian Bone 273
X. Summary 274
Acknowledgments 275

I. INTRODUCTION

Calcium (Ca) is a most efficiently regulated plasma constituent in birds. The classical Ca regulating hormones, parathyroid hormone (PTH), calcitonin (CT) and 1,25-dihydroxy vitamin D_3 (1,25-$(OH)_2D_3$), are all recognized in this class, although their actions and/or sensitivities may be different from those in mammals (Dacke, 1979, 1996). Avian skeletal metabolism is clearly amplified compared with that in mammals (Gay, 1988). Other putative Ca and bone-regulating factors such as prostaglandins (PGs; Dacke, 1989) parathyroid hormone-related peptide (PTHrP; Dacke, 1996) and calcitonin gene-related peptide (CGRP; Dacke et al, 1993a) are also present and have distinctive effects on avian Ca metabolism.

Avian Ca metabolism shares many features with mammals; the requirements of bone and Ca metabolism in growing birds and mammals are similar, but are also typified in birds by several unique characteristics related to their ability to lay large eggs with a heavily calcified eggshell (Romanoff and Romanoff, 1963). The amount of Ca per egg represents about 10% of the total body stores of Ca (Kenny, 1986), an enormous amount by any measure. The Ca metabolism of a domesticated hen compares with that of a woman in 18 months of pregnancy and lactation. In order to provide a source of Ca for eggshell calcification to supplement that from the diet, egg-laying hens uniquely possess a highly labile reservoir, medullary bone, which develops within the long bones in response to gonadal steroid activity. It is the most overtly estrogen sensitive form of vertebrate bone (Bloom et al., 1958; Dacke, 1979; Dacke et al., 1993b). A second feature which probably has influenced the evolution of bone and Ca metabolism in birds is the ability to fly which led to the development of light robust skeletons in which long bones are more hollowed out than in other vertebrates, implying a high degree of remodeling during the growth phase of the skeleton. Recent reviews of avian Ca and bone metabolism include those by Taylor and Dacke (1984), Kenny (1986), Hurwitz (1989), Gay (1996), and Dacke (1996).

II. PLASMA CALCIUM REGULATION

Skeletal growth represents a steady-state perturbation of the Ca regulatory system, involving a proportional outflow of Ca from the central pool. It changes as a function of age, genetic potential, and nutritional-environmental factors and at least some of the components of the Ca-regulating system will respond to these factors. Due to homeostatic regulation, plasma Ca concentration correlates poorly with nutritional status or rates of movement of Ca between body compartments such as soft tissue and bone. These can be studied either by use of isotopes such as ^{45}Ca, or by specific perturbations such as bolus Ca loading or chronic feeding of diets of different Ca concentrations, parathyroidectomy, or vitamin D deficiency. However, such drastic approaches result in a large departure of metabolism from normal (Hurwitz and Bar, 1996).

The feedback regulation of avian plasma Ca includes several components, with either rapid or slow response times (Figure 1). The Ca control system responds to perturbations in plasma Ca by modulating Ca flows, rapid responses being either directly associated with the Ca concentration, such as those involving the kidney, or with the action of peptide hormones, most importantly, rapid PTH release. Chronic perturbation, e.g., dietary Ca deficiency, which results in increased bone resorption, leads to a sluggish but more economic increase in Ca absorption in response to $1,25(OH)_2D_3$, production of which is also stimulated by PTH (Hurwitz and Bar, 1996). Mammals respond to an EGTA-induced hypocalcemic challenge within tens of minutes to a few hours, while one week old chickens correct such challenges within minutes, recovery being dependent upon the presence of PTH (Koch et al., 1984). Approximately 40% of an original intravenous

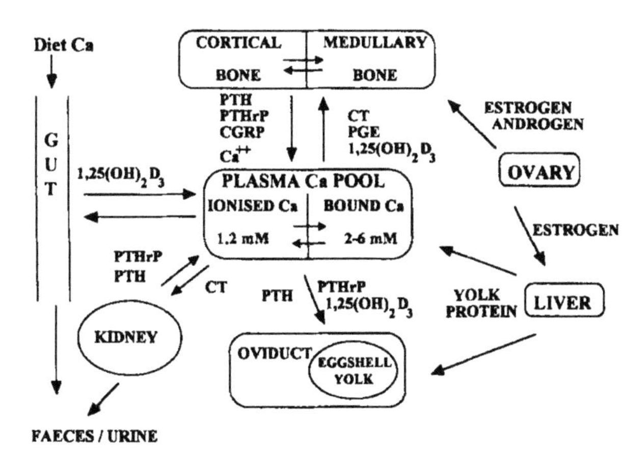

Figure 1. Model illustrating the essential organs and potential endogenous hormonal and humoral factors that regulate Ca^{2+} and bone metabolism in the egglaying bird. Adapted from Dacke (1979).

dose of ^{45}Ca in the chick clears rapidly from the plasma pool and becomes located within the skeleton by 15 minutes. We calculated that total skeletal outflux of Ca is approximately 80% of influx, and the plasma pool of Ca is cleared into the skeleton every few minutes; typically, net Ca accumulation into the chick femur is about 0.28 mole $min^{-1}g^{-1}$ wet weight (Shaw et al., 1989). Factors which modify either influx or outflux of Ca^{2+} in this system are likely to profoundly affect minute to minute plasma Ca modulation in the rapidly growing animal. Bronner and Stein (1992) calculated that the $t_{1/2}$ for ^{45}Ca uptake by the chick femur is less than 10 minutes, compared with around 30 minutes in rabbit, dog, and rat.

III. AVIAN BONE MORPHOLOGY

Compared with mammals, avian bone formation is rapid, beginning around day 6 of embryogenesis when osteoblasts first appear on the surfaces of the cartilage anlage (Anderson, 1973). Well-mineralized bones are developed in hatchlings. Rapid growth continues and, at maturity, medullary bone develops within the endosteal cavities of the endochondral skeleton (Schraer and Hunter, 1985). When considering avian bone turnover, several types of bone should be considered: cortical, medullary, and trabecular. Much work on avian bone cells has been done using the growing chick tibia from which osteoclasts can be isolated from the endosteum (Hunter et al., 1988) and osteoblasts from the periosteal surface (Gay et al., 1994). Medullary bone is the most physiologically responsive type of bone and as such is a valuable source of large, relatively pure populations of cells such as osteoclasts (Dacke et al., 1993b); however, less attention has been given to avian trabecular bone (Gay, 1996).

A. Structure and Formation of Medullary Bone

Medullary bone, a nonstructural woven bone, is found in long bones of egg-laying birds and may completely fill the marrow spaces (Figures 2 and 3; Dacke, 1979). It forms shortly before the onset of egg laying and persists throughout the egg-laying period. It is more or less continuously present in domesticated species such as chickens and Japanese quail, which lay eggs throughout the year. Medullary bone is richly vascular, with a large surface area available for mineral exchange. The mineral phase is similar to that of cortical bone consisting of a hydroxyapatite lattice (Dacke et al., 1993b). In cortical bone, the hydroxyapatite crystals are oriented with respect to the or-

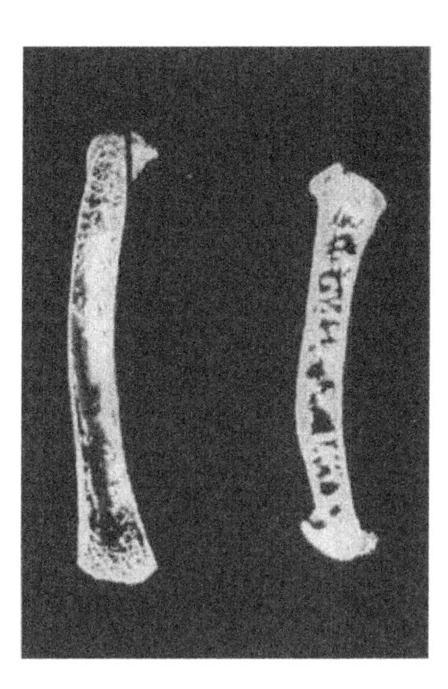

Figure 2. Longitudinal sections of femurs from chickens. The organic material has been removed with solvents, leaving the mineral elements. The cockerel bone, left, has a hollow shaft whereas the marrow cavity of an egglaying hen femur is filled with medullary bone.

ganic matrix (Neuman and Neuman, 1958), while medullary bone has apatite crystals randomly distributed throughout the matrix. It is more heavily calcified than cortical bone, although its collagen fibril content is much lower, and the apatite to collagen ratio is higher (Taylor et al., 1971). The proteoglycan component of medullary bone matrix has been investigated with emphasis on its time of appearance after bone induction by estrogen and on the nature of its glycosaminoglycan side chains (Schraer and Hunter, 1985). In chicken and quail medullary bone, these consists of keratin sulphate (Candlish and Holt, 1971; Fisher and Schraer, 1982). Proteoglycans bind Ca, and Stagni et al. (1980) extracted a proteoglycan from medullary bone with high Ca affinity and phosphatase activity which hydrolyzed ATP, GTP, and pyrophosphate. The differences between medullary and cortical bone allow for greater deposition of mineral, at the expense of strength (Dacke et al., 1993b).

The synergistic action of androgens and estrogens stimulate medullary bone formation in female birds during oogenesis. It can be induced artificially in adult males by administration of these hormones (Dacke, 1979; Miller and Bowman, 1981). Estrogen induces differentiation of endosteal cells to form osteoblasts and decreases osteoclast numbers on the endosteal surface in adult male Japanese quail (Kusuhara and Schraer, 1982; Ohashi

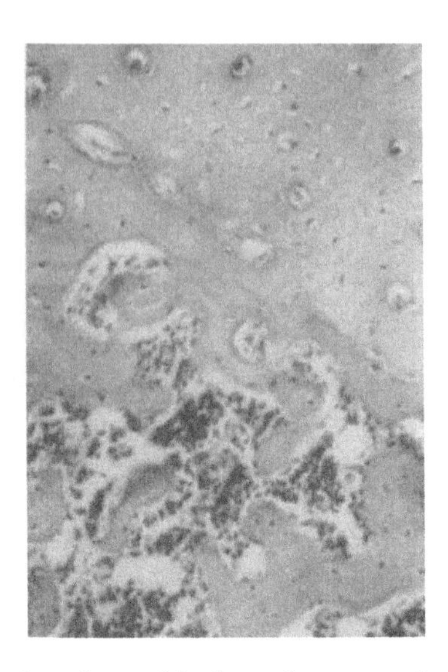

Figure 3. Transverse section of part of the femur from an egglaying Japanese quail. Part of the cortical bone (top right) is shown, while the medullary bone (bottom left) can be clearly delineated as irregular whorls of bone with large numbers of cells in the intervening spaces. Magnification: 200 x.

et al., 1987). The anti-estrogen, tamoxifen, suppresses medullary bone formation and reduces bone matrix (Ohashi et al., 1990). Ohashi et al. (1987) also reported increases in osteoclast numbers in these birds following tamoxifen treatment and suggested that the role of estrogen in the endosteal response is to accelerate osteoblast differentiation and inhibit osteoclast activity.

B. Function of Medullary Bone

Medullary bone is a labile store of Ca for eggshell formation and is metabolized 10–15 times faster than cortical bone (Hurwitz, 1965; Simkiss, 1967; Dacke et al., 1993b). This is emphasized by the high number of osteoclasts in avian medullary, compared with cortical, bone (van de Velde et al., 1984a). Eggshell formation in pigeons is accompanied by intense osteoclastic activity resulting in total removal of medullary bone, while osteoblastic activity, resulting in intensive medullary bone formation, predominates when no egg is present in the oviduct (Bloom et al., 1958). Chickens, unlike pigeons, lay eggs more or less continuously. In chickens, when bone resorption is low, formation is also low and when resorption is high, formation is high. In both phases, numbers of nuclei per osteoclast and numbers of osteoclasts are similar (van de Velde et al., 1984a). During the active period, the resorbing surface, i.e., ruffled border in each active osteoclast, increases and, although total numbers of osteoclasts remained constant, the percentage of active osteoclasts increase (Holtrop and King, 1977; Miller, 1985). van de Velde et al. (1984a) presumed that matrix formation and osteoclastic resorption are in phase such that, during eggshell formation, the organic matrix becomes low in Ca, but recalcifies during the subsequent inactive period. Hence, the total medullary bone volume of chickens does not change during the egg-laying cycle. Kusuhara (1976) reported that, during the inactive period, medullary bone contains more Ca than during the active period. Thus, the mineralization rate is high when the resorption rate is low, and vice versa. Osteoblastic and osteoclastic activities are well balanced in medullary bone, indicating a highly efficient coupling mechanism. Skeletal metabolism in chickens and Japanese quail is therefore even more intense than in the pigeon (Taylor and Dacke, 1984; Dacke et al., 1993b).

Medullary bone can supply 40% of Ca required for eggshell formation (Mueller et al., 1964, 1969). However, Taylor and Moore (1954) observed that the quantity of this bone in egg-laying hens maintained for seven days on a Ca-deficient diet is relatively unaffected, but cortical bone is depleted while showing clear signs of osteoclastic activity. Medullary bone osteoblast numbers greatly increased in Ca-deficient hens (Zambonin-Zallone and Mueller, 1969), while osteoclasts appeared temporarily exhausted (Zambonin-Zallone and Teti, 1991). Simkiss (1967) proposed that medullary and cortical bone differ in their ability to mobilize Ca. On normal diets, and also on the first day of Ca depletion when there is increased osteoclastic activity in medullary bone, medullary reserves are able to supply the Ca demand of eggshell formation. During prolonged Ca deficiency, the hen maintains the labile medullary reservoir at the expense of cortical bone. We recently re-examined this question by feeding Japanese quail hens diets deficient in either Ca or Ca and vitamin D_3 for 23 days. The total quantity of medullary versus cortical bone area in decalcified tibial sections was quantified by image analysis (Dacke et al., 1993b). Surprisingly, after 16 days on these diets, the volume of medullary bone was reduced by more than 50%, while the quantity of cortical bone remained constant. By 23 days, medullary bone volume was recovering, with some accompanying loss of cortical bone. Whether this reflects species differences in response or simply a matter of timing remains to be seen. However, if medullary bone is either resistant to, or able to recover from, a severe and continuing Ca deficiency, this presents an intriguing series of questions about the nature of mechanisms involved in this remarkable response.

C. Medullary Bone Osteoclasts

Osteoclasts destroy calcified bone by secreting proteolytic enzymes and acid into the resorption lacuna (Delaissé and Vaes, 1992). Control of osteoclast, including medullary bone osteoclast, function is the focus of much research. The functional morphology of the avian cells, i.e., changes from quiescent to active states with ruffled borders, reflects rapid changes in medullary bone turnover during the egg-laying cycle.

Avian osteoclasts are extremely large, contain up to 100 nuclei and superficially resemble mammalian osteoclasts (Figure 4; Dacke et al., 1993b). Miller (1977), using Japanese quails at different points in their egg-laying cycle, observed clear zones and ruffled borders in actively resorbing medullary bone osteoclasts, while inactive osteoclasts, although still adhering to the bone surface, did not show these features. Osteoclasts are highly motile and exist in alternating states of motility and immotility corresponding to the stages of migration and resorption. They are activated by contact with exposed bone mineral, closely adhere to the mineralized matrix via the clear zone with the aid of podosomes and are believed to recognize and bind to bone surfaces via cellular adhesion molecules (integrins) (see Dacke et al., 1993b). Immunofluorescence studies have demonstrated that vinculin and actinin are localized as a ring of dotlike structures in chick osteoclasts, such that their intracellular organization parallels the circumferential structure of the sealing zone (Marchisio et al., 1984). Podosomes are localized at the sealing zone in avian and human osteoclasts (Marchisio et al., 1984; Teti et al., 1989a) and may also be present in the clear zone of resorbing cells *in vivo* (Zambonin-Zallone et al., 1988), who also

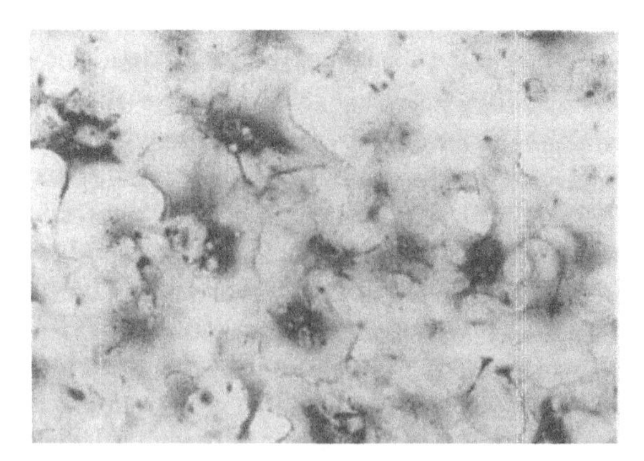

Figure 4. Light micrograph of osteoclastlike cells prepared from medullary bone of egglaying Japanese quail. The cells were settled onto glass for seven days and stained with toluidine blue. bar scale 20 μm. (From Bascal, 1993).

reported the presence of small indentations on the bone surface coinciding with podosomes on the osteoclasts. This suggests that podosomes may be capable of excavating small depressions on the bone surface. Each podosome in an avian osteoclast appears as a ventral, discrete, footlike conical structure, consisting of a short membrane protrusion (Marchisio et al., 1984) and presents a core of microfilaments associated with actinin, fimbrin, and gelsolin linked through vinculin and talin to the integrin receptor. This receptor ultimately recognizes and binds to components of the extracellular matrix (Teti et al., 1991a). Podosomes can assemble and disassemble within a few minutes (see Teti et al., 1991a); this instability provides an important point at which to regulate osteoclastic bone resorption.

Osteoclasts secrete acid, via the action of carbonic anhydrase, into the resorbing site via a polarized vacuolar proton pump, thus creating a localized acidic microenvironment (Gay et al., 1994; Gay, 1992). Inhibition of carbonic anhydrase blocks bone resorption (Delaissé and Vaes, 1992; Gay, 1992). Silver et al. (1988) demonstrated pH as low as 3.0 in the attachment zone between the cell and the base of the culture dish. This allows for dissolution of the bone mineral as well as for providing for an optimal pH for the action of lysosomal acid hydrolases, e.g., cysteine proteinases and acid phosphatases, that are secreted by the osteoclast. Removal of inorganic and organic constituents of bone results in excavations known as resorption pits on the bone surface (see Dacke et al., 1993b), following which an osteoclast moves away to resorb at another site, thus repeating the cycle.

Bone resorption generates Ca ions as well as inorganic phosphate (Pi) and, ultimately, organic components of the matrix. Silver et al. (1988), using microelectrodes, showed that osteoclasts become exposed in their acidic microcompartment to Ca^{2+} concentrations up to 40 mM. There have been several investigations of the fate of Ca^{2+} solubilized from bone mineral and its role in regulating osteoclastic bone resorption. Treatments that increase cytosolic Ca^{2+} $[Ca^{2+}]_i$, namely increased extracellular $[Ca^{2+}]$, opening of Ca^{2+} channels, Ca^{2+} ionophores, and Ca^{2+}-ATPase inhibitors, disrupt the microfilamentous core of podosomes and convert avian osteoclasts into cells devoid of the clear zone, lacking podosomes and with reduced capacity for bone resorption (Malgaroli et al., 1989; Miyauchi et al., 1990; Teti et al., 1989a).

Osteoclasts form resorption pits when settled onto slices of devitalized substratum such as bovine cortical bone, whale dentine (see Dacke et al., 1993b), or naturally occurring Ca carbonate (Guillemin et al., 1995). Oreffo et al., (1990) reported that osteoclasts harvested from medullary bones of egg-laying chickens can form resorption pits *in vitro*, although topographical details were not described. Bascal and Dacke (1992, 1996) studied isolated quail medullary bone osteoclasts settled onto devitalized bovine cortical bone slices. The resorption pits tended to be very limited in number compared with those obtained from rat osteoclasts and frequently had irregular shapes (Figure 5). We also observed shallow depressions suggestive of incomplete pit formation and/or possible points of podosome attachment (Bascal and Dacke, 1996). It is possible that bovine cortical bone slices do not represent an ideal surface for the avian cells or that the culture conditions used did not adequately reflect conditions *in vivo*.

D. Assessment of Avian Osteoclast Motility

Alterations in osteoclast motility and spread area are associated with changes in bone resorption. CT, which inhibits bone resorption in mammal-

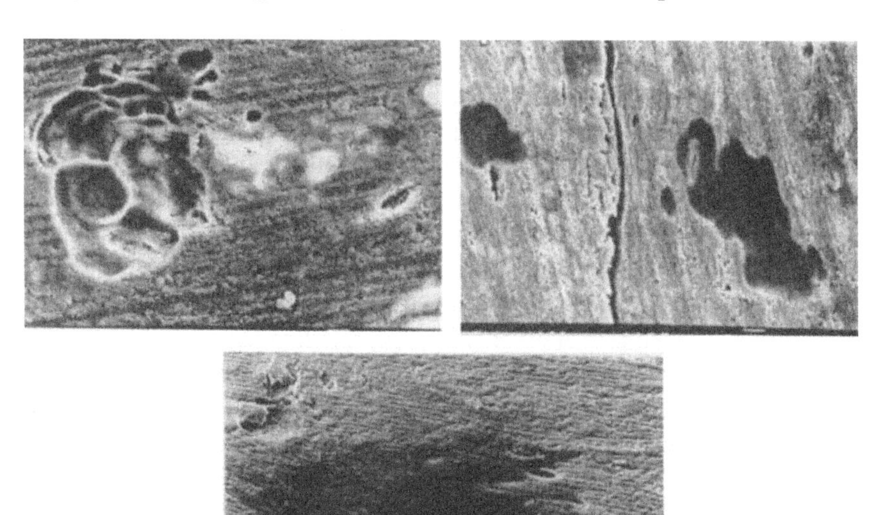

Figure 5. Scanning electron micrographs of slices of bovine cortical bone that have been exposed to (**a**) neonatal quail osteoclast like cells, on removal of the cells, resorption pits with clearly defined edges are apparent, bar scale 10 μm; (**b**) quail medullary bone osteoclastlike cells, which also produce a small number of resorption pits with clearly defined edges, bar scale 10 μm. (**c**) quail medullary bone osteoclastlike cells, which produce numerous dark patches that upon tilting at 45° reveal shallow excavations, bar scales 100 μm. (From Dacke et al., 1993b and Bascal and Dacke, 1996.) Reproduced with kind permission from the Journal of Endocrinology Ltd.

ian species, induces cellular retraction and inhibits migration of rat osteoclasts (see Chapter 17). It also reduces the spread area of chick osteoclasts in organ culture (Pandalai and Gay, 1990). Conversely, pro-absorptive stimuli, such as PTH, vitamin D_3 metabolites, and prostaglandin E_2 (PGE_2), increase osteoclast ruffled borders, clear zones, and cell size in mammalian and avian species (see below). Cytoplasmic spreading relates to functional status of the osteoclast. In rat osteoclasts, inhibition of bone resorption, e.g., by CT, is associated with loss of cytoplasmic folds from the cell surface followed by its detachment from the bone surface (Pandalai and Gay, 1990), a sequence likely to reduce the osteoclast surface area. Using chick tibia organ culture, increases in osteoclast cell spreading in response to PTH are also observed. These effects are faster in onset (between two and four minutes postexposure) than in rat osteoclasts, suggesting differences in sensitivity to PTH between avian and mammalian osteoclasts.

The cytoskeleton plays an important role in cell shape changes and, since podosomes contain cytoskeletal proteins, they would also be affected. Teti et al. (1991a) suggest that the cytoskeleton is controlled by $[Ca^{2+}]_i$ and cytosolic pH, which respond to extracellular stimuli and might act as mechanisms of signal transduction. Teti et al. (1989a) reported that low pH stimulates cultured avian osteoclasts to polarize, form podosome-rich clear zones, and increase bone resorption, while alkalization inhibited bone resorption associated with osteoclasts devoid of clear zones and podosomes. Similarly in the rat (Arnett and Dempster, 1986) and, to a lesser extent, chick (Arnett and Dempster, 1987) osteoclastic resorption *in vitro* was affected by the pH. The changes in osteoclastic function observed in response to altered intracellular pH may be mediated via changes in $[Ca^{2+}]_i$. This suggests that local acidic conditions may stimulate osteoclast adhesion to bone surfaces. Teti et al. (1989a) observed that changes in $[Ca^{2+}]_i$ are coupled to changes in intracellular pH, so that cell acidification leads to a reduction in $[Ca^{2+}]_i$

E. Osteoclastic Tartrate-Resistant Acid Phosphatase

Acid phosphatases have been implicated in bone resorption since the discovery of increased levels of enzyme activity in plasma during bone remodeling. Taylor et al. (1965) also simultaneously demonstrated cyclical changes in the activities of acid and alkaline phosphatases during eggshell calcification in hens. The tartrate-resistant form of the enzyme, or TRAP, is located biochemically and histochemically in bone and is used as a relatively specific marker for osteoclasts (Figure 6; see Dacke et al., 1993b). In

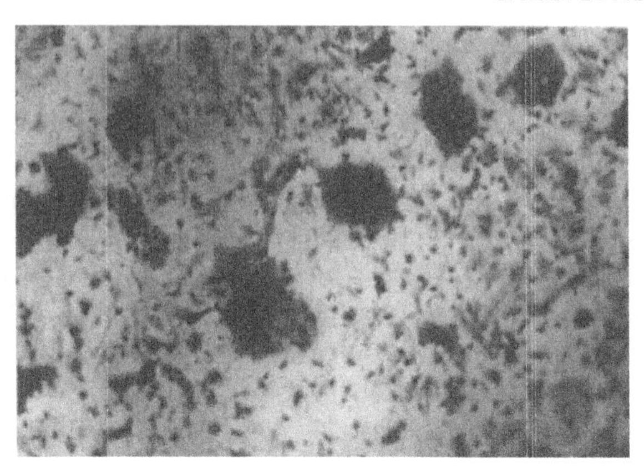

Figure 6. Light micrograph of osteoclastlike cells prepared from medullary bone of egglaying Japanese quail and settled onto slices of devitalized bovine cortical bone for seven days. The cells are stained for the presence of TRAP, which shows as a deep red brown color. bar scale 20 μm. Reproduced with kind permission of the Journal of Endocrinology Ltd.

avian, like rat, bone TRAP is localized to the ruffled border region of osteoclasts and to the bone surfaces facing the ruffled border zone (Fukushima et al., 199la). *In vitro* bone cultures have provided evidence for the potential role of calciotropic hormones in TRAP release which correlates with bone resorption. Miller (1985) provided in vivo evidence that activated medullary bone osteoclasts in egg-laying Japanese quail secrete acid phosphatase only three hours after oviposition, that is, before commencement of eggshell calcification and intensive bone resorption. The osteoclasts lacked ruffled borders but stained for acid phosphatase. There was no evidence of extracellular acid phosphatase activity associated with these cells; however, 20 minutes after PTH administration the enzyme was present in the matrix and extracellular space adjacent to the ruffled border of active osteoclasts associated with resorbing bone.

Mammalian osteoclasts *in vitro* release large quantities of TRAP. CT inhibits TRAP secretion. PGE_1 transiently inhibits TRAP secretion, while PTH and 1,25-dihydroxyvitamin D_3 have no influence in pure osteoclast cultures. PTH raises TRAP activity in osteoclasts only when co-cultured with osteoblasts, suggesting that the effects of PTH on osteoclasts are mediated via osteoblasts (see Dacke et al., 1993b). Using chicken medullary bone osteoclast cultures, Oreffo et al. (1988) observed increases in TRAP activity and inhibition of bone resorption following retinoic acid or retinol treatment.

Datta et al. (1989) and Moonga et al. (1990) proposed that $[Ca^{2+}]_i$ acts as a second-messenger in regulating TRAP secretion. However, unlike the situation in most cells, where increased $[Ca^{2+}]_i$ leads to increased exocytosis, the situation is reversed in osteoclasts. The Ca^{2+} ionophore, ionomycin, and elevated $[Ca^{2+}]_e$ both increase $[Ca^{2+}]_i$ and inhibit TRAP secretion in isolated neonatal rat osteoclasts. We recently investigated TRAP secretion in quail medullary bone osteoclast monolayers, cultured for a week. The cells stained specifically for TRAP (Figure 6) and secreted significant amounts of the enzyme. Raised $[Ca^{2+}]_e$ decreased TRAP secretion over a period of 24 hours, while ionomycin had a similar effect although low doses of the ionophore produced a small stimulatory effect. As in rodent osteoclasts, TRAP secretion was also reduced by PGE_2 and the phosphodiesterase inhibitor 1-isobutyl 3-methylxanthine, the latter implicating a role for cAMP. Raised ambient [Pi] also reduced TRAP secretion—a response which may be physiologically as important as that to $[Ca^{2+}]_i$, since local Pi levels are probably high in the vicinity of the resorbing bone surface. CT and CGRP, however, failed to influence TRAP production by these cells (Dacke et al., 1993b; Bascal and Dacke, 1996).

IV. PARATHYROID HORMONE, RELATED PEPTIDES, AND AVIAN BONE

PTH injection causes transient hypercalcemia in birds; this response is greater in laying hens than cockerels, possibly due to Ca^{2+} binding to yolk proteins or extra PTH receptors in medullary bone and oviduct (Dacke, 1979). It produces rapid and sensitive responses in immature birds. Japanese quail or chicks respond within minutes (Dacke and Kenny, 1973; Kenny and Dacke, 1974), too brief a time scale for significant osteoclastic resorption (Hurwitz, 1989b) or for changes in intestinal or renal transport mechanisms. This response has been used as a PTH bioassay method. The primary role of PTH during eggshell formation is to protect plasma Ca^{2+} levels at the expense of bone and calcifying eggshell.

Chicken PTH consists of 88 amino acids (Khosla et al., 1988). Russell and Sherwood (1989) found the nucleotide sequence of chicken pre-pro-PTH mRNA to be approximately three times the size of the equivalent mammalian mRNA. The chicken pre-pro-PTH mRNA encoded a 119 amino acid precursor peptide, and an 88 amino acid hormone which is four residues longer than all known mammalian homologues. There is significant homology of sequence in the biologically active 1–34 region with mammalian hormones, much less in the middle and carboxyl-terminal regions.

van de Velde et al (1984b) measured plasma PTH-like bioactivity during the chicken egg-laying cycle by cytochemical bioassay and found it to be elevated during eggshell calcification. Singh et al. (1986) also measured levels of PTH during the egg cycle of chickens using an *in vitro* bioassay and found them to be higher in Ca deficient hens than in those fed a high Ca diet. They were highest during the phase of shell calcification than shortly after ovulation in both groups of hens and inversely related to plasma ionized Ca levels (Figure 7), indicating a role for PTH-like activity in the regulation of Ca metabolism of the egg-laying bird. The PTH-like activity measured in these studies was probably a mixture of PTH and PTHrP and it will be useful

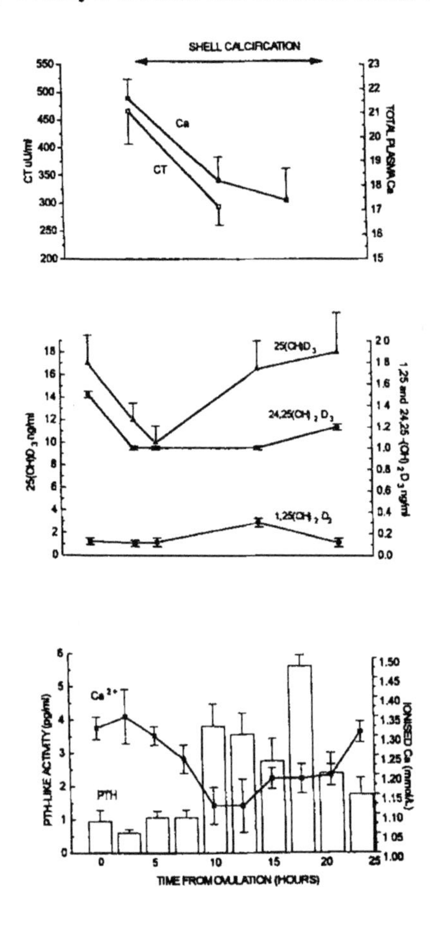

Figure 7. Changes in levels of circulating calcitonin, parathyroidlike activity, and vitamin D metabolites during the ovulation cycle of hens. CT data are from Japanese quail (after Dacke et al., 1973); PTH and vitamin D data are from chickens (Singh et al., 1986, Nys et al., 1986, Nys, 1993).

in the future to distinguish between the two circulating peptides using specific antibodies.

A. Skeletal Actions

The PTH response involves osteoclast recruitment and activation, mediated via osteoblastic receptors (Hurwitz, 1989a,b). Kenny and Dacke (1974) using acute ^{45}Ca labeling, demonstrated that the initial (0 to 30 minute) response to PTH in immature birds involved reduced plasma Ca clearance. The development of a method for temporal microwave fixation of injected radioisotopes (Shaw and Dacke, 1985; Shaw et al., 1989), enabled the demonstration that decreased plasma ^{45}Ca clearance reflects an inhibition of net skeletal Ca uptake. These responses were elicited using bPTH(1–34), but were also seen following intravenous injection of 16,16-dimethyl PGE_2 (Table 1). They are very rapid (three minute) and most apparent in long bone (Dacke and Shaw, 1987). Phosphodiesterase inhibitors, e.g., 3-isobutyl-1-methylxanthine, mimic these effects in chicks, suggesting a role for cAMP (Shaw and Dacke, 1989). Oxidation of bPTH(1–34) with hydrogen peroxide reduced Ca uptake and the concurrent cAMP activation, but not the hypercalcemia. The analogue [Nle^8, Nle^{18}, Tyr^{34}]-bPTH(1–34) gave a smaller hypercalcemic response and slightly reduced effects on plasma ^{45}Ca clearance and bone uptake suggesting that the initial hypercalcemic response to PTH in chicks does not merely reflect its acute effects on skeletal Ca uptake (Dacke and Shaw, 1988).

The stimulus for medullary bone mobilization during shell formation probably involves PTH secretion (Dacke, 1979). PTH injection into quail hens within four hours of ovulation caused eggshell thinning, but chronic ^{45}Ca mobilization from bone to eggshell suggests a dual effect of the hor-

Table 1. Acute Effects of Calcitrophic Agents on Chick Ca Metabolism

Agent	*Plasma Ca*	*Plasma ^{45}Ca*	*Bone ^{45}Ca*	*Osteoclast Cell Spread Area*
bPTH(1-34)	↑↑	↑↑	↓↓	↑↑
bPTHrP(1-34)	↑	↑	↓↓↑	—
PGE_2	↑↑↑	↑↑↑	↓↓↓	—
CT	0	—	↓↓	↓↓
CGRP	↑↑	↓↓	↑	—
Amylin	0	0	—	—
Ca^{2+}	↑↑	—	—	↓↓

Key: ↑ = increase, ↓ = decrease, 0 = no change from control value,— = not measured. See text for abbreviations. Modified from Dacke et al., 1993a.

mone on bone and oviducal Ca^{2+} transport. It had no such effect if injected later in the cycle when endogenous secretion is high (Dacke, 1976). Taylor (1970) proposed differences in cortical and medullary bone sensitivities to PTH and also in the circulating levels of PTH on high- and low-Ca diets. He suggested that normally small increases in circulating PTH levels stimulate medullary bone resorption but, under severe Ca deficiency, they greatly increase (de Bernard et al., 1980) producing substantial responses in cortical bone with only slight further effects on medullary bone. This suggestion was supported by Bannister and Candlish (1973), who measured the collagenolytic response to PTH of medullary and cortical bone in laying hens. Medullary bone was not destroyed by feeding chickens Ca-deficient diets, but its composition changed more than that of any other bone. The magnesium, sodium, potassium, and phosphorus contents of bone ash were increased, while those of Ca, CO_2 and citrate decreased (Simkiss, 1967). Hence, the medullary bone trabeculae were constituted of poorly calcified osteoid. PTH stimulates resorption in cultured avian bone (Ramp and McNeil, 1978). It destroys bone matrix associated with an increase in proteolytic, lysosomal, and acid-producing enzymes in bone, including acid phosphatase, B-glucuronidase, and carbonic anhydrase (see Dacke, 1996). These changes emphasize that medullary bone is a most labile form of bone.

PTH receptors are located on osteoblasts, but were considered absent on osteoclasts (Hurwitz, 1989a). This is now disputed (Pandala and Gay, 1990; Gay, 1996). Teti et al. (1991b) reported PTH binding to cultured medullary bone osteoclasts although it is reported not to directly affect second messenger generation in rat and avian (including medullary bone) osteoclasts (Nicholson et al., 1986; Rifkin et al., 1988). The reported elevation of cAMP levels in osteoclast-rich cultures may be attributable to contaminating osteoblasts (Ito et al., 1985; Nicholson et al., 1986). The discrepancy between motility, cAMP levels, and bone resorption observed with avian osteoclasts, but not mammalian osteoclasts, highlights our lack of understanding the relationship between these activities. Dacke et al. (1993b) recently found that PGE_2 but not CT, CGRP, and PTH, stimulated production of cAMP by quail medullary bone osteoclasts settled onto plastic.

Biochemical mechanisms underlying avian osteoclast function resemble those in mammals, but are functionally distinct from the latter class (Gay, 1988). Thus the ruffled border contains a proton pump-ATPase and an Na^+, K^+-ATPase. It also contains carbonic anhydrase closely associated with the cytoplasmic side of the membrane. Ca^{2+}-ATPase is present on the plasma membrane of the narrow side of osteoclasts, but absent in the ruffled borders; its role is presumably to direct outward flow transmembrane Ca^{2+} flux

(Akisaka et al., 1988). May et al. (1993) and Guillemin et al. (1995) demonstrated a direct effect of PTH on chick osteoclasts resulting in decreased carbonic anhydrase activity and increased acid production. The mechanism involves activation of adenylate cyclase via a G_s type protein. Stimulation of acidification by PTH and cAMP is blocked by estradiol. Estradiol was inhibitory to the same extent as CT; these effects were not additive. Estradiol-17β in micromolar, but not in nanomolar, amounts, blocked H^+ pumping in isolated plasma membrane vesicles (Gay et al., 1993). Figure 11 summarizes the actions of PTH and other calciotrophic factors on avian bone cell function.

PTH induces rapid morphological changes in avian osteoblasts and osteoclasts *in situ;* scanning electron microscopy reveals retraction of osteoblasts from a flattened to a stellate shape (Pandalai and Gay, 1990). Dramatic changes occur in avian osteoclast morphology within 10 to 60 minutes of PTH exposure both in organ culture and *in vivo* (Miller, 1978; Miller et al., 1984; Pandalai and Gay, 1990). Miller (1977, 1978) described responses of medullary bone osteoclasts in egg-laying Japanese quail *in vivo* during the inactive phase of eggshell calcification. PTH induced ruffled border formation, bounded by filamentous-rich clear zones within 20 minutes, these changes being characteristic of active bone resorbing cells found during shell calcification. Similarly PTH induced ruffled border formation in medullary bone osteoclasts maintained in culture (Sugiyama and Kusuhara, 1996). More recently, using tibiae from Ca deficient chicks, osteoclasts *in situ* were shown to increase their cell spread area by 40% within two and four minutes of PTH challenge (Pandala and Gay, 1990). This is remarkably fast compared with more classical PTH responses. Bronner (1996) speculated that an important mechanism underlying the minute to minute regulation of blood Ca levels in both birds and mammals is an ability of bone lining cells, osteoclasts as well as osteoblasts, to alter their size and shape and migrate to and from areas of the bone surface where high or low affinity binding sites for Ca^{2+} are located (Figure 8). This interesting hypothesis has not been fully tested.

PTH can rapidly change Ca transfer by osteoblasts and osteocytes. Thus PTH stimulated increases in Ca uptake by these cells have been observed, while others have reported either no response in $[Ca^{2+}]_i$ or a net Ca efflux from bone cells, at least in embryonic chick bone *in vitro* (Hurwitz, 1989b; Malgaroli et al., 1989). Ypes et al. (1988) described at least two types of voltage controlled ionic channels, using patch clamp techniques in cultured embryonic chick osteoblasts, and predicted a role for these channels in the response to PTH. Bone surfaces are widely believed to be lined with a con-

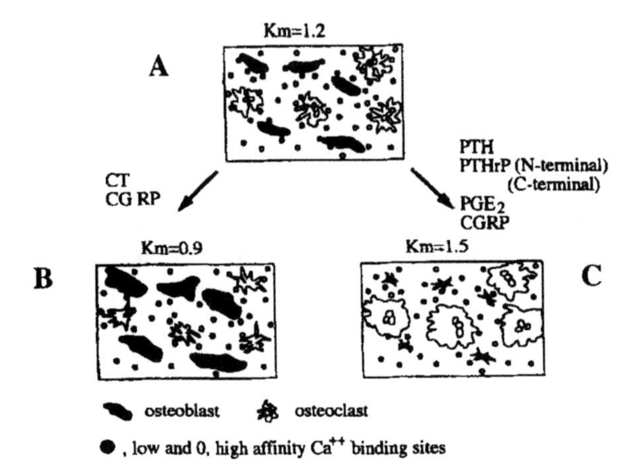

Figure 8. Diagram representing acute effect of PTH or CT on spatial relationships of osteoblasts and osteoclasts on a bone surface. Open and closed circles represent K_m values of high- affinity and low-affinity bone Ca-binding sites. (**A**) K_m = 1.2 normocalcemic situation with an equal number of high and low affinity Ca^{2+} binding sites, (**B**) K_m = 0.9 mM, hypocalcemic situation where agents such as CT cause exposure of high affinity binding sites due to osteoclast shrinkage. K_m = 1.5 hypercalcemic situation where osteoblast shrinkage has exposed low-affinity sites and associated expansion of osteoclasts has blocked high-affinity sites, leading to an average K_m of 1.5 and hypercalcemia. (**C**) Result of CT administration, where osteoclasts shrinkage has exposed high-affinity sites and consequent expansion of osteoblasts has blocked low-affinity sites, leading to an average K_m of 8 hypocalcemia. Note: high-affinity sites are considered to have an apparent K_m of 0.9 mM Ca, with low affinity sites an apparent K_m of 1.5 mM Ca. Bone mineral with a high Ca/P ratio is considered to have a relatively higher affinity for Ca binding than bone mineral with a low Ca/P ratio (see text and Table 1). (Redrawn from Dacke et al., 1993 and modified from Bronner and Stein 1992).

tinuous layer of cells and the osteoblasts become tightly adherent to the bone surface. Thus mineralization of osteoid proceeds in a distinct extracellular compartment (Gay, 1996). Osteoblasts respond to PTH by changing shape, thus cultured chick periosteal osteoblasts, in common with neonatal rat cells, increased in area within one minute of PTH treatment (Lloyd et al., 1995). A spread in cell area would cause osteoblasts arranged in a sheet to push tightly together and reduce transmembrane fluid movement. Not all studies agree with these observations, however. For example, osteoblasts cultured to confluence become stellate when treated with PTH (Miller et al., 1976).

Gay and her co-workers have begun to evaluate mechanisms of osteoblastic Ca^{2+} uptake and release at mineralizing surfaces. Plasma membrane Ca-ATPase was found by ultrastructural histochemistry of chick metaphysis to be present along the apico-lateral sides of the cell and absent along the cell surface in contact with osteoid (Akisaka et al., 1988). Plasma membrane vesicles from cultured chick periosteal osteoblasts translocated Ca^{2+} at a rate of 9.9 ± 2.3 nmol/mg prot/min (Gay and Lloyd, 1995). This is similar in magnitude to rates found in red blood cells, but is substantially less than that found in intestine and kidney, tissues which can absorb or reabsorb massive quantities of Ca^{2+}. The direction of Ca^{2+} transport by osteoblast plasma membrane vesicles was outward, as reported for other tissues (Garrahan and Rega, 1990). On the basis of location as well as direction and magnitude of pumping, the plasma membrane Ca^{2+}-ATPase in osteoblasts appears not to be involved in Ca^{2+} delivery to sites of mineralization. Its more likely role in osteoblasts is to restore $[Ca^{2+}]_i$ to basal levels following hormone treatment. Lloyd et al. (1995) used a Ca^{2+}-specific dye, Ca green C_{18} loaded into the plasma membrane of chick periosteal osteoblasts. This dye trapped emerging Ca^{2+} causing it to fluoresce. PTH stimulation resulted in peripheral cell fluorescence within 10 seconds; this was reduced by vanadate, quercetin, and trifluoperazine, indicating a Ca^{2+}-ATPase translocation mechanism. Thapsigarin, which blocks Ca^{2+} reentry into intracellular stores, prolonged the response, indicating that maintenance of $[Ca^{2+}]_i$ involves both intracellular stores and plasma membrane efflux.

B. Osteoclast Activation and the Role of PTH

Osteoblasts move aside to allow increasing numbers of osteoclasts to attach to bone surfaces during physiological and pathological bone resorption (Zheng et al., 1991). In medullary bone osteoclast numbers do not fluctuate but appear to oscillate between active and inactive states (Zambonin-Zallone and Mueller 1969; van de Velde et al., 1984b). Gay (1996) estimates that osteoclasts occupy 50% of the endosteal surface of growing chick tibia and suggests that the tightly adherent osteoblastbone lining cell layer moves apart in discrete locales to allow osteoclast attachment, a process probably locally regulated by cytokines.

PTH acts on osteoclasts indirectly probably via osteoblastic or marrow stromal cell secretions. While osteoblasts are widely held as the pivotal cell in regulating bone metabolism, at least two avian studies and one mammalian study identifies neighboring cells, not osteoblasts, as the mediator (Duong et al., 1990; Gay, 1996). Figure 11 includes a hypothetical role for the direct and indirect action of PTH on osteoclasts. PTH causes osteoblasts to extrude Ca^{2+} (Lloyd et al., 1995). Because of the close proximity of osteoblasts and osteoclasts *in vivo* it is possible that the concentration of Ca^{2+} adjacent to osteoclasts rises substantially when osteoblasts have been stimulated with PTH. Gay (1996) suggests that the inhibitory effect of Ca^{2+}, derived from PTH-stimulated osteoblasts (see below), may be overridden by the ability of osteoclasts to respond directly to PTH.

C. Parathyroid Hormone-Related Peptide

PTHrP is associated with malignancy-associated hypercalcemia in humans and also as the predominant peptide in fetal mammals; the three known forms vary in size from 139 to 173 amino acids. Its spectrum of actions in mammals, many in common with PTH, range from stimulation of osteoclastic bone resorption to enhancement of placental mineral transport (see Dacke, 1996). PTHrP is expressed in a variety of tissues in chick embryos. This molecule is highly conserved; the first 21 residues are identical with the human sequence (Schermer, 1991). It is also expressed in the isthmus and shell gland of the hen's oviduct (Thiede et al., 1991). Thiede et al. (1991) followed the cyclical expression of PTHrP in the shell gland and found peptide levels to increase as the egg moves through the anterior oviduct, gradually returning to basal levels in the 15-hour calcification period. The fluctuations in PTHrP mRNA and PTHrP levels were localized to the shell gland serosal and smooth muscle layer, suggesting a role in modulating vascular smooth muscle activity. Chicken PTHrP(1-34)NH_2 relaxed resting tension of isolated shell gland blood vessels in a dose-dependent manner. These data indicate that expression of the PTHrP gene in the avian oviduct is regulated during the egg-laying cycle and that PTHrP may function as a local modulator of shell gland smooth muscle activity. The vasorelaxant property of N-terminal fragments of PTHrP suggests a function to increase blood flow to the shell gland during egg calcification. We can also speculate PTHrP secreted from the oviduct feeds back to regulate medullary bone turnover during the egg-laying cycle.

Both PTH and PTHrP enhance cAMP and inhibit collagen synthesis in avian epiphyseal cartilage cells, an effect which is blocked by the antagonist PTH(3-34) (Pines et al., 1990). PTHrP(1-34) showed only slight PTH agonist activity in the chick hypercalcemic assay with respect to either plasma Ca levels or ^{45}Ca clearance. In femur, it substantially decreased ^{45}Ca uptake but in calvarium the opposite effect apparently occurred (Dacke et al., 1993a). Fenton et al. (1991, 1994) found that basal bone resorption by embryonic chick osteoclasts was directly inhibited by chicken and human PTHrP-(107-139) and PTHrP-(107-111). Numbers of resorption pits and total area resorbed per bone slice were reduced by PTHrP-(107-139) while resorption stimulated by hPTH-(1-34) in co-cultured chicken osteoclasts and osteoblasts was also inhibited by cPTHrP-(107-139). These results indicate C-terminal PTHrP to be a paracrine regulator of bone cell activity.

Schermer et al. (1994) studied the effects of synthetic chicken PTHrP fragments in avian (chicken renal plasma membranes and 19-day-old chick embryonic bone cells) and mammalian (canine renal plasma membranes and rat osteosarcoma [UMR-106-H5]) cells. [36-Tyr]cPTHrP(1-36)NH_2 and hPTHrP(1-34)NH_2 activated adenylate cyclase to a similar degree in chick bone cells. In UMR-106 cells and chicken renal membranes, the potency of [36Tyr]cPTHrP(1-36)NH_2 for activation of adenylate cyclase was half that of [36Tyr]hPTHrP(1-36)NH_2. Binding of ^{125}I-[36Tyr]cPTHrP(1-36)NH_2 to chick bone cells and chicken renal membranes was completely displaced by bPTH(1-34) and hPTHrP(1-34)NH_2. This does not support the concept of a distinct chicken PTHrP n-terminal receptor and suggests that PTH and PTHrP utilize the same receptor due to high homology of the amino termines of each peptide.

V. CALCITONIN, RELATED PEPTIDES, AND AVIAN BONE

A. Calcitonin

The role of CT in Ca metabolism remains obscure. Only in mammals is it shown to regulate plasma Ca levels by inhibiting osteoclastic bone resorption (Copp and Kline, 1989). Plasma Ca levels in birds are refractory to CT (Dacke, 1979). Whether or not this is due to the high circulating levels of biologically active CT (Boelkins and Kenny, 1973) causing receptor downregulation is unclear. We found that dosing heavily fasted (22 hour) chicks with salmon CT *in vivo* caused a rapid (10 minutes) but variable inhibition of net ^{45}Ca uptake into the skeleton, with the long bones being most affected (Ancill et al., 1991). This effect is similar to that of PTH and PGE_2, but its physiological significance is obscure. Other studies indicate the presence of CT receptors in birds. Thus CT infusion into 19-day chick embryos caused mild hypercalcemia (Baimbridge and Taylor, 1980). CT injected into laying hens when an egg shell was not being calcified caused hypocalcemia; but during eggshell formation was ineffective (Luck et al., 1980) This suggests that medullary bone osteoclast inhibition by CT can be overridden, a process possibly unique to medullary bone.

High levels of CT circulate in birds and are detectable in Japanese quail using bioassays. They are higher in adult males than females apart from a brief period before commence of lay (Dacke, 1979). In the quail hen, plasma CT levels rise following ovulation but fall during shell calcification (Figure 7; Dacke et al., 1972). Chronic CT deficiency after ultimobranchiolectomy in chickens does not alter serum Ca^{2+} or alkaline phosphatase levels, or the chemical composition of bone (Brown et al., 1970). However, chronic CT administration to growing chickens increases bone mass (Belanger and Copp, 1972). They reported that CT dosage affected quality and quantity of medullary bone in laying hens by inhibiting bone resorption, while cortical bone showed evidence of osteoporotic breakdown, suggesting that medullary bone is more sensitive than cortical bone to CT. Luck et al. (1980) reported that large doses of salmon CT (sCT) injected into hens 15 to 16 hours after ovulation, reduce plasma ionized, but not total, Ca levels by about 18%. CT also reduced carbonic anhydrase expression in metatarsi from chicks and their osteoclasts were smaller and rounder, with the ruffled borders being dramatically shortened; both features indicate cell inactivity (Anderson et al., 1982).

In vitro salmon CT acutely increased calvarial cell proliferation, [^{3}H]-thymidine incorporation into DNA, [^{3}H]-proline incorporation into bone matrix collagen, and [^{3}H]-hydroxyproline in intact embryonic chick calvaria and tibiae. The increased [^{3}H]-hydroxyproline incorporation was associated with increased alkaline phosphatase activity in the bones. [^{3}H]-proline incorporation in embryonic chicken calvaria also increased during three days of exposure (i.e., four hours/day) to CT. The proliferative action(s) of CT also occur in cultured neonatal mouse calvaria (Farley et al., 1988). The main target for CT is the osteoclast. Avian osteoclasts did not respond to CT in terms of post-receptor events (Miyaura et al., 1981; Ito et al., 1985; Nicholson et al., 1986). It also failed to enhance adenylate cyclase activity from the plasma membrane of medullary bone osteoclasts (Felix et al., 1983). However, cAMP responses were found in osteoclasts from chicks fed low Ca or rachitogenic diets (Eliam et al., 1988; Rifkin et al., 1988). Moreover, osteoclasts from Ca deficient chicks respond to CT *in vitro* within four minutes by a reduction in cell spread area (Pandala and Gay, 1990) and also by an inhibition of their bone resorptive activity (de Vernejoul et al., 1988). CT also causes disappearance of ruffled borders in cultured medullary bone osteoclasts (Sugiyama and Kusuhara, 1996). Unlike its actions on rat osteoclasts, CT failed to alter motility or bone resorption by freshly isolated chick osteoclasts (Arnett and Dempster, 1987). [^{131}I]-labeled sCT did not bind to avian osteoclasts, indicating the absence of receptors on these cells (Nicholson et al., 1987).

To eliminate the possibility that native chicken CT (cCT), is required to produce an inhibitory effect, Dempster et al. (1987) examined cCT effects on chick osteoclastic activity. It failed to inhibit bone resorption or motility, however it rapidly and drastically reduced the area of cultured chick osteoclasts on endosteal bone surface (Pandalai and Gay, 1990). This effect was augmented by dibutyryl cAMP, suggesting a cAMP-mediated response. Rifkin et al. (1988) observed that CT increases cAMP accumulation in chick osteoclasts; similarly, individual chick osteoclasts cultured for between three and five days responded to CT within 30 minutes by retracting. Immunocytochemical studies showed an alteration of the cytoskeletal protein elements of these cells (Hunter et al. 1989). They also demonstrated a reduction in the intracellular acidity of chick osteoclasts cultured in the presence of CT (Hunter et al., 1988; Gay et al., 1993). CT was also reported to inhibit avian osteoclastic release of ^{45}Ca from prelabeled bone in culture (de Vernejoul et al., 1988).

Generally, studies indicating positive effects of CT on avian osteoclasts, involved chicks maintained on Ca and vitamin D-deficient diets. In Ca-deficient chickens, CT increased cAMP accumulation and induced cellular retraction in isolated osteoclasts. Weak, but measurable, binding of biotinylated CT to osteoclasts has been shown (Hall et al., 1994). Osteoclasts from chickens fed normal diets showed none of these responses (Eliam et al., 1988). Additionally, plasma Ca^{2+} and CT levels were markedly reduced in Ca^{2+}-deficient chickens. This supports the concept that in normal chickens CT receptors are downregulated, whereas when fed Ca-deficient diets, the fall in circulating levels of both CT and Ca^{2+} upregulates these receptors. For this to occur, the chicks must be hypocalcemic (and possibly vitamin-D-deficient) for several weeks. The effect of diet also appears to regulate other parts of the CT receptor-effector system. Fukushima et al. (1991) compared the distribution of adenylate cyclase in osteoclasts from chicks fed normal or Ca-deficient diets. The enzyme was not detected in osteoclasts from normal chicks, while abundant activity was found in Ca deficient chicks. Isolated avian osteoclasts were found to respond to CT in several studies. CT inhibited resorption pit formation (de Vernejoul et al., 1988); focal adhesion kinase expression was suppressed by prolonged exposure to CT in osteoclasts from 18-day-old chick embryos, as well as in human osteoclasts (Berry et al., 1994). Osteoclasts on cultured medullary bone fragments lost their ruffled borders and actin filament orientation under the influence of CT (Sugiyama and Kusuhara, 1996).

B. Calcitonin Gene-Related Peptide and Amylin

CGRP, a 37 amino acid neuropeptide derived from the same gene as CT is expressed in birds mainly within the central and peripheral nervous system (see Dacke, 1996). Its structure is conserved with 90% homology between chicken and human CGRPs, compared with 50% between respective CT molecules (Zaidi et al., 1990a). CGRP presence in bone neurons coupled with its interaction with osteoclastic CT receptors (Goltzman and Mitchell, 1985) suggests a paracrine role in bone. A further member of the CT/CGRP family, amylin, a peptide from the pancreatic islet cells, is the most potent non-CT peptide, at least in mammalian assay systems (Zaidi et al., 1990a). CGRP is a potent vasodilator and is implicated in neurotransmission and neuromodulation (Zaidi et al., 1990a). Its role in bone and Ca metabolism was only recently recognized. It shares the acute hypocalcemic effects of CT in rodents, albeit at around 1,000-fold less potency, and in inhibiting bone resorption, stimulating cAMP production in mouse calvaria, and inhibiting neonatal rat osteoclastic spreading (Zaidi et al., 1990a). In the rabbit, CGRP causes transient hypocalcemia followed by sustained hypercalcemia (Tippins et al., 1984). In the same paper, a preliminary report of the *in vivo* hypercalcemic effect of the peptide in chicks was given. Bevis et al. (1990) and Ancill et al. (1990) repeated and extended these findings in chicks. The former paper gave details of comparative dose response curves for CGRP and PTH, the two peptides being approximately equipotent. In the latter we investigated the effect of CGRP on a simultaneously injected ^{45} Ca label. Intravenous injection of rat CGRP gave a rapid hypercalcemic response lasting for at least one hour which was most evident in fed chicks. Fasted chicks by contrast showed a hypophosphatemic response and also an increased plasma ^{45}Ca clearance. Subsequently, it was found that both rat and chicken CGRP caused transient increases in ^{45}Ca uptake into a variety of bone types in the chick (Ancill et al, 1991). These responses were well developed in fasted chicks but absent in fed ones and were most pronounced in calvaria and vertebrae. With low doses of CGRP, reversal of the response was noticed in calvaria, but not other bone types, while in fed animals this was the only response seen, again in calvaria. These findings indicate that CGRP may have a variety of effects on bone and Ca metabolism in the chick involving acute effects on net movement of Ca into and out of the skeleton. However, while consistent with changes in plasma ^{45}Ca clearance, they are too transient to account for them or for the hypercalcemic responses, although alternative targets such as kidney (Zaidi et al., 1990b) may be responsible. In a preliminary experiment, the effect of amylin on Ca metabolism in chicks *in vivo* appeared to be lacking (Dacke et al., 1993a).

Mixed bone cell cultures obtained from new-born chick and rodent calvaria respond to CGRP with increases in cAMP formation (Michelangeli et al., 1989). This effect was not the result of an action as a weak CT agonist, since in most instances CT had no effect. They concluded that chick, rat, and mouse bones contain osteoblast-rich cell populations that respond specifically to CGRP by a rise in cAMP.

VI. IS THERE A CALCIUM RECEPTOR IN AVIAN BONE?

Levels of plasma total and ionized Ca fall as eggshell formation proceeds (Figure 7), indicating a drain on Ca reserves not fully compensated for by bone mineral mobilization (Dacke et al., 1973). Mechanisms by which mobilized Ca reaches the bloodstream are unknown. It could be actively transported through the osteoclast cytoplasm, diffuse through a paracellular route between bone surface and osteoclast membrane or diffuse into extracellular fluid when osteoclasts migrate from resorption sites as a result of increased $[Ca^{2+}]_e$ in the microevironment (Bronner and Stein, 1992). Indirect evidence exists for a Na^+/Ca^{2+} exchanger in the basolateral membrane of osteoclasts (Baron et al., 1986), which could be responsible for Ca^{2+} extrusion entering the cell passively at the apical ruffled border membrane. The Ca^{2+}-ATPase in chick osteoclast basolateral membranes (Akisaka et al., 1988) may also be involved in transcellular movement and Ca^{2+} extrusion from these cells.

Several studies suggest that the raised $[Ca^{2+}]_e$, found in the resorbing microenvironment, can inhibit osteoclastic bone resorption (Malgaroli et al., 1989; Zaidi et al., 1989; Miyauchi et al., 1990); such changes occur locally as a consequence of bone resorption (Silver et al., 1988). In isolated rat osteoclasts, increased $[Ca^{2+}]_e$ triggers an acute elevation in $[Ca^{2+}]_i$, followed by marked retraction of the osteoclast margin and loss of secretory and bone resorptive activities (see also Chapter 16). These observations led to the suggestion that changes in the $[Ca^{2+}]_e$ are monitored by a unique sensor or receptor for Ca^{2+} (Moonga et al., 1990; Zaidi et al., 1991). Datta et al. (1989) observed that increased $[Ca^{2+}]_e$ dramatically reduced fresh rat osteoclast size, accompanied by reduced TRAP release and inhibition of bone resorption, while Miyauchi et al. (1990) showed podosome expression to be decreased in avian osteoclasts cultured on glass surfaces exposed to a high $[Ca^{2+}]_e$ (Figure 9). This potentially represents a physiologically important negative feedback control mechanism, in which osteoclastic activity would generate $[Ca^{2+}]_e$ locally, limiting further activity. Thus osteoclast contraction would expose the pit to extracellular fluid leading to diffusion of Ca^{2+} from the microenvironment. The reduced $[Ca^{2+}]_e$ would allow the osteoclast to recommence resorbing or migrate to a new site. Teti et al. (1991a) hypothesized that the increase

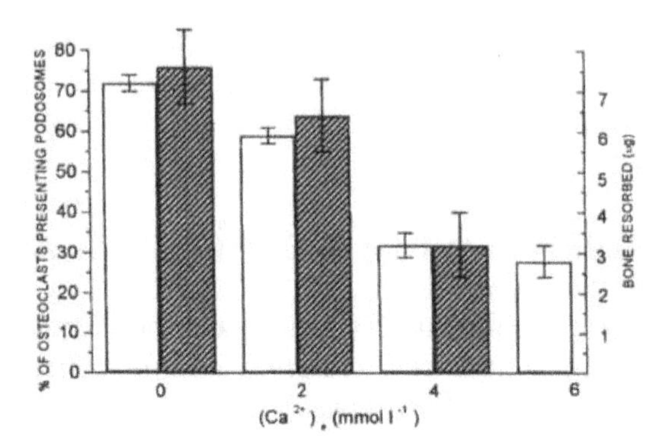

Figure 9. Effects of raised $[Ca^{2+}]_e$ on podosome expression (clear columns) and resorption of labeled bone particles (hatched columns) by cultured chicken medullary bone osteoclasts. (Redrawn from Miyauchi et al., 1990).

in $[Ca^{2+}]_i$ following increased $[Ca^{2+}]_e$ activates gelsolin which, in turn, depolymerizes actin filaments, causing rearrangement of adhesion proteins located in the podosomes and detachment of the osteoclast from the bone.

Freshly prepared individual osteoclasts from quail medullary bone, unlike those from neonatal rats, do not exhibit a rise in $[Ca^{2+}]_i$ in response to increased $[Ca^{2+}]_e$ (Bascal et al., 1992). Medullary bone osteoclasts, when cultured away from bone substrata for several days, recover an ability to respond to elevated $[Ca^{2+}]_e$ ranging from 5 to 40 mM with dose-dependent increases in $[Ca^{2+}]_i$, but neither the fresh nor cultured cells showed such a response to CT (Arkle et al., 1994). Freshly isolated quail medullary osteoclasts are also refractory to $[Ca^{2+}]_e$ in terms of changes in cell spread area. However they do respond to ionomycin (a Ca^{2+} ionophore) with a modest reduction in cell spread area. This suggests that fresh quail cells retain intracellular mechanisms necessary for elaboration of the aforementioned responses, but lack receptors for detecting changes in $[Ca^{2+}]_e$. However, when cultured away from the microenvironment of the bone in medium containing 1.25 mM-Ca^{2+} for seven days, their responsiveness to increases in $[Ca^{2+}]_e$ is restored. This restoration included a prompt rise in $[Ca^{2+}]_i$ and a 50% reduction in cell spread area, a response similar to that in fresh neonatal rat osteoclasts (Bascal et al., 1993; Bascal et al., 1994). The latter findings suggest that the putative Ca^{2+} receptors on freshly isolated quail medullary bone osteoclasts are normally downregulated but reappear after several days in culture. Thus, during the eggshell calcification cycle, when resorption of medullary bone prevails and raised local Ca^{2+} levels are generated by intense osteoclastic activity, the osteoclasts become insensitive to inhibitory factors such as elevated $[Ca^{2+}]_e$ (Figure 10). A

a

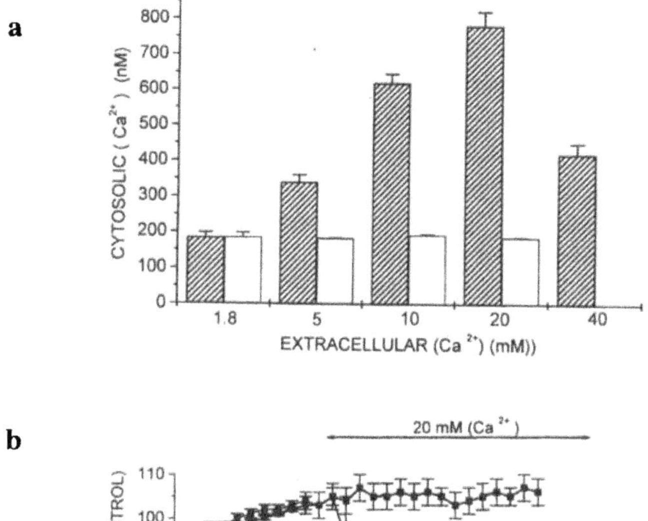

b

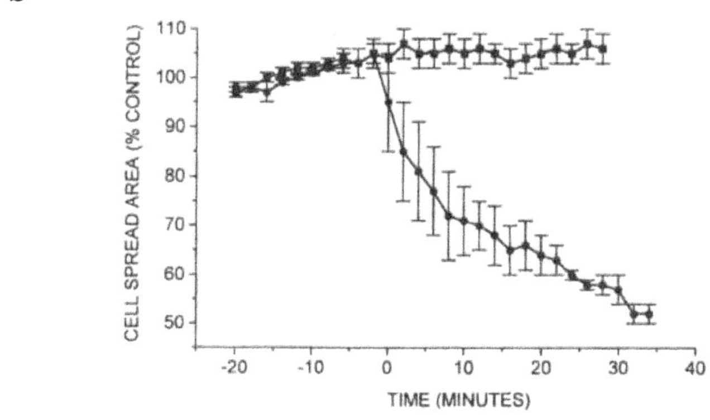

Figure 10. Response of quail medullary bone osteoclasts to an elevated $[Ca^{2+}]_e$. Fresh cells show no response to this stimulus, either in terms of cytosolic Ca $[Ca^{2+}]_i$ (**a**. open columns) or cell spread area (**b**. filled squares). In cells cultured for about seven days, 20mM $[Ca^{2+}]_e$ caused a dramatic cell retraction to around 50% of control values (**b**. filled circles) but unlike rat cells, those cultured from the quail did not recover spontaneously. When $[Ca^{2+}]_i$ levels were measured in monolayers of cultured cells, $[Ca^{2+}]_e$ elevations produced prompt increases in these values in a manner dependent upon the cat ion concentration (**a**. hatched columns). Data redrawn from Bascal et al., 1992, 1994; Dacke et al., 1993b; Zaidi et al., 1996.

complete recovery of osteoclasts in terms of their lost sensitivity to Ca^{2+}, we believe, represents the fact that we removed these cells from the microenvironment of the long bone. These findings provided the first clues that $[Ca^{2+}]_e$ was being recognized by the osteoclast by a receptorlike entity. Expression of this putative receptor could be controlled, at least in quail osteoclasts, by the level of increased bone turnover required during egg lay.

Although the transduction process for this putative receptor in avian osteoclasts is unknown, a comparison may be drawn with PTH-secreting cells, where it is thought to be linked to phosphatidylinositol turnover and generation of an inositol trisphosphate ($InsP_3$)-dependent Ca^{2+} signal (Brown, 1991). Thus, osteoclasts may monitor changes in $[Ca^{2+}]_c$ by means of a Ca^{2+} receptor but its coupling is unclear. It may be due to Ca^{2+} influx into the osteoclast via a special type of receptor-operated (or Ca^{2+}-activated) Ca^{2+} channel or to intracellular redistribution of Ca^{2+} consequent on the activation of a Ca^{2+}-binding protein or receptor. Alternatively, a greater proportion of Ca^{2+} channels may be open in the actively resorbing osteoclast (Zambonin-Zallone and Teti, 1991).

Zambonin-Zallone and colleagues have characterized Ca^{2+} channels and their regulatory role in isolated avian and mammalian osteoclasts. They showed that osteoclasts possessed both Ca^{2+}-operated (CaOCC) (Malgaroli et al., 1989) and dihydropyridine-sensitive voltage-dependent Ca^{2+} channels (VDCC) (Teti et al., 1989b; Miyauchi et al., 1990), although VDCCs may not be involved in $[Ca^{2+}]_c$-stimulated elevation of $[Ca^{2+}]_i$ (Datta et al., 1989). Increases in $[Ca^{2+}]_i$, together with the opening of the CaOCCs, also stimulates Ca^{2+} release from intracellular stores, leading to a transient spike of $[Ca^{2+}]_i$ followed by a sustained phase at a higher than basal level (Miyauchi et al., 1990). CaOCCs appear to operate, while VDCCS are downregulated when the cells are incubated in the presence of bone fragments.

Zaidi and colleagues have since characterized the activation properties of the putative Ca^{2+} receptor in rat osteoclasts. A classical pattern of drug-receptor interaction emerged in which there is evidence of concentration-dependent activation as well as use-dependent inactivation. Occupancy of this putative triggers a distinct $[Ca^{2+}]_i$ release mechanism. They suggested that a ryanodine receptor is involved, akin to mechanisms described for voltage-sensing in striated muscle. They have recently demonstrated a novel form of the type II ryanodine receptor that is localized to the osteoclast surface membrane.

Apart from Ca^{2+}, membrane potential and ambient pH can also control osteoclast Ca^{2+} receptor function. These cells are exposed to pH values as low as 4.0 and can rest at either one of the two preferred membrane potentials of -15 and -70 mV, respectively. Either a low pH or a hyperpolarized membrane potential markedly enhanced Ca^{2+} receptor sensitivity . Thus, the Ca^{2+} receptor may regulate osteoclast activity not only by sensing changes in $[Ca^{2+}]_e$ and $[H^+]_e$, but also by monitoring simultaneous alterations in the membrane potential (Zaidi et al., 1996).

VII. ARE THERE ESTROGEN RECEPTORS IN AVIAN BONE?

In recent years evidence has accumulated for an estrogen receptor (ER) in bone cells. Avian medullary bone, the most estrogen dependent of all vertebrate bone types, presents a unique model in which to investigate this receptor. Medullary bone forms in estrogen-dosed male birds within a matter of days, a process which is blocked by the simultaneous administration of anti-estrogenic compounds such as tamoxifen (Williams et al., 1991). Upon cessation of estrogen treatment medullary bone resorbs just as rapidly. Studies on estradiol-17β (E_2) interaction with bone cells show some inconsistencies. For osteoblasts, evidence for a classic nuclear/cytoplasmic ER is quite disparate. Saturable binding site assays indicate both low, ~ 200 sites per nucleus (Komm et al., 1988), and high, 1,600 sites per nucleus, levels of high affinity receptor (Eriksen et al., 1988). Specific binding of ^{3}H-estradiol to discrete sites along the endosteum of E_2-induced male Japanese quail was reported by Hunter and colleagues (1988; Turner et al., 1993); these sites are presumably where medullary bone will form. In Japanese quail medullary bone from both females and E_2-induced males, osteoblasts, preosteoblasts, and bone lining cells were positive for ER by immunostaining and binding of fluorescently-tagged estrogen (Ohashi et al., 1990). However, in sections of mammalian bone from donors at the correct stage of oestrus, Braidman et al., (1995) failed to find ER in osteoblasts by immunostaining. Both avian (Ohashi et al., 1991a,b) and mammalian (Braidman et al., 1995) osteocytes exhibit very high levels of ER by immunostaining. Since osteocytes derive from osteoblasts, Gay (1996) suggested that under certain conditions, osteoblasts would also express ER as, for example, at the onset of medullary bone formation. A nuclear/cytoplasmic E_2 receptor was also demonstrated in medullary bone osteoclasts by the high affinity nuclear binding assay, by cDNA probing for mRNA, and by Western blotting (Oursler et al., 1991). E_2 was also found to rapidly reduce acidification (Gay et al., 1993) leading to the proposal that a plasma membrane E_2 receptor exists in avian osteoclasts (Brubaker and Gay, 1994). An E_2 bovine serum albumin complex conjugated to fluorescein (E_2-BSA-Fl) has been detected on osteoclast surfaces using confocal microscopy (Brubaker and Gay, 1994). Rapid responses to complexed E_2 were noted: namely, changes in cell shape and reduced pH. To achieve a maximal effect 2 μM E_2-BSA-F1 was needed; 1 μM E_2 was adequate to block binding of the complex.

VIII. THE AVIAN VITAMIN D SYSTEM

The control of vitamin D_3 metabolism in birds has been reviewed by Norman (1987), Hurwitz (1989a), Norman and Hurwitz (1993), Nys (1993), and Dacke (1996). Like mammals, birds metabolize vitamin D_3 to 25-OH-D_3 and 1,25-$(OH)_2D_3$ in their liver and kidneys, respectively. The avian kidney, like that in mammals, synthesizes and secretes 1,25-$(OH)_2D_3$. Circulating vitamin D_3 metabolite levels in Japanese quail and egg-laying chickens have been determined. Increases in intestinal Ca absorption occurring after sexual maturity and during eggshell formation are related to enhanced circulating 1,25-$(OH)_2D_3$ levels (Sedrani and Taylor, 1977, Castillo et al., 1979) and its accumulation in the intestinal mucosa (Bar and Hurwitz, 1979). Increases in 25-$(OH)D_3$-1-hydroxylase activity can be induced by injecting estrogen into immature birds (Baksi and Kenny, 1977). Abe et al. (1979) reported that plasma concentrations of 25-$(OH)D_3$ and 1,25-$(OH)_2D_3$ but not 24,25-$(OH)_2D_3$ in egg-laying hens fluctuated during the eggshell calcification cycle (Figure 7). These results were confirmed by Nys et al. (1986) who also demonstrated that hens laying shell-less eggs do not show the cyclical fluctuation in 1,25-$(OH)_2D_3$ levels. Using Ca deficient birds, Bar and Hurwitz (1979) demonstrated that the stimulatory effect of estrogen on renal 25-$(OH)D_3$-hydroxylase is eliminated, suggesting that increased 1,25-$(OH)_2D_3$ production results from increased Ca induced by estrogens.

Vitamin D deficient diets result in medullary bone resorption in laying hens while in nonlaying birds osteodystrophy results (Wilson and Duff, 1991). 1,25-$(OH)_2D_3$ facilitates bone formation by inducing biosynthesis of osteocalcin (bone γ-carboxy-glutamic acid protein). The function of this small (MW, 5,500) noncollagenous vitamin D binding protein in skeletal mineralization is obscure, although it is a specific product of osteoblasts during bone formation. It has been purified from chicken bone and sequenced (Nys, 1993). It binds Ca and shows affinity for hydroxyapatite, suggesting its involvement in the mineral dynamics of bone (Hauschka et al., 1989). Osteocalcin is released into the circulation and provides a convenient index of bone turnover, reflecting new osteoblast formation rather than release of matrix protein during bone resorption (Nys, 1993). 1,25-$(OH)_2D_3$ stimulates osteocalcin synthesis by binding to promotor elements and enhancing osteocalcin gene transcription. However unlike intestinal calbindin, substantial osteocalcin synthesis occurs in vitamin D-deficient chicks (Lian et al., 1982). Medullary bone matrix formation is induced by sex steroids regardless of vitamin D status, although it only becomes fully mineralized when both vitamin D_3 and the sex steroids are present (Takahashi et al., 1983). Nys (1993) reports that changes in blood osteocalcin levels parallel those of 1,25-$(OH)_2D_3$ in laying hens, rise in hens fed low Ca diets, and decrease in hens laying shell-less eggs. It is possible that increased osteocalcin levels in response to estrogen reflect increased vitamin D receptor expression by osteoblasts (Liel et al., 1992). However osteoclasts from medullary bone as well as from rat bone appear to be devoid of 1,25-$(OH)_2D_3$ receptors and the effects of the metabolite are considered to be mediated via the osteoblasts (Merke et al., 1986). Harrison and Clark (1986) succeeded in growing medullary bone in organ culture from egg-laying hens. They demonstrated a response in these cultures to 1,25-$(OH)_2D_3$ by a dose dependent inhibition of $[^3H]$-proline uptake.

IX. PROSTAGLANDINS AND AVIAN BONE

The role of PGs and other eicosanoids in vertebrate (including avian) Ca and bone metabolism has been reviewed by Dacke (1989, 1996). PG effects on mammalian bone cells mimic those of PTH in that they stimulate cAMP production, cause transient increases in Ca^{2+} influx, activate carbonic anhydrase, release lysosomal enzymes, and may inhibit collagen synthesis. They also elicit morphological responses in osteoclasts and osteoblasts similar to those with other osteolytic agents (Dacke, 1989). The stable PGE_2 analogue, 16,16-dimethyl PGE_2, is profoundly hypercalcemic in chicks (Kirby and Dacke, 1983). Indomethacin, a PG synthesis inhibitor, produces hypocalcemia in egg-laying chickens (Hammond and Ringer, 1978) and quail (Dacke and Kenny, 1982). In chickens this is accompanied by delayed oviposition and thicker eggshells. PGE_2 and other eicosanoids rank alongside PTH and 1,25-$(OH)_2D_3$ as powerful stimulators of bone resorption (Dacke, 1989). The effects of PGE_2 on AMP production were studied in osteoclast-rich cultures derived from avian medullary bone and long bones of newborn rats. PGE_2 increased cAMP production in both types of osteoclasts suggesting essentially similar mechanisms (Nicholson et al., 1986; Arnett and Dempster, 1987). In addition to PTH, CGRP, and possibly CT, PGs can acutely influence skeletal uptake of ^{45}Ca (Table 1). We previously published a model (Shaw et al., 1989) in which the rapid effects of PTH and PGE_2 on skeletal ^{45}Ca uptake could be explained in terms of cAMP mediated inhibitions of outwardly directed Ca^{2+} pumps located in the membranes of bone lining cells. An alternative model proposed by Bronner (1996) (Figure 8) would involve rapid changes in shape and location of bone lining cells.

X. SUMMARY

In this chapter, evidence for the presence and functional significance of receptors in avian bone for PTH, CT and associated peptides, for the novel Ca^{2+} receptor, and for an estrogen receptor is reviewed. New evidence is emerging for a direct action of PTH on osteoclasts, much of this evidence being derived from studies of cells from avian bone, which may be more responsive to PTH than those from mammals. Nuclear estrogen receptors are found in osteoblasts, osteocytes, and osteoclasts and new evidence indicates the possible presence of a plasma membrane estrogen receptor in osteoclasts; this has yet to be confirmed. It is apparent that a variety of nonclassical factors can influence bone and Ca metabolism in birds as well as in mammals. Avian responses are often more explicit than the equivalent ones in mammals, for example the hypercalcemic responses to PGs. Other factors such as PTHrP and CGRP are represented in birds and the next few years should provide fertile ground for new research on their role in bone and Ca metabolism in general, to which avian models are likely to make an important contribution. While birds appear to be refractory to CT except during extreme Ca deficiency, they do exhibit sensitive responses to CGRP. Levels of PTH-like bioreactivity in the circulation of egg-laying birds rise as eggshell formation proceeds. This suggests that PTH and/ or PTHrP in birds may play an important role in regulating the supply of Ca for the eggshell, or possibly have other functions such as regulating oviduct smooth muscle function. Interaction of PTHrP with the PTH receptor of bone cells is an emerging area of investigation. Their interactions with gonadal steroids in forming and maintaining avian medullary bone may prove a particularly rewarding area for future studies. Much recent evidence indicating the presence of a Ca^{2+} receptor in bone has been gleaned from studies in freshly harvested medullary bone osteoclasts, in which the putative receptor was demonstrated to be downregulated, but becomes upregulated after several days culture away from a bone substratum. It seems likely that these may function as part of a local system to regulate osteoclast activity in this bone type, possibly allowing for the very rapid switching that is required to change from a state of net mineral deposition to mobilization at different points in the hen's egg-laying cycle. However, the precise nature of the signals that control this function is still largely obscure and requires much additional research.

Our understanding of mechanisms controlling medullary bone formation and turnover is still sparse. With a cell cycle time about 25% that of cortical bone, it represents an extremely active remodeling system. PTH and or PTHrP probably play important roles, as does the vitamin D_3 system. CT appears to have less in-

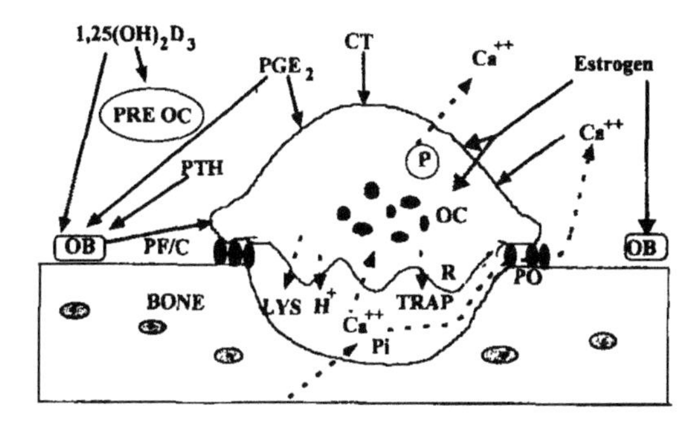

Figure 11. Diagrammatic representation of sites of regulation of avian osteoclastic bone resorption in birds. Inorganic phosphate (Pi), osteoclast (OC), osteoblast (OB), tartrate-resistant acid phosphatase (TRAP), pump (P), 1,25-$(OH)_2D_3$, CT, PTH, PGE_2, paracrine and cytokine factors (PF/C), podosomes (PO), lysosymes (LYS), ruffled border (R), solid lines = controlling factors, dotted lines = secretions/ion fluxes.

fluence, except possibly during times of extreme Ca deficiency, although the high circulating levels of this hormone in birds remain an enigma. Evidence is emerging for an important regulatory role for Ca^{2+} and possibly Pi as well as for PGs. These studies indicate that avian medullary bone osteoclasts have important differences from mammalian osteoclasts particularly with respect to responses to CT. However, avian cortical bone osteoclasts seem to be quite mammalian-like. Determining what the differences are at the signal pathway level will be of considerable value. Medullary bone represents a potentially rapidly responding model for studies of effects of anti-osteoporotic drugs such as bisphosphonates. Preliminary studies indicate that the bisphosphonate alendronate can protect structural bone and inhibit medullary bone formation if given to hens before the commence of egglay. When given during egglay the drug reduces medullary bone volume and, at higher doses, eggshell quality (Thorp et al., 1993).

ACKNOWLEDGMENTS

The author is grateful to Mrs. Gill Whitaker who assisted with preparation of the figures.

REFERENCES

Abe, E., Tanabe, R, Suda, T., and Yoshiki, S. (1979). Circadian rhythm of 1,25-dihydroxyvitamin D_3 production in egg-laying hens. Biochem. Biophys. Res. Comm. 88, 500-507.

Akisaka, T., Yamamoto, T. and Gay, C. (1988). Ultracytochemical investigation of calcium-activated adenosine triphosphatase (Ca^{++}-ATPase) in chick tibia. J. Bone Min. Res. 3, 19-25.

Ancill, A.K., Bascal, Z.A., Whitaker, G., and Dacke, C.G. (1990). Effects of rat and chicken calcitonin gene-related peptides (CGRP) upon calcium metabolism in chicks. Regul. Peptides 30, 231-238.

Ancill, A.K., Bascal, Z.A., Whitaker., G., and Dacke, C.G. (1991). Calcitonin gene-related peptide promotes transient radiocalcium uptake into chick bone in vivo. Exp. Physiol. 76, 146.

Anderson, H. (1973). Calcium-accumulating vesicles in the intercellular matrix of bone. Hard Tissue. Growth, Repair, and Remineralization. Ciba Foundation Symposium 11, pp. 213-246. Elsevier, Amsterdam.

Anderson, R., Schraer, H., and Gay, C. (1982). Ultrastructural immunocytochemical localization of carbonic anhydrase in normal and calcitonin-treated chick osteoclasts. Anat. Rec. 204, 9-20.

Arkle, S., Wormstone, I. M., Bascal, Z.A., and Dacke, C.G. (1994). Estimation of intracellular calcium activity in confluent monolayers of primary cultures of quail medullary bone osteoclasts. Exp. Physiol. 79, 975-982.

Arnett, T. R., and Dempster, D.W. (1986) Effect of pH on bone resorption by rat osteoclasts in vitro. Endocrinol. 119, 119-124.

Arnett, T.R., and Dempster, D.W. (1987). A comparative study of disaggregated chick and rat osteoclasts in vitro: Effects of calcitonin and prostaglandins. Endocrinol. 120, 602-608.

Baimbridge, K. and Taylor, T. (1980). Role of calcitonin in calcium homeostasis in the chick embryo. J. Endocr. 85, 171-185.

Bannister, D.W. and Candlish, J.K. (1973). The collagenolytic activity of avian medullary bone: effect of laying status and parathyroid extract. Brit. Poult. Sci. 14, 121-125.

Bar, A. and Hurwitz, S. (1979). The interaction between dietary calcium and gonadal hormones in their effect on plasma calcium, bone, 25-hydroxycholecalciferol-1-hydroxylase and duodenal calcium-binding protein, measured by a radio immunoassay in chicks. Endocrinol. 104, 1455-1460.

Baron, R., Neff, L., Roy, C., Boisvert, A., and Caplan, M. (1986). Evidence for a high and specific concentration of (Na^+,K^+)ATPase in the plasma membrane of the osteoclast. Cell 46, 311-320.

Bascal, Z.A (1993). Studies on the biology and pharmacology of avian medullary bone osteoclasts in vitro. PhD Thesis, University of Portsmouth, Portsmouth, UK.

Bascal, Z.A., Moonga, B.S., Dacke, C.G., and Zaidi, M..(1992). Osteoclasts from medullary bone of egg-laying Japanese quail do not express the putative calcium 'receptor'. Exp. Physiol. 77, 501-504.

Bascal, Z.A., Alam, A.S.M.T., Zaidi, M., and Dacke, C.G. (1994). Effect of raised extracellular calcium on cell spread area in quail medullary bone osteoclasts. Exp. Physiol. 79, 15-24.

Bascal, Z.A. and Dacke, C.G. (1992). Resorption pit formation by quail medullary bone osteoclasts in vitro. Abstacts of the Fifth International Symposium on Avian Endocrinology. AFRC Institute of Animal Physiology and Genetics, Edinburgh.

Bascal, Z.A. and Dacke, C.G. (1996). Ability of quail medullary osteoclasts to form resorption pits and stain for tartrate-resistant acid phosphatase. In: *The Comparative*

Endocrinology of Calcium Regulation. (Dacke, C.G., Caple, I. Danks, J., and Flik, G., Eds.). pp. 143-148. J. Endocrinol. (Bristol).

Baksi, S.N., and Kenny, A.D. (1977). Vitamin D_3 metabolism in immature Japanese quail: Effects of ovarian hormones. Endocrinol. 101, 1216-1220.

de Bernard B., Stagni, N., Camerroto R., Vittur, F., Zanetti, M., Zambonin-Zallone, A., and Teti, A. (1980). Influence of depletion on medullary bone of laying hens. Calcif. Tiss. Int. 32, 221-228.

Belanger, L.F. and Copp, D.H (1972). The skeletal effects of prolonged calcitonin administration in birds under various conditions. In: Calcium, Parathyroid Hormone, and the Calcitonins. Proc. Fourth Parathyroid Conference. (Talmage, R.V. and Munson, P.L., Eds.), pp. 41-50. Exerpta Medica, Amsterdam.

Bevis, P.J.R., Zaidi, M., and MacIntyre, I. (1990). A dual effect of calcitonin gene-related peptide on plasma calcium levels in the chick. Biochem. Biophys. Res. Comm. 169, 846-850.

Berry, V., Rathod, H., Pulman, L.B. and Datta, H.K. (1994). Immunofluorescent evidence for the abundance of focal adhesion kinase in the human and avian osteoclasts and its downregulation by calcitonin. J. Endocr. 141, R11-R15.

Bloom, M.A., Domm, L.V., Nalbandov, A.V., and Bloom, W. (1958). Medullary bone of laying chickens. Am. J. Anat. 102, 411-453.

Boelkins, J.N. and Kenny, A.D. (1973). Plasma calcitonin levels in Japanese quail. Endocrinology 92, 1754-1760.

Braidman, I., Davenport, L., Carter, D., Selby, P., Mawer, E., and Freemont, A. (1995). Preliminary in situ identification of estrogen target cells in bone. J. Bone Min. Res. 10, 74-80.

Bronner, F. (1996). Calcium metabolism in birds and mammals. In: *Comparative Endocrinology of Calcium Regulating Hormones.* (Dacke, C.G., Danks, J., Flik, G., and Caple, I., Eds.) pp. 131-135. J. Endocrinol. Bristol, England.

Bronner, F. and Stein, W.D. (1992). Modulation of bone calcium-binding sites regulates plasma calcium: A Hypothesis. Calcif. Tiss. Int. 50, 483-489.

Brown, D.M., Perey, D.Y.E., and Jowsey, J. (1970). Effects of ultimobranchiolectomy on bone composition and mineral metabolism in the chicken. Endocrinol. 87, 1282-1291.

Brown, E.M. (1991). Extracellular Ca^{2+}-sensing regulation of parathyroid cell function and role of Ca^{2+} and other ions as extracellular (first) messengers. Physiol. Revs. 71, 371-411.

Brubaker, K. and Gay, C. (1994). Specific binding of estrogen to osteoclast surfaces. Biochem. Biophys. Res. Comm. 200, 899-907.

Candlish, J. and Holt, F. (1971). The proteoglycans of fowl cortical and medullary bone. Comp. Biochem. Physiol. 40B, 283-293.

Castillo, L., Tanaka, Y., Wineland, M.J., Jowsey, J.O., and Deluca, H.F. (1979). Production of 1,25-dihydroxyvitamin D_3 and formation of medullary bone in the egg laying hen. Endocrinol. 104, 1598-1606.

Copp, D.H. and Kline, L.W. (1989). Calcitonin. In: Vertebrate Endocrinology Fundamentals and Biomedical Implications. (Pang, P.K.T. and Schreibman, M.P., Eds.), Vol. 3, pp. 79-103. Academic Press, New York.

Dacke, C.G. (1979). Calcium regulation in sub-mammalian vertebrates. Academic Press, London.

Dacke, C.G. (1976). Parathyroid hormone and eggshell calcification in Japanese quail. J. Endocrinol. 71, 239-243.

Dacke, C.G., (1989). Eicosanoids, steroids and miscellaneous hormones. In: Vertebrate Endocrinology: Fundamentals and Biomedical Implications, (Pang, P.K.T. and Schreibman, M. P. Eds,), Vol. 3, pp. 171-210. Academic Press, New York.

Dacke, C.G. (1996; In Press). The parathyroids, calcitonin, and vitamin D. In: Sturkie's Avian Physiology, Fifth Ed., (Causey Whittow, G., Ed.), Academic Press, Orlando.

Dacke, C.G., Boelkins J.N., Smith, W.K., and Kenny, A.D. (1972). Plasma calcitonin levels in birds during the ovulation cycle. J. Endocr. 54, 369-370.

Dacke, C.G., Musacchia, X.J., Volkert, W.A., and Kenny, A.D. (1973). Cyclical fluctuations in the levels of blood calcium pH and CO_2 in Japanese quail. Comp. Bioch. Physiol. 44A, 1267-1275.

Dacke, C.G. and Kenny, A.D. (1973). Avian bioassay for parathyroid hormone. Endocrinol. 92, 463-470.

Dacke, C.G. and Kenny, A.D. (1982) Prostaglandins: Are they involved in avian calcium homeostasis? In: Aspects of Avian Endocrinology: Practical and Theoretical Implications (Scanes, C.G., Ottinger, M., Balthazart, J.A., Cronshaw, J., and Chester Jones, I., Eds.) Grad. Studies Texas Tech University, 26, 255-262.

Dacke, C.G. and Shaw, A J. (1988). Effects of synthetic bovine parathyroid hormone (1-34) and its analogues on ^{45}Ca uptake and adenylate cyclase activation in bone and plasma calcium levels in the chick. Quart. J. Exp. Physiol. 73, 573-584.

Dacke., C.G. and Shaw, A.J. (1987). Studies of the rapid effects of parathyroid hormone and prostaglandins on ^{45}Ca uptake into chick and rat bone in vivo. J. Endocr. 115, 369-377.

Dacke, C.G., Ancill, A.K., Whitaker, G., and Bascal, Z.A. (1993a). Calcitrophic peptides and rapid calcium fluxes into chicken bone in vivo. In: Avian Endocrinology. (Sharp, P.J., Ed.), pp. 239-248. J. Endocrinol. Ltd., Bristol.

Dacke, C.G., Arkle, S., Cook, D., Wormstone, I., Jones, S., Zaidi, M., and Bascal, Z. (1993b). Medullary bone and avian calcium regulation. J. Exp. Biol. 184, 63-84.

Datta, H.K., MacIntyre, I., and Zaidi, M. (1989). The effect of extracellular calcium elevation on morphology and function of isolated osteoclasts. Biosci. Rep. 9, 747-751.

de Vernejoul, M., Horowitz, M., Demignon, J., Neff, L., and Baron, R. (1988). Bone resorption by isolated chick osteoclasts in culture is stimulated by murine spleen cell supernatant fluids (osteoclast-activating factor) and inhibited by calcitonin and prostaglandin E_2. J. Bone Miner. Res. 3, 69-80.

Delaissé, J. and Vaes, G. (1992). Mechanism of mineral solubilization and matrix degradation in osteoclastic bone resorption. Biology and Physiology of the Osteoclast. (Rifkin, B.R. and Gay, C.V. Eds). pp. 289-314, Boca Raton, CRC Press, Inc.

Dempster, D.W., Murrils, F.J., Horbert, W.R., and Arnett, T.R. (1987). Biological activity of chicken calcitonin: Effects on neonatal rat and embryonic chick osteoclasts. J. Bone. Min. Res. 2, 443-448.

Duong, L., Grasser, W., De Haven, P. and Sato, M. (1990). Parathyroid hormone receptors identified on avian and rat osteoclasts. J. Bone Min. Res. 5, S203.

Eliam, M., Baslé, M., Bouizar, Z., Bielakoff, J., Moukhtar, M., and de Vernejoul, M. (1988). Influence of blood calcium on calcitonin receptors in isolated chick osteoclasts. J. Endocr. 119, 243-248.

Eriksen, E., Colvard, D., Berg, N., Graham, M., Mann, K., Spelsberg, T., and Riggs, B. (1988). Evidence of estrogen receptors in normal human osteoblastlike cells. Science 241, 84-86.

Farley, J.R., Tarbaux, N.M., Hall, S.L., Linkhart, T.A., and Baylink, D.J. (1988). The anti-bone-resorptive agent calčitonin also acts in vitro to directly increase bone formation and bone cell proliferation. Endocrinol. 123, 159-167.

Felix, R., Rizzoli, R., Fleisch, H, Zambonin-Zallone, A., Teti, A., and Primavera, M. (1983). Isolation, culture and hormonal response of chicken medullary bone osteoclasts. Calcif. Tiss. Int. 35, 641.

Fenton, A., Kemp, B., Kent, G., Moseley, J., Zheng, M., Rowe, D., Britto, J., Martin, T. and Nicholson, G. (1991). A carboxyl terminal peptide from the parathyroid hormone-related peptide inhibits bone resorption by osteoclasts. Endocrinol. 129, 1762-1768.

Fenton, A.J., Martin, T.J., and Nicholson, G.C. (1994). Carboxyl-terminal parathyroid hormone-related protein inhibits bone resorption by isolated chicken osteoclasts. J. Bone Min. Res. 9, 515-519.

Fisher, L. and Schraer, H. (1982). Keratan sulfate proteoglycan isolated from the estrogen-induced medullary bone in Japanese quail. Comp. Biochem. Physiol. 72B, 227-232.

Fukushima, O., Yamamoto, T., and Gay, G.V. (1991) Ultrastructural localization of tartrate-resistant acid phosphatase (purple acid phosphatase) activity in chicken osteoclasts. J. Histochem. Cytochem. 39, 1207-1213.

Garrahan, P. and Rega, A. (1990) Plasma membrane calcium pump. In: Intracellular Calcium Regulation (Bronner, F. Ed.) pp. 271-303. Wiley-Liss, New York.

Gay, C.V. (1988). Avian bone resorption at the cellular level. CRC Critical Revs. in Poult. Biol. 1, 197-210.

Gay, C. (1992). Osteoclast ultrastructure and enzyme histochemistry: Functional implications. In: Biology and Physiology of the Osteoclast. (Rifkin, B.R. and Gay, C.V., Eds.), pp. 129-150. CRC Press, Boca Raton.

Gay, C.V. (1996). Avian bone turnover and the role of bone cells. In: Comparative Endocrinology of Calcium Regulating Hormones. (Dacke, C.G., Danks, J., Flik, G., and Caple, I. Eds.), pp. 113-121. J. Endocr. Ltd., Bristol, England.

Gay, C., Kief, N., and Bekker, P. (1993). Effect of estrogen on acidification in osteoclasts. Biochem. Biophys. Res. Comm. 192, 1251-1259.

Gay, C., Lloyd, Q., and Gilman, V. (1994). Characteristics and culture of osteoblasts derived from avian long bone. In Vitro Cell. Dev. Biol. 30A, 379-383.

Gay, C. and Lloyd, Q. (1995). Characterization of calcium efflux by osteoblasts derived from long bone periosteum. Comp. Biochem. Physiol. 111, 267-271.

Goltzman, D. and Mitchell, J. (1985). Interaction of calcitonin and calcitonin gene-related peptide at receptor sites in target tissues. Science 227, 1343-1345.

Guillemin, G., Hunter, S.J., and Gay, C.V. (1995). Resorption of natural calcium carbonate by avian osteoclasts in vitro. Cells and Materials: Scanning Electron Microscopy International. 5, 157-165.

Hall, M., Kief, N., Gilman, V., and Gay, C, (1994). Surface binding and clearance of calcitonin by avian osteoclasts. Comp. Biochem. Physiol. 108A, 59-63.

Hammond, R.W. and Ringer, R.K. (1978). Effcct of indomethacin on the laying cycle, plasma calcium, and shell thickness in the laying hen. Poult. Sci. 57, 1141.

Harrison, J.R and Clark, N.B. (1986). Avian medullary bone in organ culture: Effects of vitamin D metabolites on collagen synthesis. Calcif. Tiss. Int. 39, 35-43.

Hauschka, P.V., Lian, J.B., Cole, D.E., and Gundberg, C.M. (1989). Osteocalcin and matrix Gla protein; vitamin K-dependent protein in bone. Physiol. Revs. 69, 990-1034.

Holtrop, M. and King, G.K. (1977). The ultrastructure of the osteoclast and its functional implications. Clin. Orthop. 123, 177-196.

Hunter, S., Schraer, H., and Gay, C. (1988). Characterization of isolated and cultured chick osteoclasts: The effects of acetazolamide, calcitonin, and parathyroid hormone on acid production. J. Bone Min. Res. 3, 297-303.

Hunter, S.J., Schraer, H., and Gay, C.V. (1989). Characterisation of the cytoskeleton of isolated chick osteoclasts; effects of calcitonin. J. Histochem. Cytochem. 37, 1529-1537.

Hurwitz, S. (1965). Calcium turnover in different bone segments of laying fowl. Am. J. Physiol. 208, 203-207.

Hurwitz, S. (1989a). Calcium homeostasis in birds. Vitamins and Hormones. 45, 173-221.

Hurwitz, S. (1989b). Parathyroid hormone. In. Vertebrate Endocrinology: Fundamentals and Biomedical Implications. (Pang, P.K.T. and Schreibman, M.P., Eds.), Vol. 3, pp. 45-77, Academic Press, New York.

Hurwitz , S. and Bar, A. (1996). Response of the calcium regulating system to growth. In: The Comparative Endocrinology of Calcium Regulating Hormones, (Dacke, C.G., Danks, J., Flik, G. and Caple, I. Eds.) J. Endocr. Ltd., Bristol.

Ito, M.B., Schraer, H., and Gay, C.V. (1985). The effects of calcitonin, parathyroid hormone, and prostaglandin E_2 on cyclic AMP levels of isolated osteoclasts. Comp. Biochem. Physiol. 81A, 653-657.

Kenny, A.D. (1986). Parathyroid and ultimobranchial glands. In: Avian Physiology. (Sturkie, P.D. Ed.), 4th ed., pp. 466-478. Springer-Verlag, New York.

Kenny, A.D. and Dacke, C.G. (1974). The hypercalcaemic response to parathyroid hormone in Japanese quail. J. Endocrinol. 62, 15-23.

Khosla, S., Demay, M., Pines, M., Hurwitz, S., Potts, J.T., and Kronenberg, H.M. (1988). Nucleotide sequence of cloned cDNAs encoding chicken preproparathyroid hormone. J. Bone Min. Res. 3, 689-698.

Kirby, G.C. and Dacke, C.G. (1983). Hypercalcaemic responses to 16,16-dimethyl prostaglandin E_2, a stable prostaglandin E_2 analogue, in chicks. J. Endocr. 99, 115-122.

Komm, B., Terpening, C., Benz, D., Graeme, K., Gallegos, A., Korc, M., Greene, G., O'Malley, B., and Haussler, M. (1988). Estrogen binding, receptor mRNA, and biologic response in osteoblastlike osteosarcoma cells. Science 241, 81-84.

Koch, J., Wideman, RF., and Buss, E.G. (1984). Blood ionic calcium response to hypocalcemia in the chicken induced by ethylene glycol-bis-(B-aminoethylether)-N,N'-tetraacetic acid: Role of the parathyroids. Poult. Sci. 63, 167-171.

Kusuhara, S. (1976). Histochemical and microradiographical studies of medullary bones in laying hens. Jap. J. Zootech. Sci. 47, 141-146.

Kusuhara, S and Schraer, H. (1982). Cytology and autoradiography of estrogen-induced differentiation of avian endosteal cells. Calcif. Tissue Int. 34, 352-358.

Lian, J.B., Glimcher, M.J., Roufosse, A.H., Hauscha, P.V., Gallop, P.M., Cohen-Solal, L., and Reit, B. (1982). Alterations in the γ-carboxyglutamic acid and osteocalcin concentrations in vitamin D-deficient chick bone. J. Biol. Chem. 257, 4999-5003.

Liel, Y., Kraus, S., Levy, J., and Shany, S. (1992). Evidence that estrogens modulate activity and increase the number of 1,25-dihydroxyvitamin D receptors in osteoblastlike cells. Endocrinol. 130, 2597-2601.

Limm, S.K., Gardella, T., Thompson, A., Rosenberg, J., Keutmann, H., Potts, J., Kronenberg, H., and Nussbaum, S. (1991). Full-length chicken parathyroid hormone.

Biosynthesis in Escherichia coli and analysis of biologic activity. J. Biol. Chem. 266, 3709-3714.

Lloyd, Q., Kuhn, M., and Gay, C. (1995). Characterization of calcium translocation across the plasma membrane of primary osteoblasts using a lipophilic calcium-sensitive fluorescent dye, calcium green C_{18}. J. Biol. Chem. 270, 22445-22451.

Luck, M., Sommerville, B., and Scanes, C. (1980). The effect of eggshell calcification on the response of plasma calcium, activity to parathyroid hormone, and calcitonin in the domestic fowl (*Gallus domesticus*). Comp. Biochem. Physiol. 65A, 151-154.

Malgaroli, A., Meldolesi, J., Zambonin-Zallone, A., and Teti, A. (1989). Control of cytosolic free calcium in rat and chicken osteoclasts. J. Biol. Chem. 264, 14342-14347.

Marchisio, P.C., Cirillo, D., Naldini, L., Primavera, M.V., Teti, A., and Zambonin-Zallone, A. (1984). Cell-substratum interaction of cultured avian osteoclasts is mediated by specific adhesion structures. J. Cell Biol. 99, 1696-1705.

May, L., Gilman, V., and Gay, C. (1993). Parathyroid hormone regulation of avian osteoclasts. In: Avian Endocrinology. (Sharp, P.J., Ed.), pp. 227-238, J. Endocr. Ltd., Bristol.

Merke, J., Klaus G., Hugel, U., Waldherr, R., and Ritz, E. (1986). No 1,25-dihydroxyvitamin D_3 receptors on osteoclasts of calcium-deficient chickens despite demonstrable receptors on circulating monocytes. J. Clin. Invest. 77, 312-314.

Michelangeli, V.P., Fletcher, A.E., Allan, E.H., Nicholson, G.C., and Martin, T.J. (1989). Effects of calcitonin gene-related peptide on cyclic AMP formation in chicken, rat, and mouse bone cells. J. Bone Miner. Res. 4, 269-272.

Miller, S.C. (1977). Osteoclast cell-surface changes during the egg-laying cycle in Japanese quail. J. Cell Biol. 75, 104-118.

Miller, S.C. (1978). Rapid activation of the medullary bone osteoclast cell surface by parathyroid hormone. J. Cell Biol. 76, 615-618.

Miller, S.C. (1985). The rapid appearance of acid phosphatase activity at the developing ruffled border of parathyroid hormone activated medullary bone osteoclasts. Calcif. Tissue Int. 37, 526-529.

Miller, S.C. and Bowman, B.M. (1981). Medullary bone osteogenesis following estrogen administration to mature male Japanese quail. Develop. Biol. 87, 52-63.

Miller, S.C., Bowman, B.M., and Myers, R.L. (1984). Morphological and ultrastructural aspects of the activation of avian medullary bone osteoclasts by parathyroid hormone. Anat. Rec. 208, 223-231.

Miller, S.C., Wolf, A., and Arnaud C. (1976). Bone cells in culture: Morphologic transformation by hormones. Science 192, 1340-1342.

Miyauchi, A., Hruska, K.A., Greenfield, E.M., Duncan, R., Alverez, J., Barattolo R, Colucci S., Zambonin-Zallone, A., Teitlelbaum, S.L., and Teti, A. (1990). Osteoclast cytosolic calcium regulated by voltage-gated calcium channels and extracellular calcium controls podosome assembly and bone resorption. J. Cell Biol. 111, 2543-2552.

Miyaura, C., Nagata, N., and Suda, T. (1981). Failure to demonstrate the stimulatory effect of calcitonin on cyclic AMP accumulation in avian bone in vitro Endocrinol. Jap. 28, 403-408.

Moonga, B.S., Moss, D.W., Patchell, A., and Zaidi, M. (1990). Intracellular regulation of enzyme secretion from rat osteoclasts and evidence for a functional role in bone resorption. J. Physiol. 429, 29-45.

Mueller, W.J., Brubaker, R.L., and Caplain, M.D. (1969). Eggshell formation and bone resorption in egglaying hens. FASEB J. 28, 1851-1855.

Mueller, W.J., Schraer, R., and Shraer, H. (1964). Calcium metabolism and skeletal dynamics of laying pullets. J. Nutr. 84, 20-26.

Neumann, W.F. and Neumann, M.W. (Eds.) (1958). The chemical dynamics of bone mineral. pp. 137-168, University of Chicago Press, Chicago.

Nicholson, G.C., Livesey, S.A., Moseley, J.M., and Martin, T.J. (1986). Actions of calcitonin, parathyroid hormone, and prostaglandin E_2 on cyclic AMP formation in chicken and rat osteoclasts. J. Cell Biochem. 31, 229-241.

Nicholson, G.C., Moseley. J.M., Sexton, P.M., and Martin, T.J. (1987). Chicken osteoclasts do not possess calcitonin receptors. J. Bone Min. Res. 2, 53-59.

Norman, A. and Hurwitz, S. (1993). The role of the vitamin D endocrine system in avian bone biology. J. Nutr. 123, 310-316

Norman, A.W. (1987). Studies on the vitamin D endocrine system in the avian. J. Nutr. 117, 797-807.

Nys, Y. (1993). Regulation of plasma 1,25-$(OH)_2D_3$, of osteocalcin, and of intestinal and uterine calbindin in hens. In: Avian Endocrinology. (Sharp, P.J., Ed.), pp. 345-357. J. Endocr. Ltd, Bristol.

Nys, Y., N'Guyen, T.M., Williams, J., and Etches, R.J. (1986). Blood levels of ionized calcium, inorganic phosphorus, 1,25-dihydroxycholecalciferol and gonadal hormones in hens laying hard-shelled or shell-less eggs. J. Endocr. 111, 151-157.

Ohashi, T., Kusuhara, S., and Ishida, K. (1987). Effects of oestrogen and anti-oestrogen on the cells of the endosteal surface of male Japanese quail. Br. Poult. Sci. 28, 727-732.

Ohashi, T, Kusuhara, S., and Ishida, K. (1990). Histochemical identification of oestrogen target cells in the medullary bone of laying hens. Brit. J. Poult. Sci. 31, 221-224.

Ohashi T, Kusuhara S, and Ishida, K. (1991a). Immunoelectron microscopic demonstration estrogen receptors in osteogenic cells of Japanese quail. Histochem. 96, 41-44.

Ohashi, T., Kusuhara, S., and Ishida, K. (1991b). Estrogen target cells during the early stage of medullary bone osteogenesis: Immunohistochemical detection of estrogen receptors in osteogenic cells of estrogen-treated male Japanese quail. Calcif. Tiss. Int. 49, 124-127.

Orrefo, R.C.O., Teti, A., Triffit, J.T., Francis, M.J.O., Carano, A., and Zambonin-Zallone, A. (1988). Effect of vitamin A on bone resorption: Evidence for direct stimulation of isolated chicken osteoclasts by retinol and retinoic acid. J. Bone Min. Res. 3, 203-210.

Orrefo, R.C.O., Bonewald, L., Kukita, A., Garrett, I.R., Seyden, S.M., Rosen, D., and Mundy, G.R. (1990). Inhibitory effects of the bone-derived growth factors, osteoinductive factor, and transforming growth factor on isolated osteoclasts. Endocrinol. 126, 3069-3075.

Oursler, M., Osdoby, P., Pyfferoen, J., Riggs, B., and Spelsberg, T. (1991). Avian osteoclasts as estrogen target cells. Proc. Nat. Acad. Sci. 88, 6613-6617.

Pandalai, S. and Gay, C.V. (1990). Effects of parathyroid hormone, calcitonin, and dibutyryl-cyclic AMP on osteoclast area in cultured chick tibiae. J. Bone Min. Res. 5, 701-705.

Pines, M., Granot, I., and Hurwitz, S. (1990). Cyclic AMP-dependent inhibition of collagen synthesis in avian epiphysial cartilage cells: Effect of chicken and human parathyroid hormone and parathyroid hormone-related peptide. Bone Min. 9, 23-34.

Ramp, W.K. and McNeil, R.W.L. (1978). Selective stimulation of net calcium efflux from chick embryo tibiae by parathyroid hormone in vitro. Calcif. Tissue Res. 25, 227-232.

Rifkin, B.R., Auszmann, J.M., Kleckner, A.P., Vernillo, A.T., and Fine, A.S. (1988). Calcitonin stimulates cAMP accumulation in chick osteoclasts. Life Sci. 42, 799-804.

Romanoff, A.L. and Romanoff, A.J. (1963). The Avian Egg. 2nd ed. Wiley, New York.

Russell, J. and Sherwood, L.M. (1989). Nucleotide sequence of the DNA complementary to avian (chicken) prepoparathyroid hormone mRNA and the deduced sequence of the hormone precursor. Mol. Endocrinol. 3, 325-331.

Schermer, D.T., Chan, S.D.H., Bruce, R., Nissenson, R.A., Wood, W.I., and Strewler, G.J. (1991). Chicken parathyroid hormone-related protein and its expression during embryologic development J. Bone Min. Res. 6, 149-155.

Schermer, D.T., Bradley, M.S., Bambino, T.H., Nissenson, R.A., and Strewler, G.J. (1994). Functional properties of a synthetic chicken parathyroid hormone-related protein 1-36 fragment. J. Bone Min. Res. 9, 1041-1046.

Schraer, H. and Hunter, S. (1985). The development of medullary bone: A model for osteogenesis. Comp. Biochem. Physiol. 82A, 13-17.

Sedranni, S. and Taylor, T.G. (1977). Metabolism of 25-hydroxycholecalciferol in Japanese quail in relation to reproduction. J. Endocr. 72, 405-406.

Shaw, A.J. and Dacke, C.G. (1985). Evidence for a novel inhibition of calcium uptake into chick bone in response to bovine parathyroid hormone (1-34) or 16,16-dimethyl prostaglandin E_2 in vivo. J. Endocr. 105, R5-R8.

Shaw, A.J., Whitaker, G., and Dacke, C.G. (1989). Kinetics of rapid ^{45}Ca uptake into chick skeleton in vivo: Effects of microwave fixation Quarterly J. Exp. Physiol. 74, 907-915.

Shaw, A.J. and Dacke, C.G. (1989). Cyclic nucleotides and the rapid inhibitions of bone ^{45}Ca uptake in response to bovine parathyroid hormone and 16,16-dimethyl prostaglandin E_2 in chicks. Calcif. Tissue Int. 44, 209-213.

Silver, I.A., Murrils, R.J., and Etherington, D.J. (1988). Microelectrode studies on the acid microenvironment beneath adherent macrophages and osteoclasts. Exp. Cell Res. 175, 266-276.

Simkiss, K. (1967). Calcium in reproductive physiology. Chapman and Hall, London.

Singh, R., Joyner, C.J., Peddie, M.J., and Taylor, T.G. (1986). Changes in the concentrations of parathyroid hormone and ionic calcium in the plasma of laying hens during the egg cycle in relation to dietary deficiencies of calcium and vitamin D. Gen. Comp. Endocrinol. 61, 20-28.

Stagni, N., De Bernard, B., Liut, G.F., Vittur, F., and Zanetti, M. (1980). Ca^{2+}-binding glycoprotein in avian bone induced by estrogen. Conn. Tissue Res. 7, 121-125.

Sugiyama, T. and Kusuhara, S. (1996; In press). Morphological changes of osteoclasts on hen medullary bone during the egg-laying cycle and their regulation. In: The Comparative Endocrinology of Calcium Regulating Hormone. (Dacke, C.G., Danks, J., Flik, G., and Caple, I. Eds.), J. Endocrinol. Ltd., Bristol.

Takahashi, N., Shinki, T., Abe, E., Morinchi, N., Yamaguchi, A., Yoshiki,S., and Suda, T. (1983). The role of vitamin D in the medullary bone formation in egg-laying Japanese quails and in immature chicks treated with sex hormone. Calcif. Tiss. Int. 35, 465-471.

Taylor, T.G. (1970). The role of the skeleton in eggshell formation. Annls. Biol. Anim. Biochem. Biophys. 10, 83-91.

Taylor, T.G., Williams, A., and Kirkley, J. (1965). Cyclic changes in the activities of plasma acid and alkaline phosphatases during eggshell calcification in the domestic fowl. Can. J. Physiol. 43, 451-457.

Taylor, T.G., Simkiss, K., and Stringer, D.A. (1971). The skeleton: Its structure and metabolism. In: Physiology and Biochemistry of the Domestic Fowl. (Freeman, B.M., Ed.), pp. 125-170. Academic Press, London.

Taylor, T.G. and Dacke, C.G. (1984). Calcium metabolism and its regulation. In: Physiology and Biochemistry of the Domestic Fowl, (Freeman, B.M., Ed), Vol. 5, pp. 125-170. Academic Press, London.

Taylor, T.G. and Moore, J.M. (1954). Skeletal depletion in hens laying on a low calcium diet. Br. J. Nutr. 8, 112-124.

Teti, A., Blair, H.C., Schlesinger, P., Grano, M., Zambonin-Zallone, A., Kahn, A.J., TeitelBaum, S.J., and Hruscka, K.A. (1989a). Extracellular protons acidify osteoclasts, reduce cytosolic calcium, and promote expression of cell-matrix attachment structures. Am. Soc. Clin. Invest. 84, 773-780.

Teti, A., Grano, M.S., Argentino, L., Bartollo, R., Miyauchi, A., Teitelbaum, S.L., Hruska, K.A., and Zambonin-Zallone, A. (1989b). Voltage-dependent calcium channel expression in isolated osteoclasts. Boll. Soc. It. Biol. Sper. 65, 1115-1118.

Teti, A., Rizzoli, R., and Zambonin Zallone, A. (1991a). Parathyroid hormone binding to cultured avian osteoclasts. Biochem. Biophys. Res. Comm. 174, 1217-1222.

Teti, A., Grano, M., Colucci, S., and Zambonin Zallone, A. (1991b). Osteoblastic control of osteoclast bone resorption in a serum-free co-culture system. Lack of effect of parathyroid hormone. J. Endocrin. Invest. 15, 63-68.

Thiede, M.A., Harm, S.C., McKee, R.L., Grasser, W.A., Duong, L.T., and Leach, R.M. (1991). Expression of the parathyroid hormone-related protein gene in the avian oviduct: Potential role as a local modulator of vascular smooth muscle tension and shell gland motility during the egg-laying cycle. Endocrinol. 129, 1958-66.

Thorp, B.H., Wilson, S., Rennie, S., and Solomons, S. (1993). The effect of a bisphosphonate on bone volume and eggshell structure in the hen. Avian Pathol. 22, 671-682.

Tippins, J.R., Morris, H.R., Panico, M., Etienne, T., Bevis, P., Girgis, S., MacIntyre, I., Azria, M., and Attinger, M. (1984). The myotropic and plasma-calcium modulating effects of calcitonin gene-related peptide (CGRP). Neuropeptides 4, 425-434.

Turner, R., Bell, N., and Gay, C. (1993). Evidence that estrogen binding sites are present in bone cells and mediate medullary bone formation in Japanese quail. Poult. Sci. 72, 728-740.

van de Velde, J., Vermeiden, J., Touw, J., and Veldhuijzen, J. (1984a). Changes in activity of chicken medullary bone cell populations in relation to the egg-laying cycle. Metab. Bone Disease and Related Res. 5, 191-193.

van de Velde, J.P., Loveridge, N., and Vermeiden, J.P. (1984b). Parathyroid hormone responses to calcium stress during eggshell calcification. Endocrinol. 115, 1901-1904.

Williams, D.C., Paul, D.C., and Herring, J.R. (1991). Effects of antiestrogenic compounds on avian medullary bone formation. J. Bone Min. Res. 6, 1249-1256.

Wilson, S. and Duff, S.R.I. (1991). Effects of vitamin or mineral deficiency on the morphology of medullary bone in laying hens. Res. Vet. Sci. 50, 216-221.

Ypes, D.L., Ravesloot, J.H., Buisman, H.P., and Nijweide, P.J. (1988). Voltage activated ionic channels and conductance's in embryonic chick osteoblast cultures. J. Membrane Biol. 101, 141-150.

Zaidi, M., Moonga, B., Bevis, J.R., Bascal, Z.A., and Breimer, M.D. (1990a). The calcitonin gene peptides: Biology and clinical relevance. Crit. Revs. Clin. Lab. Sci. 28, 109-174.

Zaidi, M., Datta, H.K., and Bevis, P.J.R. (1990b). Kidney: A target organ for calcitonin gene-related peptide. Exp. Physiol. 75, 27-32.

Zaidi, M., Bascal, Z. A., Adebanjo, O.A., Arkle, S.A., and Dacke, C.G. (1996). Ca^{2+}-sensing receptor in avian osteoclasts. In: *The Comparative Endocrinology of Calcium Regulation.* (Dacke, C.G., Danks, J., Flik, G., and Caple, I. Eds.), pp. 123-130. J. Endocr. Ltd., Bristol.

Zaidi, M., Datta, H.K., Patchell, A., Moonga, B.S., and MacIntyre, I. (1989) >Calcium-activated= intracellular calcium elevation: A novel mechanism of osteoclast regulation. Biochem. Biophys. Res. Comm. 163, 1461-1465.

Zaidi, M., Kerby, J., Huang, C.L.-H., Alam, A.S.M.T., Rathod, H., Chambers, T.J., and Moonga, B.S. (1991). Divalent cations mimic the inhibitory effects of extracellular ionized calcium on bone resorption by isolated rat osteoclasts: Further evidence for a "calcium receptor". J. Cell. Physiol. 149, 422-427.

Zambonin-Zallone, A. and Mueller, W.J. (1969). Medullary bone of laying hens during calcium depletion and replication. Calcif. Tissue Res. 4, 136-146.

Zambonin-Zallone, A., Teti, A. (1991). Isolation and behavior of cultured osteoclasts. In: Bone. Vol. 2: The Osteoclast. (Hall, B.K., Ed.), pp. 87-118. CRS Press, Ann Arbor, MI.

Zambonin-Zallone, A., Teti, A., Carano, A., and Marchisio, P.C. (1988). The distribution of podosomes in osteoclasts cultured in bone laminae: Effect of retinol. J. Bone Miner. Res. 3, 517-523.

Zheng, M., Nicholson, G., Wharton, A., and Papadimitriou, J. (1991). What's new in osteoclast ontogeny? Pathology, Research, and Practice 187, 117-125.

Printed and bound by CPI Group (UK) Ltd, Croydon, CR0 4YY
08/05/2025
01865012-0001